Physics Olympiad

Problems and Solutions

Physics Olympiad
Problems and Solutions

Mihail Sandu

Romanian Physics Olympiad Committee, Romania

World Scientific

NEW JERSEY · LONDON · SINGAPORE · BEIJING · SHANGHAI · HONG KONG · TAIPEI · CHENNAI · TOKYO

Published by

World Scientific Publishing Europe Ltd.
57 Shelton Street, Covent Garden, London WC2H 9HE
Head office: 5 Toh Tuck Link, Singapore 596224
USA office: 27 Warren Street, Suite 401-402, Hackensack, NJ 07601

Library of Congress Control Number: 2025017036

British Library Cataloguing-in-Publication Data
A catalogue record for this book is available from the British Library.

PHYSICS OLYMPIAD
Problems and Solutions

ISBN 978-1-80061-727-8 (hardcover)
ISBN 978-1-80061-715-5 (paperback)
ISBN 978-1-80061-716-2 (ebook for institutions)
ISBN 978-1-80061-717-9 (ebook for individuals)

For any available supplementary material, please visit
https://www.worldscientific.com/worldscibooks/10.1142/Q0505#t=suppl

Desk Editors: Eshak Nabi Akbar Ali/Gabriel Rawlinson/Shi Ying Koe
Translation from Romanian by Daniela Berciu

Typeset by Stallion Press
Email: enquiries@stallionpress.com

Foreword

The book is a collection of theoretical and experimental problems proposed by Prof. Mihail Sandu over two decades of physics competitions between Hungary, Moldova, and Romania. These competitions were designed to prepare participants for the International Physics Olympiad (IPhO). The problems, along with their detailed solutions, serve as an invaluable resource for high school students looking to enhance their problem-solving skills and deepen their understanding of physics.

Physics Olympiads occupy a unique space in science education. They not only reward knowledge but also nurture creativity, resilience, and a passion for discovery. Many students who participate in these competitions go on to pursue degrees in physics, engineering, and related fields, often becoming accomplished researchers and innovators.

Mihail Sandu's contributions to physics education are truly unparalleled. He is the author of nearly one hundred problem books on physics and astronomy and has been a dedicated member of Romania's National Committees for the Physics, Astronomy, and Astrophysics Olympiads for decades. In recognition of his lifelong commitment to the field, the International Astronomical Union recently named a small planet in his honor: (670740) Mihailsandu. This tribute reflects the profound impact of his work on students and, by extension, on the broader scientific community.

This book is an outstanding resource for students, teachers, and physics enthusiasts alike. The problems offer an opportunity to explore the beauty and rigor of physics, while the detailed solutions inspire critical and creative thinking. I am confident that this collection will prove invaluable to anyone preparing for Physics Olympiads or seeking to deepen their knowledge of the subject, just as Mihail Sandu's books have shaped and inspired my early experiences with physics. I also hope it will remind readers of the joy and wonder that lie at the heart of scientific inquiry.

Andrei Constantin
Royal Society Dorothy Hodgkin Fellow,
Department of Physics, University of Oxford,
and Physics & Mathematics Tutor
at Mansfield College Oxford

About the Author

Professor Mihail Sandu has many decades of experience teaching highschool physics at the Technological High School of Tourism, Călimăneşti, Romania, as well as delivering undergraduate physics and mathematics courses as an associate professor at the Lucian Blaga University of Sibiu, Romania. Beginning in 1978, he was actively involved in developing and approving the Romanian pre-university physics curricula as a member of the National Committee of Physics, Ministry of Education and Research. Prof. Sandu has been a member of the Romanian National Physics Olympiad Committee since 1997, organizing teams and competitions, providing training and instruction, and proposing problems for National and International Olympiads. He has authored many physics and astronomy textbooks at the high-school and unviersity levels, alongside several volumes of Physics, Astronomy and Astrophysics Olympiad problems.

Contents

Chapter 1

International Pre-Olympic Physics Contest 1999, Târgovişte, Romania

Problem 1. Rarefied Gas

To achieve the best possible vacuum in a sufficiently large container with volume V and containing an ideal gas at pressure p_1, the walls of the container are kept at temperature T, except for one sector with a very small surface area ΔS. This lower part of the container is kept at a temperature much lower than T. The gas molecules that hit the specified container sector adhere to it, and the gas liquifies.

Given: μ, the molar mass of the gas, and R, the universal constant of ideal gases. Consider the volume of the liquid resulting from the liquefaction of the gas to be negligible. *Determine* how long it takes the pressure of the gas in the container to become p_2.

It is known that $e^{-x} = 1 - x$, if $x \ll 1$.

Solution

Using the equation of state for ideal gases, we calculate the mass of the gas that adheres, through liquefaction, to the lower sector in the considered time interval.

It results that:

$$p_1 V = \frac{m_1}{\mu} RT;$$

$$p_2 V = \frac{m_2}{\mu} RT;$$

$$\Delta m = m_1 - m_2 = \frac{\mu(p_1 - p_2)V}{RT}.$$

On the other hand, according to the simplified model of an ideal gas, the molecules move in three mutually perpendicular directions with the thermal velocity v_T. Thus, of the total number of molecules in the container, only 1/6 will move vertically down.

Take n to be the concentration of the molecules in the vessel. After a time interval $\Delta t = t_2 - t_1$, the molecules located in the cylindrical column with section area ΔS and height $v_T \cdot \Delta t$ will reach a sector of the wall with surface area ΔS.

The number of these molecules is

$$\Delta N = \frac{1}{6} n \cdot \Delta S \cdot v_T \cdot \Delta t,$$

where $n = \frac{N}{V}$ represents the concentration of molecules in the container, so that the number of molecules adhering to the wall surface per unit time decreases according to the law

$$\frac{dN}{dt} = -\frac{1}{6} \frac{N}{V} \cdot v_T \cdot \Delta S.$$

From this, it follows that:

$$\frac{dN}{N} = -\frac{1}{6} \frac{\Delta S}{V} \cdot v_T \cdot dt; \quad \int_{N_1}^{N_2} \frac{dN}{N} = -\frac{1}{6} \frac{\Delta S}{V} \cdot v_T \int_{t_1}^{t_2} dt;$$

$$\ln N_2 - \ln N_1 = -\frac{1}{6} \frac{\Delta S}{V} \cdot v_T \cdot (t_2 - t_1);$$

$$\ln \frac{N_2}{N_1} = -\frac{1}{6} \frac{\Delta S}{V} \cdot v_T \cdot \Delta t;$$

$$\frac{N_2}{N_1} = e^{-\frac{1}{6} \cdot \frac{\Delta S}{V} \cdot v_T \cdot \Delta t} \approx 1 - \frac{1}{6} \cdot \frac{\Delta S}{V} \cdot v_T \cdot \Delta t;$$

$$N_2 = N_1 - N_1 \frac{1}{6} \cdot \frac{\Delta S}{V} \cdot v_T \cdot \Delta t;$$

$$N_1 - N_2 = N_1 \frac{1}{6} \cdot \frac{\Delta S}{V} \cdot v_T \cdot \Delta t;$$

$$\frac{m_1}{m_0} - \frac{m_2}{m_0} = \frac{m_1}{m_0} \cdot \frac{1}{6} \cdot \frac{\Delta S}{V} \cdot v_T \cdot \Delta t,$$

where m_0 is the mass of one molecule;

$$m_1 - m_2 = \frac{1}{6} \cdot m_1 \cdot \frac{\Delta S}{V} \cdot v_T \cdot \Delta t;$$

$$p_1 V = v_1 RT = \frac{m_1}{\mu} RT, \quad m_1 = \frac{\mu p_1 V}{RT};$$

$$m_1 - m_2 = \frac{1}{6} \cdot \frac{\mu p_1 V}{RT} \cdot \frac{\Delta S}{V} \cdot v_T \cdot \Delta t;$$

$$m_1 - m_2 = \frac{1}{6} \cdot \frac{\mu p_1}{RT} \cdot \Delta S \cdot v_T \cdot \Delta t;$$

$$p_1 V = v_1 RT = \frac{m_1}{\mu} RT; \quad m_1 = \frac{\mu p_1 V}{RT};$$

$$m_1 - m_2 = \frac{1}{6} \cdot \frac{\mu p_1 V}{RT} \cdot \frac{\Delta S}{V} \cdot v_T \cdot \Delta t;$$

$$m_1 - m_2 = \frac{1}{6} \cdot \frac{\mu p_1}{RT} \cdot \Delta S \cdot v_T \cdot \Delta t;$$

$$p_2 V = v_2 RT = \frac{m_2}{\mu} RT; \quad m_2 = \frac{\mu p_2 V}{RT};$$

$$m_1 - m_2 = \frac{\mu p_1 V}{RT} - \frac{\mu p_2 V}{RT};$$

$$m_1 - m_2 = \frac{\mu (p_1 - p_2) V}{RT}; \quad m_1 - m_2 = \frac{1}{6} \cdot \frac{\mu p_1}{RT} \cdot \Delta S \cdot v_T \cdot \Delta t;$$

$$\frac{1}{6} \cdot \frac{\mu p_1}{RT} \cdot \Delta S \cdot v_T \cdot \Delta t = \frac{\mu (p_1 - p_2) V}{RT};$$

$$\frac{1}{6} \cdot \frac{p_1}{RT} \cdot \Delta S \cdot v_T \cdot \Delta t = \frac{(p_1 - p_2) V}{RT};$$

$$\frac{1}{6} \cdot p_1 \cdot \Delta S \cdot v_t \cdot \Delta t = (p_1 - p_2) V;$$

$$\Delta t = \frac{6\,(p_1 - p_2)\,V}{p_1 \cdot \Delta S \cdot v_T} V_T = \sqrt{\frac{3RT}{\mu}};$$

$$\Delta t = \frac{6\,(p_1 - p_2)\,V}{p_1 \cdot \Delta S} \cdot \sqrt{\frac{\mu}{3RT}};$$

$$[\Delta t]_{SI} = \frac{\frac{N}{m^2} \cdot m^3}{\frac{N}{m^2} \cdot m^2 \cdot \frac{m}{s}};$$

$$m_1 - m_2 = \frac{1}{6} \cdot \frac{\mu p_1 V}{RT} \cdot \frac{\Delta S}{V} \cdot v_T \cdot \Delta t;$$

$$m_1 - m_2 = \frac{1}{6} \cdot \frac{\mu p_1}{RT} \cdot \Delta S \cdot v_T \cdot \Delta t;$$

$$p_2 V = v_2 RT = \frac{m_2}{\mu} RT; \quad m_2 = \frac{\mu p_2 V}{RT};$$

$$m_1 - m_2 = \frac{\mu p_1 V}{RT} - \frac{\mu p_2 V}{RT} = \frac{\mu\,(p_1 - p_2)\,V}{RT};$$

$$\frac{1}{6} \cdot \frac{\mu p_1}{RT} \cdot \Delta S \cdot v_T \cdot \Delta t = \frac{\mu\,(p_1 - p_2)\,V}{RT};$$

$$\frac{1}{6} \cdot p_1 \cdot \Delta S \cdot v_T \cdot \Delta t = (p_1 - p_2)\,V;$$

$$\Delta t = \frac{6\,(p_1 - p_2)\,V}{p_1 \cdot \Delta S \cdot v_T}; \quad v_T = \sqrt{\frac{3RT}{\mu}};$$

$$\Delta t = \frac{6\,(p_1 - p_2)\,V}{p_1 \cdot \Delta S} \cdot \sqrt{\frac{\mu}{3RT}};$$

$$m_1 - m_2 = \frac{1}{6} \cdot \frac{\mu p_1 V}{RT} \cdot \frac{\Delta S}{V} \cdot v_T \cdot \Delta t;$$

$$m_1 - m_2 = \frac{1}{6} \cdot \frac{\mu p_1}{RT} \cdot \Delta S \cdot v_T \cdot \Delta t;$$

$$p_2 V = v_2 RT = \frac{m_2}{\mu} RT; \quad m_2 = \frac{\mu p_2 V}{RT};$$

$$m_1 - m_2 = \frac{\mu p_1 V}{RT} - \frac{\mu p_2 V}{RT} = \frac{\mu\,(p_1 - p_2)\,V}{RT}$$

$$= \frac{1}{6} \cdot \frac{\mu p_1}{RT} \cdot \Delta S \cdot v_T \cdot \Delta t;$$

$$m_1 - m_2 = \frac{1}{6} \cdot \frac{\mu p_1}{RT} \cdot \Delta S \cdot v_T \cdot \Delta t;$$

$$\frac{1}{6} \cdot \frac{\mu p_1}{RT} \cdot \Delta S \cdot v_T \cdot \Delta t = \frac{\mu \, (p_1 - p_2) \, V}{RT};$$

$$\frac{1}{6} \cdot \frac{p_1}{1} \cdot \Delta S \cdot v_T \cdot \Delta t = \frac{(p_1 - p_2) \, V}{1};$$

$$\Delta t = \frac{6 \, (p_1 - p_2) \, V}{p_1 \cdot \Delta S \cdot v_T}; \quad v_T = \sqrt{\frac{3RT}{\mu}};$$

$$\Delta t = \frac{6 \, (p_1 - p_2) \, V}{p_1 \cdot \Delta S} \cdot \sqrt{\frac{\mu}{3RT}}.$$

Problem 2. Gas Cylinder

A horizontal cylindrical metal vessel with length L_0 and cross-sectional area S contains an ideal gas at pressure p_0 when the gas's temperature along the cylinder's axis is kept constant and equal to $T_0 = 0^0\text{C}$. The temperature of the gas and the cylinder walls is made to vary along the longitudinal axis of the cylinder according to the law

$$T = ax^2 + bx + c,$$

so that, at the ends of the cylinder, the temperature is t_0, and halfway between the ends, the temperature is $T_{\max}$. The gas temperature is the same at any point in a given cross-section of the cylinder. The coefficient of linear expansion of the metal from which the vessel is made is $\alpha \ll 1$. The expansion of the cylinder is only longitudinal.

(a) *Determine* the pressure p of the gas within the vessel, and calculate the displacement Δx_{CM} of the center of mass, measured from one end of the cylinder.

Two identical holes with very small surface areas are opened simultaneously at one of the vessel's ends and in the middle of its length.

(b) *Compare* the numbers of molecules that escape into the surrounding vacuum through the two holes per unit time if the temperature distribution remains as $T(x)$, given that:

$$\int \frac{dx}{ax^2 + bx + c} = \frac{1}{\sqrt{\Omega}} \ln \left| \frac{2ax + b - \sqrt{\Omega}}{2ax + b + \sqrt{\Omega}} \right|;$$

$$\Omega = b^2 - 4ac > 0;$$

$$\int \frac{x dx}{ax^2 + bx + c} = \frac{1}{2a} \ln \left| ax^2 + bx + c \right|$$

$$- \frac{b}{2a\sqrt{\Omega}} \ln \left| \frac{2ax + b - \sqrt{\Omega}}{2ax + b + \sqrt{\Omega}} \right|.$$

$$\Omega > 0.$$

Solution

(a) Corresponding to the normal initial conditions (p_0, T_0), the number N of the gas molecules in the vessel is determined as follows:

$$p_0 V_0 = v R T_0 = \frac{N}{N_{A0}} p_0 V_0 N 0_A;$$

$$N = \frac{p_0 V_0 N_A}{R T_0} = \frac{p_0 S_0 L_0 N_A}{R T_0} = \frac{p_0 S_0 L_0}{\frac{R}{N_A} T_0};$$

$$\frac{R}{N_A} = k = \text{Boltzmann's constant};$$

$$N = \frac{p_0 L_0 S_0}{k T_0}.$$

If the temperature of the gas varies along the vessel, according to the law

$$T = ax^2 + bx + c,$$

under the specified conditions it follows that:

$$a = -\frac{4(T_{\max} - T_0)}{L^2}; \quad b = \frac{4(T_{\max} - T_0)}{L}; \quad c = T_0;$$

$$T(x) = -\frac{4(T_{\max} - T_0)}{L^2} x^2 + \frac{4}{L} (T_{\max} - T_0) x + T_0.$$

So, using the equation of the new equilibrium state of the ideal gas, it results that:

$$pV = vRT = \frac{N}{N_A}; \quad p = \frac{N}{V} \cdot \frac{R}{N_A} \cdot T; \quad \frac{N}{V} = n; \quad \frac{R}{N_A} = k;$$

$$p = nkT; \quad n(x); \quad T(x);$$

$$p = n(x) \cdot k \cdot T(x);$$

$$n(x) = \frac{p}{k \cdot T(x)} = \frac{p}{k \cdot (ax^2 + bx + c)};$$

$$n(x) = \frac{p}{k \cdot \left[-\frac{4}{L^2} \cdot (T_{max} - T_0) \cdot x^2 + \frac{4}{L} \cdot (T_{max} - T_0) \cdot x + T_0\right]};$$

$$\bar{n} = \frac{1}{L} \cdot \int n(x) \cdot dx = \frac{p}{kL} \int \frac{dx}{ax^2 + bx + c} = \frac{N}{V} = \frac{S_0 L_0 p_0}{kT_0 V};$$

$$N = \frac{p_0 L_0 S_0}{kT_0};$$

$$V = S_0 L;$$

$$Q(x) = \int \frac{dx}{ax^2 + bx + c} = \frac{1}{\sqrt{\delta}} \ln \left| \frac{2ax + b - \sqrt{\delta}}{2ax + b + \sqrt{\delta}} \right|;$$

$$\delta = b^2 - 4ac > 0;$$

$$a = -\frac{4(T_{max} - T_0)}{L^2}; \quad b = \frac{4(T_{max} - T_0)}{L}; \quad c = T_0;$$

$$\delta = \frac{16(T_{max} - T_0)^2}{L^2} + \frac{16(T_{max} - T_0)T_0}{L^2};$$

$$\delta = \frac{16(T_{max} - T_0)}{L^2}(T_{max} - T_0 + T_0); \quad \delta = \frac{16(T_{max} - T_0)}{L^2} \cdot T_{max};$$

$$\sqrt{\delta} = \frac{4}{L} \cdot \sqrt{T_{max}(T_{max} - T_0)};$$

$$2ax + b - \sqrt{\delta} = -\frac{8(T_{max} - T_0)}{L^2}x + \frac{4(T_{max} - T_0)}{L}$$

$$-\frac{4}{L} \cdot \sqrt{T_{max}(T_{max} - T_0)};$$

$$2ax + b - \sqrt{\delta} = \frac{4}{L}\Big[-2\left(T_{\max} - T_0\right)x + \left(T_{\max} - T_0\right)L$$

$$- L \cdot \sqrt{T_{\max}\left(T_{\max} - T_0\right)}\,\Big];$$

$$2ax + b + \sqrt{\delta} = -\frac{8\left(T_{\max} - T_0\right)}{L^2}x + \frac{4\left(T_{\max} - T_0\right)}{L}$$

$$+ \frac{4}{L} \cdot \sqrt{T_{\max}\left(T_{\max} - T_0\right)};$$

$$2ax + b + \sqrt{\delta} = \frac{4}{L}\Big[-2\left(T_{\max} - T_0\right)x + \left(T_{\max} - T_0\right)L$$

$$+ L \cdot \sqrt{T_{\max}\left(T_{\max} - T_0\right)}\,\Big];$$

$$F(x) = \int \frac{dx}{ax^2 + bx + c} = \frac{1}{\sqrt{\delta}}\ln\left|\frac{2ax + b - \sqrt{\delta}}{2ax + b + \sqrt{\delta}}\right|;$$

$$F(x) = \frac{L}{4\sqrt{T_{\max}\left(T_{\max} - T_0\right)}}$$

$$\cdot \ln\left|\frac{-2\left(T_{\max} - T_0\right)x + \left(T_{\max} - T_0\right)L - L\sqrt{T_{\max}\left(T_{\max} - T_0\right)}}{-2\left(T_{\max} - T_0\right)x + \left(T_{\max} - T_0\right)L + L\sqrt{T_{\max}\left(T_{\max} - T_0\right)}}\right|;$$

$$F(L) = \frac{L}{4\sqrt{T_{\max}\left(T_{\max} - T_0\right)}}$$

$$\cdot \ln\left|\frac{-2\left(T_{\max} - T_0\right)L + \left(T_{\max} - T_0\right)L - L\sqrt{T_{\max}\left(T_{\max} - T_0\right)}}{-2\left(T_{\max} - T_0\right)L + \left(T_{\max} - T_0\right)L + L\sqrt{T_{\max}\left(T_{\max} - T_0\right)}}\right|;$$

$$F(L) = \frac{L}{4\sqrt{T_{\max}\left(T_{\max} - T_0\right)}}$$

$$\cdot \ln\left|\frac{-\left(T_{\max} - T_0\right)L - L\sqrt{T_{\max}\left(T_{\max} - T_0\right)}}{-\left(T_{\max} - T_0\right)L + L\sqrt{T_{\max}\left(T_{\max} - T_0\right)}}\right|;$$

$$F(L) = \frac{L}{4\sqrt{T_{\max}\left(T_{\max} - T_0\right)}}$$

$$\cdot \ln\left|\frac{-\left(T_{\max} - T_0\right) - \sqrt{T_{\max}\left(T_{\max} - T_0\right)}}{-\left(T_{\max} - T_0\right) + \sqrt{T_{\max}\left(T_{\max} - T_0\right)}}\right|;$$

$$F(0) = \frac{L}{4\sqrt{T_{\max}(T_{\max} - T_0)}}$$

$$\cdot \ln\left|\frac{(T_{\max} - T_0)L - L\sqrt{T_{\max}(T_{\max} - T_0)}}{(T_{\max} - T_0)L + L\sqrt{T_{\max}(T_{\max} - T_0)}}\right|;$$

$$F(0) = \frac{L}{4\sqrt{T_{\max}(T_{\max} - T_0)}}$$

$$\cdot \ln\left|\frac{(T_{\max} - T_0) - \sqrt{T_{\max}(T_{\max} - T_0)}}{(T_{\max} - T_0) + \sqrt{T_{\max}(T_{\max} - T_0)}}\right|;$$

$$F(0) = \frac{L}{4\sqrt{T_{\max}(T_{\max} - T_0)}}$$

$$\cdot \ln\left|\frac{\sqrt{T_{\max} - T_0} \cdot \sqrt{T_{\max} - T_0} - \sqrt{T_{\max}(T_{\max} - T_0)}}{\sqrt{T_{\max} - T_0} \cdot \sqrt{T_{\max} - T_0} + \sqrt{T_{\max}(T_{\max} - T_0)}}\right|;$$

$$F(0) = \frac{L}{4\sqrt{T_{\max}(T_{\max} - T_0)}} \cdot \ln\left|\frac{\sqrt{T_{\max} - T_0} - \sqrt{T_{\max}}}{\sqrt{T_{\max} - T_0} + \sqrt{T_{\max}}}\right|;$$

$$\bar{n} = \frac{1}{L} \cdot \int n(x) \cdot dx = \frac{p}{kL}\int \frac{dx}{ax^2 + bx + c} = \frac{N}{V} = \frac{S_0 L_0 p_0}{kT_0 V};$$

$$F(x) = \int \frac{dx}{ax^2 + bx + c} = \frac{1}{\sqrt{\delta}}\ln\left|\frac{2ax + b - \sqrt{\delta}}{2ax + b + \sqrt{\delta}}\right|;$$

$$\bar{n} = \frac{p}{kL}[F(x = L) - F(x = 0)]\int \frac{dx}{ax^2 + bx + c} = \frac{N}{V} = \frac{S_0 L_0 p_0}{kT_0 V};$$

$$\bar{n} = \frac{p}{kL}[F(x = L) - F(x = 0)]; \quad \bar{n} = \frac{p}{kL} \cdot \Delta F; \quad \bar{n} = \frac{S_0 L_0 p_0}{kT_0 V};$$

$$\frac{p}{kL} \cdot \Delta F = \frac{p}{L} \cdot \Delta F = \frac{S_0 L_0 p_0}{T_0 V};$$

$$V = SL; \quad S \approx S_0; \quad V = S_0 L;$$

$$\frac{p}{L} \cdot \Delta F = \frac{S_0 L_0 p_0}{T_0 S_0 L}; \quad p \cdot \Delta F = \frac{L_0 p_0}{T_0};$$

$$p = \frac{L_0 p_0}{T_0 \Delta F}; \quad \Delta F = F_{x=L} - F_{x=0};$$

$$F_{x=L} = \frac{L}{4\sqrt{T_{\max}(T_{\max} - T_0)}}$$

$$\cdot \ln\left|\frac{-(T_{\max} - T_0) - \sqrt{T_{\max}(T_{\max} - T_0)}}{-(T_{\max} - T_0) + \sqrt{T_{\max}(T_{\max} - T_0)}}\right|;$$

$$F_{x=0} = \frac{L}{4\sqrt{T_{\max}(T_{\max} - T_0)}} \cdot \ln\left|\frac{\sqrt{T_{\max} - T_0} - \sqrt{T_{\max}}}{\sqrt{T_{\max} - T_0} + \sqrt{T_{\max}}}\right|.$$

In Figure 2.1, the cylindrical vessel is represented in its final state (when its length is L) with the specified temperature distribution along it, divided into n identical sectors, each with length $d = L/n$, so that the temperature along each sector can be considered constant.

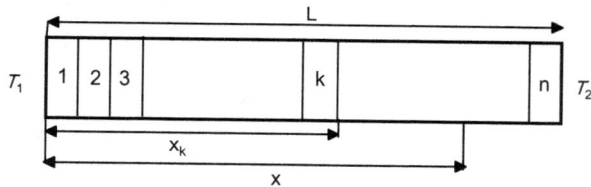

Fig. 2.1

With the temperature variation along the cylinder being as specified, it follows that at distance x from the end of the cylinder, the temperature is

$$T_x = -\frac{4}{L^2}(T_{\max} - T_0) \cdot x^2 + \frac{4}{L}(T_{\max} - T_0) \cdot x + T_0.$$

Under the specified conditions, the temperature of sector k is

$$T_k = -\frac{4}{L^2}(T_{\max} - T_0) \cdot x_k^2 + \frac{4}{L}(T_{\max} - T_0) \cdot x_k + T_0;$$

$$t_k + 273.15\ ^0\mathrm{C} = -\frac{4}{L^2}(T_{\max} - T_0) \cdot x_k^2$$

$$+ \frac{4}{L}(T_{\max} - T_0) \cdot x_k + t_0 + 273.15\ ^0\mathrm{C};$$

$$t_k = -\frac{4}{L^2}\left(T_{\max} - T_0\right) \cdot x_k^2 + \frac{4}{L}\left(T_{\max} - T_0\right) \cdot x_k + t_0;$$

$$L = nd; \quad x_k = kd;$$

$$t_k = -\frac{4}{n^2}\left(T_{\max} - T_0\right) \cdot k^2 + \frac{4}{n}\left(T_{\max} - T_0\right) \cdot xk + t_0.$$

If at temperature t_k the length of sector k is $d_k = d$, then the length of the same sector at temperature $t_0 = 0\,^{\circ}\mathrm{C}$ was

$$d_{0k} = \frac{d}{1 + \alpha t_k}.$$

It results that:

$$\sum d_{0k} = d \cdot \sum \frac{1}{1 + \alpha t_k} = \frac{L}{n}\sum \frac{1}{1 + \alpha t_k} L_0;$$

$$\frac{1}{1 + \alpha t_k} = (1 + \alpha t_k)^{-1} \approx 1 - \alpha t_k;$$

$$\sum \frac{1}{1 + \alpha t_k} = \sum (1 - \alpha t_k) = n - \alpha \sum t_k;$$

$$L_0 = \frac{L}{n}\left[n - \alpha \sum \left(-\frac{4}{n^2}\left(T_{\max} - T_0\right)k^2 + \frac{4}{n}\left(T_{\max} - T_0\right)k + t_0\right)\right];$$

$$L_0 = L + \alpha \frac{L}{n}\frac{4\left(T_{\max} - T_0\right)}{n^2}\sum k^2 - \alpha \frac{L}{n}\frac{4\left(T_{\max} - T_0\right)}{n}\sum k + \frac{L}{n}\alpha t_0;$$

$$\sum k^2 = 1^2 + 2^2 + 3^2 + \cdots + n^2 = \frac{n\left(n + 1\right)\left(2n + 1\right)}{6};$$

$$\sum k = 1 + 2 + 3 + \cdots + n = \frac{n\left(n + 1\right)}{2};$$

$$L_0 = L\left(1 - \frac{2}{3}\alpha(T_{\max} - T_0)\right); \quad L = \frac{L_0}{1 - \frac{2}{3}\alpha(T_{\max} - T_0)}.$$

The final coordinates of the center of mass of the gas in the vessel are:

$$x_{\mathrm{CM}} = \frac{\int x \cdot dm}{m} = \frac{S}{N}\int x \cdot n_x \cdot dx = \frac{S \cdot p}{N \cdot k}\int \frac{x \cdot dx}{ax^2 + bx + c};$$

$$\int \frac{x \cdot dx}{ax^2 + bx + c} = \frac{1}{2a} \ln \left| ax^2 + bx + c \right|$$

$$-\frac{b}{2a\sqrt{\delta}} \cdot \ln \left| \frac{2ax + b - \sqrt{\delta}}{2ax + b + \sqrt{\delta}} \right| = G(x);$$

$$\frac{1}{2a} \ln \left| ax^2 + bx + c \right| = H(x); \quad G(x) = H(x) - \frac{b}{2a} F(x);$$

$$x_{CM} = \frac{pS}{Nk} \cdot \Delta G = \frac{pS}{Nk}(G(L) - G(0)) = \frac{pS}{Nk} \cdot \Delta G;$$

$$\Delta G = \Delta H - \frac{b}{2a} \cdot \Delta F;$$

$$\Delta H = 0; \quad \Delta G = -\frac{b}{2a} \cdot \Delta F;$$

$$p = \frac{p_0 L_0}{T_0 \cdot \Delta F}; \quad N = \frac{p_0 L_0 S N_A}{R T_0};$$

$$x_{CM} = \frac{\Delta G}{\Delta F}; \quad x_{CM} = -\frac{b}{2a}; \quad x_{CM} = -\frac{L}{2}; \quad x_{0,CM} = \frac{L_0}{2};$$

$$\Delta x_{CM} = \frac{L}{2} - \frac{L_0}{2}; \quad L = \frac{L_0}{1 - \frac{2}{3}\alpha(T_{\max} - T_0)};$$

$$L = L_0 \left(1 + \frac{2}{3}\alpha(T_{\max} - T_0) \right);$$

$$\Delta x_{CM} = \frac{1}{3} L_0 \alpha (T_{\max} - T_0).$$

(b) Since the number of molecules crossing one unit area of the section per unit time is

$$d = \frac{\Delta N}{\Delta S \cdot \Delta t} = \frac{1}{6} n \cdot v_T,$$

it results that:

$$d_{x=0} = \frac{1}{6} n_{x=0} \cdot v_{T_0} = \frac{1}{6} \frac{p_0}{k T_0} \cdot \sqrt{\frac{3 R T_0}{\mu}};$$

$$d_{x=L/2} = \frac{1}{6} n_{x=L/2} \cdot v_{T_{\max}} = \frac{1}{6} \frac{p_0}{k T_{\max}} \cdot \sqrt{\frac{3 R T_{\max}}{\mu}};$$

$$\frac{d_{x=L/2}}{d_{x=0}} = \frac{p}{p_0} \cdot \sqrt{\frac{T_0}{T_{max}}} = \frac{L_0}{T_0 \cdot \Delta F} \cdot \sqrt{\frac{T_0}{T}};$$

$$p = \frac{L_0 p_0}{T_0 \Delta F}; \quad \Delta F = \frac{L_0 p_0}{T_0 p};$$

$$\frac{d_{x=L/2}}{d_{x=0}} = \frac{L_0}{T_0 \cdot \frac{L_0 p_0}{T_0 p}} \cdot \sqrt{\frac{T_0}{T}} = \frac{p}{p_0} \cdot \sqrt{\frac{T_0}{T}};$$

$$p = \frac{L_0 p_0}{T_0 \Delta F}; \quad \frac{p}{p_0} = \frac{L_0}{T_0 \Delta F}; \quad \frac{d_{x=L/2}}{d_{x=0}} = \frac{L_0}{T_0 \Delta F} \cdot \sqrt{\frac{T_0}{T}};$$

$$\Delta F = \frac{L_0 p_0}{T_0 p}; \quad \frac{d_{x=L/2}}{d_{x=0}} = \frac{L_0}{T_0 \frac{L_0 p_0}{T_0 p}} \cdot \sqrt{\frac{T_0}{T}};$$

$$\frac{d_{x=L/2}}{d_{x=0}} = \frac{L_0 T_0 p}{T_0 L_0 p_0} \cdot \sqrt{\frac{T_0}{T}}; \quad \frac{d_{x=L/2}}{d_{x=0}} = \frac{p}{p_0} \cdot \sqrt{\frac{T_0}{T}}.$$

Problem 3. Rutherford's Experiment

The equilateral hyperbola in Figure 3.1 represents the trajectory of a particle α at normal incidence on a very thin gold (Au) sheet, as a result of its interaction with a fixed nucleus, in Rutherford's experiment.

(a) *Determine* the minimum distance between particle α and the fixed nucleus, the velocity of particle α at that moment, and the scattering angle of particle α. It should be considered that, crossing the Au sheet, particle α interacts with a single nucleus.

   Given: m, the mass of particle α; v_0, the velocity of particle α outside the Au sheet; q_1, the electric charge of particle α; q_2, the electric charge of the Au nucleus; b, the collision parameter; and ε_0, the absolute permittivity of a vacuum. A non-relativistic approach should be taken.

(b) For which values of the collision parameter does particle α not cross the Au sheet? *Estimate* the radius of the Au nucleus.

(c) If the electron in the quantized model of a hydrogen atom is a homogeneous sphere with mass m, then *evaluate* the minimum radius that the electron can have.

   Given: h, Planck's constant, and s, the quantum spin number.

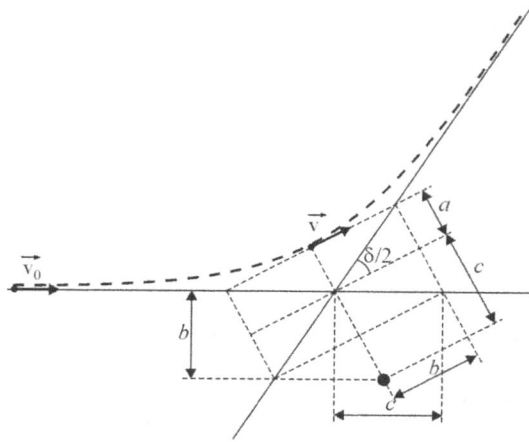

Fig. 3.1

Solution

(a) From the conservation laws of energy and kinetic momentum, it results that:

$$\begin{cases} \frac{mv_0^2}{2} = \frac{mv^2}{2} + \frac{1}{4\pi\epsilon_0}\frac{h_1 h_2}{r_{\min}}, \\ v_0 b = v r_{\min} \end{cases}$$

where $r_{\min} = a + c$;

$$v = \sqrt{\frac{q_1^2 q_2^2}{16\pi\epsilon_0^2 v_0^2 b^2 m^2 + v_0^2}} - \frac{q_1 q_2}{4\pi\epsilon_0 v_0 bm};$$

$$r_{\min} = \frac{v_0 b}{v}.$$

Considering the properties of the hyperbola, highlighted by the notation in Figure 3.1, it results that:

$$\operatorname{tg}\frac{\theta}{2} = \frac{a}{b}; \quad c^2 = a^2 + b^2.$$

As a result, we have:

$$v = \frac{v_0 b}{a + c};$$

$$\frac{mv_0^2}{2} = \frac{mv_0^2 b^2}{(a+c)^2} + \frac{1}{4\pi\varepsilon_0}\frac{q_1 q_2}{a+c};$$

$$\frac{mv_0^2}{2}\left[1 - \frac{b^2}{(a+c)^2}\right] = \frac{1}{4\pi\varepsilon_0}\frac{q_1 q_2}{(a+c)};$$

$$mv_0^2 a = \frac{q_1 q_2}{4\pi\varepsilon_0}; \quad a = \frac{q_1 q_2}{4\pi\varepsilon_0 mv_0^2};$$

$$\mathrm{tg}\,\frac{\theta}{2} = \frac{q_1 q_2}{4\pi\varepsilon_0 mv_0^2 b}.$$

(b) If particle α does not cross the Au sheet, then the scattering angle must be $\theta \geq 90°$. (If particle α crosses the Au sheet, then $\theta < 90°$.)

It results that:

$$\frac{\theta}{2} \geq 45°; \quad \mathrm{tg}\,\frac{\theta}{2} \geq \mathrm{tg}\,45° = 1;$$

$$\frac{q_1 q_2}{4\pi\varepsilon_0 mv_0^2 b} \geq 1; \quad 0 \leq b \leq \frac{q_1 q_2}{4\pi\varepsilon_0 mv_0^2}.$$

The limit condition, $\theta = 90°$, as shown in Figure 3.2, is determined by the dimensions of the fixed nucleus.

It results that

$$R = \frac{q_1 q_2}{4\pi\epsilon_0 mv_0^2}.$$

(c) In the quantized model of the hydrogen atom, the electron possesses its own intrinsic angular momentum (spin) $\vec{S}$, which is quantized according to the following law:

$$S = \sqrt{s(s+1)}\,\frac{h}{2\pi}.$$

If $\vec{\omega}$ is the angular velocity of the rotational motion of a sphere around the axis of symmetry, then

$$\vec{S} = I\vec{\omega},$$

where $I = \frac{2}{5}mR^2$ is the moment of inertia of the sphere in relation to the axis passing through its center.

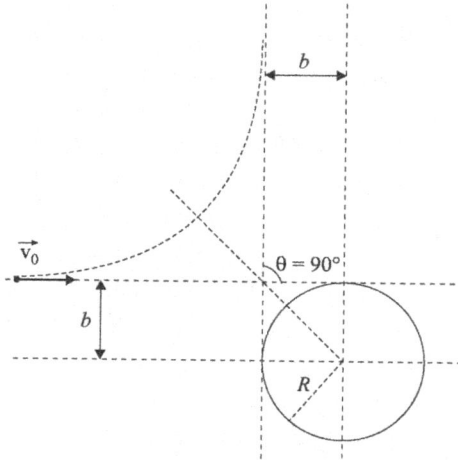

Fig. 3.2

It results that

$$\sqrt{s(s+1)}\frac{h}{2\pi} = \frac{2}{5}mR^2 \cdot \frac{v}{R}.$$

For a point on the equator of the electron sphere, we have:

$$\omega = v/R; \quad \sqrt{s(s+1)}\frac{h}{2\pi} = \frac{2}{5}mR^2 \cdot \frac{v}{R} \cdot s.$$

According to the theory of special relativity, we have $v \leq c$.
It results that:

$$\sqrt{s(s+1)}\frac{h}{2\pi} = \frac{2}{5}mRv;$$

$$v = \frac{5h\sqrt{s(s+1)}}{4\pi mR} \leq c;$$

$$R \geq \frac{5h\sqrt{s(s+1)}}{4\pi mc}.$$

Problem 4. An Artificial Satellite of Earth

An artificial satellite revolves around the Earth (considered a fixed material point), outside the atmosphere, in an ellipse-shaped orbit. As shown in Figure 4.1, the Earth is positioned in one of the ellipse's foci, so that the satellite's rotation period is T, and the areolar speed is Ω.

It is known that: $r = \frac{p}{1+e\cos\theta}$, where $p = \frac{b^2}{a}$ and $e = \sqrt{1 - \frac{b^2}{a^2}}$, with the ellipse's semi-axes (a and b) being unknown; and $r^2 = \frac{d\theta}{dt}\sqrt{pKM}$, where K (the constant of universal attraction) and M (the Earth's mass) are known.

(a) *Determine* the time dependence of the elements of the position vector of the satellite in relation to the Earth, $r = f(t)$ and $\theta = g(t)$, if at the moment t_0 the satellite passes to perigee. Use the substitutes $a - r = ae\cos u$ and $n = \frac{1}{a}\sqrt{\frac{KM}{a}}$.

(b) *Determine* the geometric interpretation of the angle u in the circle whose center is at the ellipse's center and whose radius is equal to the major semi-axis of the ellipse.

(c) Any satellite is inserted into the desired elliptical orbit by injecting it into one of the orbit's points; it is only necessary for the satellite to reach the altitude of the chosen point and to give it the required speed in the direction of the tangent to the trajectory at that point.

To arrive at the base elliptical orbit shown, which is known to reach perigee at a high altitude, the satellite is first raised

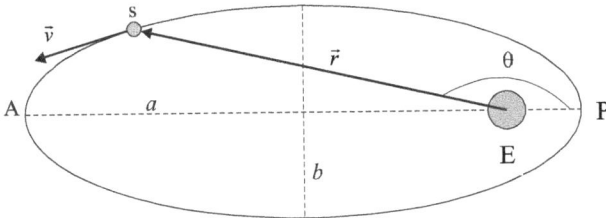

Fig. 4.1

to a certain point of an elliptical parking orbit, whose apogee has a low altitude. Then, after a few rotations on this orbit, the satellite is passed into the base orbit.

Propose, from an energetic point of view, the optimal injection point on the parking orbit and transfer point from the parking orbit to the base orbit. Give a qualitative justification.

Solution

(a)
(1) Dependence $r = f(t)$

From the equation of the ellipse, by differentiation, it results that:

$$r = \frac{p}{1 + e \cos \theta}; \quad \frac{1}{r} = \frac{1}{p} + \frac{e}{p} \cos \theta;$$

$$-\frac{1}{r^2} dr = -\frac{e}{p} \sin \theta \, d\theta;$$

$$\cos \theta = \frac{p}{e} \left(\frac{1}{r} - \frac{1}{p} \right); \quad \sin \theta = \sqrt{1 - \frac{p^2}{e^2} \left(\frac{1}{r} - \frac{1}{p} \right)^2};$$

$$\frac{p}{r^2} dr = \sqrt{e^2 - \left(\frac{p}{r} - 1 \right)^2} \, d\theta;$$

$$r^2 d\theta = \frac{p \, dr}{\sqrt{e^2 - \left(\frac{p}{r} - 1 \right)^2}}.$$

On the other hand, knowing that $r^2 \dot{\theta} = C$, it results that:

$$r^2 \frac{d\theta}{dt} = C; \quad dt = \frac{r^2 d\theta}{C};$$

$$dt = \frac{p}{C} \frac{r \, dr}{\sqrt{e^2 r^2 - (p - r)^2}};$$

$$p = a(1 - e^2);$$

$$e^2r^2 - (p - r)^2 = e^2r^2 - [a(1 - e^2) - r]^2$$

$$= e^2r^2 - a^2(1 - e^2)^2 + 2a(1 - e^2)r - r^2$$

$$= -r^2(1 - e^2) - a^2(1 - e^2)^2 + 2a(1 - e^2)r$$

$$= (1 - e^2)[2ar - r^2 - a^2(1 - e^2)]$$

$$= (1 - e^2)[a^2e^2 - (a^2 - 2ar + r^2)]$$

$$= (1 - e^2)[a^2e^2 - (a - r)^2];$$

$$dt = \frac{p}{C} \frac{r dr}{\sqrt{1 - e^2}\sqrt{a^2e^2 - (a - r)^2}};$$

$$C = \sqrt{pKM}; \quad p = a(1 - e^2);$$

$$dt = \sqrt{\frac{a}{KM}} \frac{r dr}{\sqrt{a^2e^2 - (a - r)^2}};$$

$$a - r = ae \cos u,$$

where u is the *eccentric anomaly*;

$$r = a(1 - e \cos u);$$

$$dr = ae \sin u du;$$

$$dt = a\sqrt{\frac{a}{KM}}(1 - e \cos u)du.$$

By integrating from the value $u = 0$, which corresponds to the passing of the satellite to the perigee, where $r_{\min} = a(1 - e)$, we get:

$$t - t_0 = a\sqrt{\frac{a}{KM}} \int_0^u (1 - e \cos u)du,$$

where t_0 is the moment of the passing of the satellite to the perigee;

$$u - e \sin u = \frac{1}{a}\sqrt{\frac{KM}{a}}(t - t_0); \quad \Omega = \frac{L}{2m}; \quad \Omega = \frac{b}{2}\sqrt{\frac{KM}{a}};$$

$$n = \frac{1}{a}\sqrt{\frac{KM}{a}}; \quad n = \frac{2\pi}{T}; \quad T = 2\pi\sqrt{\frac{a^3}{KM}},$$

where n is the *average movement* of the satellite;

$$u - e \sin u = n(t - t_0),$$

which is Kepler's equation, where $n(t - t_0)$ is the *average anomaly*.

By solving the previous equation, the eccentric anomaly of the satellite, u, is determined at the moment t, and then the time dependence of r will be determined:

$$r = a(1 - e \cos u).$$

(2) Dependence $\theta = g(t)$

Let us now determine the time dependence of the other polar coordinate, θ, called the *true anomaly*. For this, we will first establish the relationship between θ and u.

Comparing expressions

$$r = \frac{p}{1 + e \cos \theta} = \frac{a\left(1 - e^2\right)}{1 + e \cos \theta}$$

and

$$r = a(1 - e \cos u),$$

it results that:

$$1 - e \cos u = \frac{1 - e^2}{1 + e \cos \theta};$$

$$1 + e \cos \theta = \frac{1 - e^2}{1 - e \cos u};$$

$$\cos\theta = \frac{\cos u - e}{1 - e\cos u};$$

$$1 - \cos\theta = \frac{(1+e)(1-\cos u)}{1 - e\cos u};$$

$$1 - \cos u = \frac{r}{a}; \quad 1 - \cos\theta = 2\sin^2\frac{\theta}{2}; \quad 1 - \cos u = 2\sin^2\frac{u}{2};$$

$$\sin^2\frac{\theta}{2} = \frac{a\,(1+e)\sin^2\frac{u}{2}}{r};$$

$$1 + \cos\theta = \frac{(1-e)(1+\cos u)}{1 - e\cos u};$$

$$1 - e\cos u = \frac{r}{a}; \quad 1 + \cos\theta = 2\cos^2\frac{\theta}{2}; \quad 1 + \cos u = 2\cos^2\frac{u}{2};$$

$$\cos^2\frac{\theta}{2} = \frac{a\,(1-e)\cos^2\frac{u}{2}}{r};$$

$$\sqrt{r}\sin\frac{\theta}{2} = \sqrt{a(1+e)}\sin\frac{u}{2}; \quad \sqrt{r}\cos\frac{\theta}{2} = \sqrt{a(1-e)}\cos\frac{u}{2};$$

$$\operatorname{tg}\frac{\theta}{2} = \sqrt{\frac{1+e}{1-e}}\operatorname{tg}\frac{u}{2}.$$

(b) To establish the geometric significance of the eccentric anomaly, u, using the drawing in Figure 4.2 (which represents, along with the ellipse of the satellite's orbit around the Earth, the circle whose center aligns with the center of the ellipse and whose radius is equal to the semi-major axis of the ellipse), it results that:

$$\text{OS}'' = a; \quad \text{OS}' = \text{OE} + \text{ES}' = a \cdot \cos\angle(\text{S}'\text{OS}'');$$

$$\text{OE} = a - r_{\min} = a - \frac{p}{1+e} = a - \frac{a\,(1-e^2)}{1+e} = ae;$$

$$\text{ES}' = r\cos\theta;$$

$$ae + r\cos\theta = a\cos\angle(\text{S}'\text{OS}'').$$

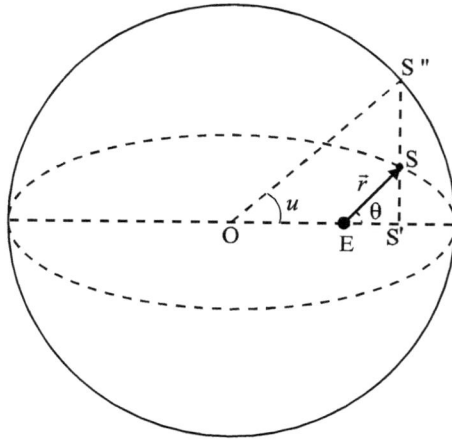

Fig. 4.2

Using the equation of the ellipse in polar coordinates, it results that:

$$r = \frac{p}{1 + e\cos\theta} = \frac{a\left(1 - e^2\right)}{1 + e\cos\theta};$$

$$1 + e\cos\theta = \frac{a\left(1 - e^2\right)}{r}; \quad e\cos\theta = \frac{a\left(1 - e^2\right)}{r} - 1;$$

$$\cos\theta = \frac{a\left(1 - e^2\right)}{er} - \frac{1}{e} = \frac{1}{r}\left[\frac{a\left(1 - e^2\right)}{e} - \frac{r}{e}\right];$$

$$\cos\theta = \frac{1}{r}\left(\frac{a}{e} - ae - \frac{r}{e}\right);$$

$$r\cos\theta + ae = \frac{a - r}{e};$$

$$ae + r\cos\theta = a\cos\left(S'\overset{3}{O}S''\right);$$

$$a - r = ae \cdot \cos\left(S'\overset{3}{O}S''\right);$$

$$a - r = ae\cos u;$$

$$u = \angle(S'OS'').$$

(c) A satellite that will revolve around the Earth in an elliptical orbit, whose elements (semi-axes, minimum altitude, maximum altitude, etc.) have been fixed according to the satellite's mission, can be inserted into the elliptical orbit at any point. It only needs to be raised to the altitude of the chosen "injection" point and to be given there the necessary speed (called the "injection speed") in the direction of the tangent to the trajectory at that point.

A carrier rocket helps transport the satellite from the Earth to the altitude of the injection point on the elliptical orbit so that on this sector of the trajectory (active trajectory), the movement of the ensemble is the result of the composition of the forces of reaction (traction) with the forces of gravitational attraction.

At the injection point, the satellite detaches from the carrier rocket. The higher the altitude of this point, the lower the injection speed that will have to be given to the satellite, as shown in Figure 4.3.

The first part of the active trajectory taken by the launch vehicle is approximately vertical to the launch site. It then curves according to a program established so that when the engine of the last reactive stage of the rocket stops working, both the injection point and the speed with which the satellite can continue moving on the ellipse have been reached.

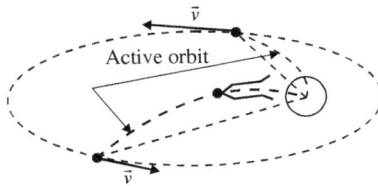

Fig. 4.3

If the satellite must be placed (injected) into an elliptical orbit with a minimum velocity, it must be raised to the apogee altitude. Could this also lead to the conclusion that the optimal solution from the energetic point of view is to insert the satellite into orbit via "injection" at the apogee, as it requires the minimum injection speed there? The injection of the satellite at the apogee, however, would require its transportation over the maximum distance, which would mean that the energy consumption exceeds the energy savings due to the injection point's advantages.

***Conclusion*:** The optimal solution from the energetic point of view is injection at the perigee, which assumes the maximum injection speed but offers the advantage of energy economy, because the satellite must be transported over the minimum distance.

This is how it is achieved in the case of elliptical orbits with a low perigee (Figure 4.4). The satellite is lifted with the help of a carrier rocket until the perigee. Then, it is injected into the elliptical orbit.

However, if the elliptical orbit on which the satellite will have to revolve (base orbit) has a high perigee, then, from the active orbit, the satellite is first inserted through "injection" into an elliptical orbit with a low perigee, whose apogee coincides with the perigee of the "base" orbit. This is called "parking" or "transfer".

The injection of the satellite into the "parking" orbit is done at its perigee (see point p in Figure 4.4), by giving the satellite the speed $\vec{v}_{max}$ corresponding to the parameters of the "parking" ellipse.

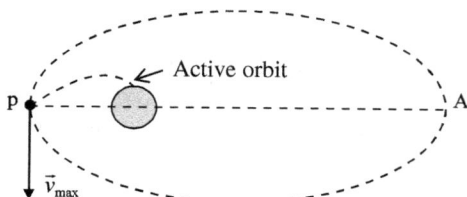

Fig. 4.4

Problem 5. Contact of Two Spherical Molecules

When two rigid spherical molecules, each with diameter d, come into close proximity, they cannot overlap. As a result, each molecule can be represented as a point surrounded by a protective sphere — a region that the center of another molecule cannot enter. This protective sphere defines a forbidden volume for the movement of other molecules: the space around one molecule that is inaccessible to the center of another due to their finite sizes.

Consequently, when many such rigid spherical molecules are introduced into a container, part of the container's volume becomes inaccessible to the centers of other molecules. This reduction in accessible volume arises from the finite size of the particles.

The average excluded volume per mole of molecules, accounting for all pairwise interactions, is known as the co-volume. This quantity corresponds to the parameter b in the van der Waals equation and represents the total volume effectively unavailable for molecular motion due to repulsive interactions. It should not be confused with the actual molar volume occupied by the molecules themselves.

(a) *Determine* the *co-volume* b of the gas in a vessel if there are N_A identical molecules, each a rigid sphere with a radius r.

(b) One of the equations for real gases is the van der Waals equation:

$$\left(p + \frac{a\nu}{V\nu^2}\right)(V - \nu b) = \nu RT,$$

where a and b are coefficients that determine the pressure and volume corrections. The dependence $p = f(V)$, whose graph is represented in Figure 5.1, for different values of T, can be extracted from it.

The graph has a point of inflection at point C, which is associated with the substance's critical state.

Determine, using two methods, the parameters of the critical state (p_c, V_c, T_c) and then *evaluate* the radius of the gas molecule.

(c) *Write* the van der Waals equation according to the reduced parameters (the reduced form of the van der Waals equation): $p_r = p/p_c$, $V_r = V/V_c$, $T_r = T/T_c$.

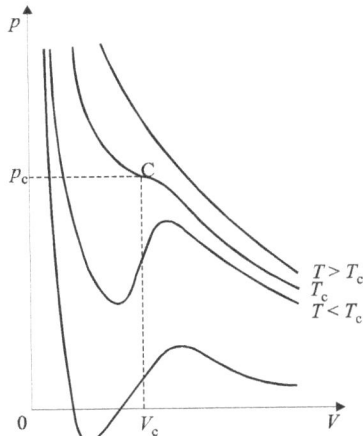

Fig. 5.1

Solution

(a) Let us consider the molecules as rigid spheres with radius r and diameter $d = 2r$. When two molecules collide, the spheres touch, and the distance between their centers equals d.

The distance between the centers of the molecules cannot be smaller than d. Then, we can consider the molecules reduced to material points located in the centers of the spheres and imagine that each material point has around it a sphere of protection with the radius d, which the center of another molecule cannot penetrate.

The volume of the protective sphere is a forbidden volume and has the value $V_{sp} = 4\pi d^3/3 = 32\pi r^3/3$. For a single molecule, the forbidden volume is $V_{sp}/2 = 16\pi r^3/3$.

Therefore, the forbidden (unavailable) volume for those N_A gas molecules placed in the container, i.e., the *co-volume* of the gas, $b = 16N_A\pi r^3/3$, represents the real volume taken up by those N spherical molecules.

Conclusion: The volume available for agitating the molecules is $V-b$, where V is the volume of the vessel.

(b)
Method 1
The critical point is an inflection point of the critical isotherm, and the tangent at this point is horizontal.

It results that:

$$\left(p + \frac{a\nu^2}{V^2}\right)(V - \nu b) = \nu RT;$$

$$p = \frac{\nu RT}{V - \nu b} - \frac{a\nu^2}{V^2};$$

$$\left(\frac{dp}{dV}\right)_C = \frac{-\nu RT_C}{(V_C - \nu b)^2} + \frac{2a\nu^2 V_C}{V_C^4} = 0;$$

$$\left(\frac{dp}{dV}\right)_C = \frac{-\nu RT_C}{(V_C - \nu b)^2} + \frac{2a\nu^2}{V_C^3} = 0;$$

$$\frac{\nu RT_C}{(V_C - \nu b)^2} = \frac{2a\nu^2}{V_C^3};$$

$$\left(\frac{d^2p}{dV^2}\right)_C = \frac{\nu RT_C 2(V_C - \nu b)}{(V_C - \nu b)^4} - \frac{2a\nu^2 3V_C^2}{V_C^6}0;$$

$$\frac{\nu RT_C}{(V_C - \nu b)^3} = \frac{3a\nu^2}{V_C^4};$$

$$V_C = 3\nu b; \quad T_C = \frac{8a}{27bR}; \quad p_C = \frac{a}{27b^2}.$$

Method 2

$$\left(p + \frac{a\nu^2}{V^2}\right)(V - \nu b) = \nu RT;$$

$$V^3 - \nu\left(b + \frac{RT}{p}\right)V^2 + \frac{a\nu^2}{p}V - \frac{ab\nu^3}{p} = 0.$$

For the critical point, the equation is written as follows:

$$V^3 - \nu\left(b + \frac{RT_C}{p_C}\right)V^2 + \frac{a\nu^2}{p_C}V - \frac{ab\nu^3}{p_C} = 0;$$

$$(V - V_C)^3 = 0.$$

From this, by identifying the coefficients, it results that:

$$\nu\left(b + \frac{RT_C}{p_C}\right) = 3V_C; \quad \frac{a\nu^2}{p_C} = 3V_C^2; \quad \frac{ab\nu^3}{p_C} = V_C^3;$$

$$V_C = 3\nu b; \quad T_C = \frac{8a}{27bR}; \quad p_C = \frac{a}{27b^2}; \quad b = \frac{RT_C}{8p_C};$$

$$b = 4N_A\frac{4\pi r^3}{3};$$

$$r = \frac{1}{4}\sqrt[3]{\frac{3RT_C}{2\pi N_A p_C}}.$$

(c)

$$\left(p + \frac{a\nu^2}{V^2}\right)(V - \nu b) = \nu RT;$$

$$b = \frac{V_C}{3\nu};$$

$$R = \frac{8}{3\nu}\frac{p_C V_C}{T_C};$$

$$a = \frac{3p_C V_C^2}{\nu^2};$$

$$\left(p + \frac{\nu^2}{V^2}\frac{3p_C V_C^2}{\nu^2}\right)\left(V - \nu\frac{V_C}{3\nu}\right) = \nu T\frac{8}{3\nu}\frac{p_C V_C}{T_C};$$

$$p_C\left(\frac{p}{p_C} + 3\frac{V_C^2}{V^2}\right)\frac{V_C}{3}\left(3\frac{V}{V_C} - 1\right) = \frac{8}{3}p_C V_C\frac{T}{T_C};$$

$$p_r = \frac{p}{p_C};$$

$$V_r = \frac{V}{V_C};$$

$$T_r = \frac{T}{T_C};$$

$$\left(p_r + \frac{3}{V_r^2}\right)(3V_r - 1) = 8T_r.$$

Problem 6. The Reunion of a Space Station with a Satellite

A space station, coupled to a satellite, revolves in a circular orbit with radius $1.25R$ around the Earth, where R is the Earth's radius. At a certain moment, the satellite is catapulted from the station in the direction of the tangent to the circle. The satellite will continue its movement on an elliptical orbit, with the apogee at a distance of $10R$ from the center of the Earth. The station and the satellite are considered as material points.

If m is the satellite's mass and M is the space station's mass, then determine the value of the ratio m/M for which the satellite meets the station after one satellite rotation around the Earth.

Given: $10^{2/3} = 4.64$, $11^{2/3} = 4.94$.

Solution

In Figure 6.1, the circular trajectory of the system before disconnection and the elliptical trajectories of the system elements after their disconnection are represented: 1 — the trajectory of the satellite with mass m; 2 — the trajectory of the carrier rocket with mass M.

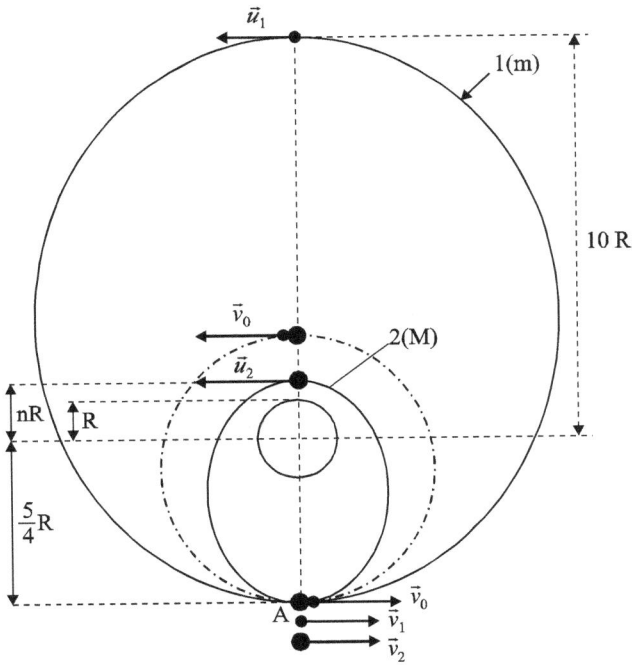

Fig. 6.1

If T_1 and T_2 are the rotation periods of the satellite and the rocket, respectively, then the reunion of the elements at point A, after a rotation of the satellite, is possible if

$$\frac{T_1}{T_2} = k > 1,$$

where k must be an integer.

With the semi-axes of the two ellipses being

$$a_1 = \frac{1}{2}\left(\frac{5}{4}R + 10R\right) = \frac{45}{8}R, \quad a_2 = \frac{1}{2}\left(\frac{5}{4}R + nR\right) = \frac{5+4n}{8}R,$$

from Kepler's third law, it results that:

$$\frac{T_1^2}{T_2^2} = \frac{a_1^3}{a_2^3}; \quad \frac{T_1}{T_2} = \left(\frac{45}{5+4n}\right)^{\frac{3}{2}} = k.$$

The disconnection of the system elements proceeds according to the law of conservation of momentum:

$$(M+m)\vec{v}_0 = m\vec{v}_1 + M\vec{v}_2, (M+m)v_0 = mv_1 + Mv_2.$$

Here, v_0 is the speed of the system on the circular path:

$$v_0 = 2\sqrt{\frac{g_0 R}{5}}; g_0 = K\frac{M^2}{R^2}, \text{ where } M_E \text{ is the Earth's mass.}$$

The evolution of the elements after disconnection follows the laws of conservation of mechanical energy and kinetic momentum.

At a certain moment, the gravitational potential energies of the satellite–Earth and rocket–Earth systems are as follows:

$$E_{p1} = mg_0\frac{R^2}{r_1}; \quad E_{p2} = -Mg_0\frac{R^2}{r_2};$$

$$\frac{mv_1^2}{2} - mg_0\frac{R^2}{\frac{5R}{4}} = \frac{mu_1^2}{2} - mg_0\frac{R^2}{10R};$$

$$\frac{Mv_2^2}{2} - Mg_0\frac{R^2}{\frac{5}{4}R} = \frac{Mu_2^2}{2} - Mg_0\frac{R^2}{nR};$$

$$\frac{5}{4}v_1 = 10u_1; \quad \frac{5}{4}v_2 = nu_2; \quad v_1 = \frac{4v_0}{3}; \quad v_2 = 2v_0\sqrt{\frac{2n}{4n+5}};$$

$$n = 5\left(1 - \frac{m}{3M}\right)^2 : 4\left(2 - \left(1 - \frac{m}{3M}\right)^2\right).$$

By imposing the condition that the carrier rocket does not fall to Earth $(n < 1)$, it results that:

$$\frac{m}{M} < 3 - 2\sqrt{2};$$

$$\frac{m}{M} = 3 - \sqrt{2\left(9 - k^{2/3}\right)}; \quad k < 5^{3/2} = 11.2;$$

$$\frac{m}{M} > 0; \quad k > 9.5;$$

$$9.5 < k < 11.2.$$

Because k must be an integer, its possible values are $k_1 = 10$ and $k_2 = 11$.

It results that:

$$\left(\frac{m}{M}\right)_1 = 3 - \sqrt{2(9 - 10^{2/3})} = 0.047;$$

$$\left(\frac{m}{M}\right)_2 = 3 - \sqrt{2(9 - 11^{2/3})} = 0.150.$$

Chapter 2

International Pre-Olympic Physics Contest 2001, Craiova, Romania

Problem 1. Satellite Outside the Earth's Atmosphere

A carrier rocket transports a satellite outside the Earth's atmosphere. It is launched at a point Q, at the distance r from the center of the Earth, so that the movement with respect to the Earth (considered fixed) follows a parabola arc, as shown in Figure 1.1, with the parameter p, having the Earth in the focus of the parabola.

(a) *Determine* the elements of the vector, representing the velocity, $\vec{v}$, of the satellite with respect to the Earth at the point of injection on the parabolic trajectory.

   Given: K, the constant of universal attraction; and M, the mass of the Earth.

   The direction of the axis of the parabola is fixed, and it is known.

(b) *Write* the equations that would allow one to establish the time dependences of the satellite's plane polar coordinates θ and r, knowing that the trajectory of a material point in the field of the central gravitational force, in general, is a conic whose equation, in plane polar coordinates, is

$$r = \frac{p}{1 + e \cos \theta},$$

   where e is the numerical eccentricity of the conic.

(c) Considering that the satellite is injected into the trajectory at point A, representing the vertex of the parabola, *determine* the distance traveled by it to point Q, at which $\theta = 2\pi/3$, knowing that:

$$\int \frac{dx}{\cos^3 x} = \frac{\sin x}{2\cos^2 x} + \frac{1}{2}\ln\left|\text{tg}\left(\frac{\pi}{4} + \frac{x}{2}\right)\right|;$$

$$\text{tg}\frac{5\pi}{12} = 2 + \sqrt{3}; \quad \ln(2 + \sqrt{3}) = 1.3.$$

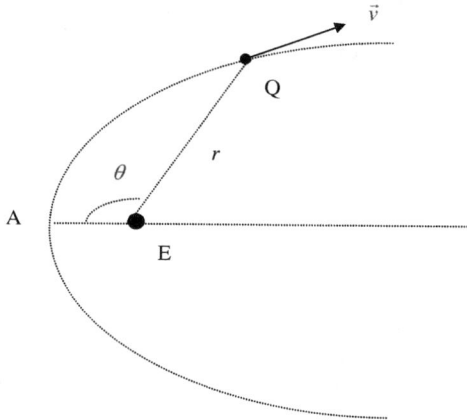

Fig. 1.1

Solution

A *parabola* is the geometric locus of points in a plane at equal distances from a fixed point called the focus and a fixed line called the directrix.

A satellite evolves on a parabolic trajectory, with the Earth in its focus. The satellite must escape from the terrestrial gravitational field and reach somewhere very far away, and its speed with respect to the Earth should be null.

Suppose a satellite is prepared, by calculation, to escape from the terrestrial gravitational field on the parabola represented in Figure 1.2, whose focus is the Earth. Its equation in Cartesian coordinates (x, y) is $y^2 = 2px$, for which the parameter of the parabola, $p = 2r_{min}$, is known.

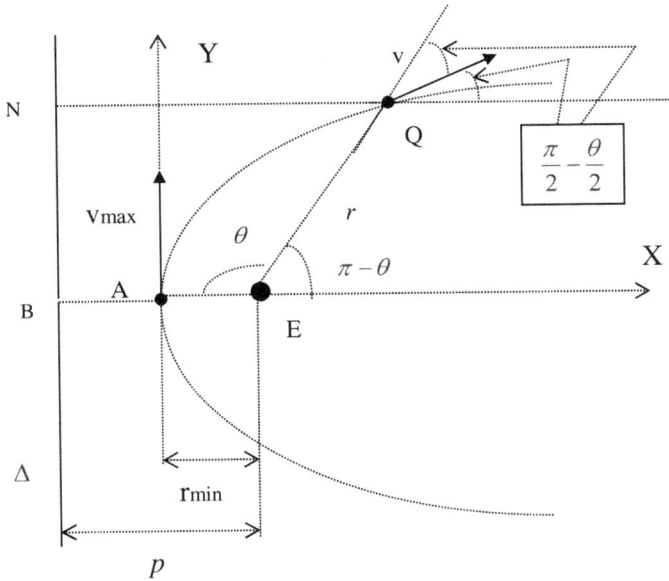

Fig. 1.2

To achieve such an escape, the satellite is first raised with the help of a carrier rocket to the altitude of the chosen injection point (Q), and there, it is given the velocity $\vec{v}$, called the injection velocity, on a tangent to the parabola, so that the total mechanical energy of the satellite–Earth system is

$$E = \frac{mv^2}{2} - K\frac{mM}{r} = 0.$$

When the escape is successful, and the satellite arrives very far from the Earth ($r \to \infty$), it is at rest with respect to the Earth ($v_\infty = 0$).

The optical property of the parabola is demonstrated: all light rays emitted from the focus of a concave parabolic mirror, after reflection, become parallel to the principal optical axis, and reciprocally, the incident light rays parallel to the principal optical axis are reflected through the focus.

As a result, the tangent to the parabola at point Q is the bisector of the angle EQN.

Considering the definition of the parabola, it follows that:

$$QE = QN;$$

$$r = EB + QP \cos(\pi - \theta);$$

$$r = 2r_{\min} + r(-\cos\theta);$$

$$2r_{\min} = r(1 + \cos\theta);$$

$$r_{\min} = r\cos^2\frac{\theta}{2}.$$

If the satellite injection were to be done at the point where $r = r_{\min}$, the injection speed would have to be $v = v_{\max}$, so that we would have:

$$E = \frac{m}{v_{\max}^2 2\frac{mM}{r_{\min}}} v_{\max} = \sqrt{\frac{2KM}{r_{\min}}};$$

$$r_{\min}v_{\max} = rv\sin\left(\frac{\pi}{2} - \frac{\theta}{2}\right) = rv\cos\frac{\theta}{2};$$

$$r\cos^2\frac{\theta}{2}v_{\max} = rv\cos\frac{\theta}{2};$$

$$v_{\max} = \frac{v}{\cos\frac{\theta}{2}}.$$

From the equation of the trajectory (conic) written in polar coordinates, for $e = 1$ (parabola), it results that:

$$r = \frac{p}{1 + \cos\theta}; \quad r = \frac{p}{2\cos^2\frac{\theta}{2}} = \frac{p}{2}\left(1 + \text{tg}^2\frac{\theta}{2}\right).$$

For the evolution on the parabola, taking place under the action of the central force of gravitational attraction, we have:

$$r^2\theta = C = \sqrt{pKM};$$

$$r^2 d\theta = C dt; \quad \frac{p^2}{4}\left(1 + \text{tg}^2\frac{\theta}{2}\right)^2 d\theta = C dt;$$

$$\text{tg}\frac{\theta}{2} = u; \quad \frac{1}{2}\frac{d\theta}{\cos^2\frac{\theta}{2}} = du;$$

$$d\theta = 2\cos^2\frac{\theta}{2}du = \frac{2}{1+\text{tg}^2\frac{\theta}{2}}du;$$

$$d\theta = \frac{2}{1+u^2}du;$$

$$\frac{p^2}{2C}(1+u^2)du = dt;$$

$$t - t_0 = \frac{p^2}{2C}\int_0^u (1+u^2)du,$$

where t_0 is the time at which the satellite passes through the point corresponding to r_{min}, for which $\theta = 0$ and $u = 0$. Thus, t is the moment when the satellite's polar coordinates are r and θ:

$$t - t_0 = \frac{p^2}{2C}\left(u + \frac{u^3}{3}\right);$$

$$\text{tg}\frac{\theta}{2} + \frac{1}{3}\text{tg}^3\frac{\theta}{2} = \frac{2\sqrt{pKM}}{p^2}(t - t_0),$$

from the solution of which we can deduce $\theta = f(t)$.

Then, from the equation of the parabola, $r = \frac{p}{2\cos^2\frac{\theta}{2}}$, we deduce the dependence $r = f(t)$.

The distance traveled by the satellite between the two points on the parabolic trajectory is:

$$S = \int v\,dt = v_{max}\int \cos\frac{\theta}{2}dt;$$

$$dt = \frac{r^2 d\theta}{C}; \quad r^2 = \frac{p^2}{4\cos^4\frac{\theta}{2}}; \quad C = \sqrt{pKM};$$

$$v_{max} = 2\sqrt{\frac{KM}{p}};$$

$$dt = \frac{p^2}{4\cos^4\frac{\theta}{2}\sqrt{pKM}}d\theta;$$

$$S = \frac{p}{2} \int_0^{2\pi/3} \frac{d\theta}{\cos^3 \frac{\theta}{2}}; \quad x = \frac{\theta}{2};$$

$$S = p \int_0^{\frac{\pi}{3}} \frac{dx}{\cos^3 x} = \left[\frac{\sin x}{2\cos^2 x} + \frac{1}{2}\ln \left| \text{tg} \left(\frac{\pi x}{4} + \frac{x}{2} \right) \right| \right]_0^{\frac{\pi}{3}};$$

$$S = p \left(\sqrt{3} + \frac{1}{2}\ln \frac{\text{tg} \frac{5\pi}{12}}{\text{tg} \frac{\pi}{4}} \right);$$

$$S = p \left[\sqrt{3} + \frac{1}{2}\ln \left(2 + \sqrt{3} \right) \right];$$

$$S \approx 2.4 \cdot p.$$

Problem 2. Space Probe Leaving the Solar System

In a cosmic project, two options for launching a space probe from Earth so that the probe leaves the solar system are discussed. In the first variant (I), it is proposed that the probe be launched with a sufficiently high speed to leave the solar system directly. In the second variant (II), the probe would first approach Mars (a planet further from the Sun than the Earth) and then, thanks to this planet, change its direction of movement and reach the necessary speed to leave the solar system.

It will be assumed that the probe moves so that, at each point of its trajectory, it is either only under the action of the Sun's gravitational field or only under Mars's gravitational field. At any point, the Sun's or Mars's gravitational field is more intense.

(a) *Establish* the elements of the vector $\vec{v}_a$, representing the minimum speed that the probe should have at the moment of launch in relation to the Earth, so that it can leave the solar system in the first variant (I). It will be admitted that the Earth's orbit in relation to the Sun is circular.

(b) Considering that the probe would be launched in the previously established direction but at a speed whose modulus should be $v_b < v_a$, *determine* the modulus of the components (normal, $\vec{v}_n$, and tangential, $\vec{v}_t$) of the speed vector of the probe in relation to the Sun, corresponding to the moment when the probe's trajectory will intersect the trajectory of Mars in relation to the Sun. The orientations of the $\vec{v}_n$ and $\vec{v}_t$ components are considered relative to Mars's circular orbit in relation to the Sun. At that moment, the probe will be considered very far from Mars.

(c) Let us now admit that, intersecting the orbit of Mars, the probe enters the region where the intensity of the gravitational field of the planet Mars is greater than the intensity of the Sun's gravitational field. Corresponding to this moment, *determine* the relationship between the speed of the probe in relation to the planet Mars, $\vec{v}$, and the components of the speed of the probe in relation to the Sun, $\vec{v}_n$ and $\vec{v}_t$, determined at point (b), when Mars is far away from the probe. During the interaction with the probe, the exact position of the planet Mars is not important. The component $\vec{v}_t$ and the vector $\vec{v}_M$ representing the speed of the planet Mars can be considered two colinear vectors.

Determine the minimum speed that the space probe should have at the moment of launching, with respect to the Earth, $v_{b,min}$, so that the probe can leave the solar system via the second variant (II). It is useful to specify that from point (a), we know the value and optimal orientation of the probe's speed upon launch from Earth so that the probe leaves the solar system directly.

Given: the speed of the Earth in relation to the Sun, $v_p = 30\,\text{km/s}$; and the ratio of the radii of the circular orbits of the planets Earth and Mars, respectively, r_E/r_M. The air resistance, the rotational movement of the Earth, and the energy consumed to detach from the Earth are neglected. The orbits of the planets Earth and Mars are coplanar.

Solution

(a) In accordance with the notation in Figure 2.1 (S — Sun, E — Earth, s — space probe, $\vec{v}_E$ — the speed of the Earth with respect to the Sun, $\vec{v}_a$ — the speed of the probe in relation to the Earth at the time of launch, r_E — the radius of the circular orbit of the Earth), the results for variant I are:

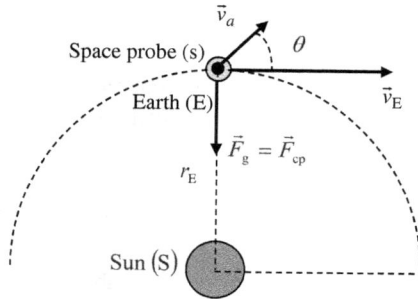

Fig. 2.1

$$K\frac{M_{\text{E}}M_{\text{S}}}{r_{\text{E}}^2} = \frac{M_{\text{E}}v_{\text{E}}^2}{r_{\text{E}}}; \quad K\frac{M_{\text{S}}}{r_{\text{E}}} = v_{\text{E}}^2,$$

where M_{E} is the mass of the Earth, M_{S} is the mass of the Sun, and K is the constant of universal attraction;

$$v_{\text{E}}^2 = K\frac{M_{\text{S}}}{r_{\text{E}}};$$

$$E_i = E_{p,i} + E_{k,i},$$

representing the total energy of the space probe–Sun system, at the time of the launch of the space probe;

$$E_i = -K\frac{m_{\text{s}}M_{\text{S}}}{r_{\text{E}}} + \frac{m_{\text{s}}v_{\text{s}}^2}{2},$$

where m_{s} is the mass of the space probe, and $\vec{v}_{\text{s}}$ is the speed of the space probe with respect to the Sun at the time of the launch of the space probe;

$$\vec{v}_{\text{s}} = \vec{v}_{\text{P}} + \vec{v}_a;$$

$$v_{\text{s}}^2 = v_{\text{E}}^2 + v_a^2 + 2v_{\text{E}}v_a\cos\theta,$$

whose maximum value will be achieved if the launch is in the direction $\theta = 0$ (on the tangent to the trajectory of the Earth around the Sun, in the direction of the movement of the Earth's revolution), so that the vectors $\vec{v}_a$ and $\vec{v}_{\text{E}}$ have identical orientations;

$$v_{\text{s}}^2 = v_{\text{E}}^2 + v_a^2 + 2v_{\text{E}}v_a = (v_a + v_{\text{E}})^2; \quad v_{\text{s}} = v_{\text{E}} + v_a;$$

$$E_i = -K \frac{m_s M_S}{r_E} + \frac{m_s (v_a + v_E)^2}{2};$$

$$E_i = -K \frac{m_s M_S}{r_E} + \frac{m_s (v_a + v_E)^2}{2};$$

$$E_i = -m_s v_E^2 + \frac{m_s (v_a + v_E)^2}{2};$$

$$E_f = E_{pf} + E_{cf},$$

representing the total energy of the probe–Sun system, after the probe has left the solar system;

$$E_f = 0;$$

$$E_i = E_f,$$

according to the law of conservation of the energy of the system,

$$-m_s v_E^2 + \frac{m_s (v_a + v_E)^2}{2} = 0;$$

$$v_a^2 + 2 v_E v_a - v_E^2 = 0;$$

$$v_a = (\sqrt{2} - 1) v_E,$$

representing the required speed of the probe, relative to the Earth, at the moment of launch so that, according to variant I, the probe, when launched from the Earth, exits the gravitational field of the Sun and leaves the solar system;

$$v_a \approx 12.4 \ \text{km/s};$$

$$v_s = v_E + v_a;$$

$$v_s = \sqrt{2} v_E = \sqrt{2 \frac{K M_s}{r_E}},$$

representing the required speed of the probe, with respect to the Sun, at the moment of launch from the Earth in the direction of the tangent to the Earth's orbit, in the direction of the movement of the Earth's revolution, so that, according to variant I, the probe, when

launched from the Earth, exits the gravitational field of the Sun and
leaves the solar system, on a trajectory in the shape of a parabola,
with the Sun at the focus, as indicated in Figure 2.2, the minimum
distance from the Sun being r_E.

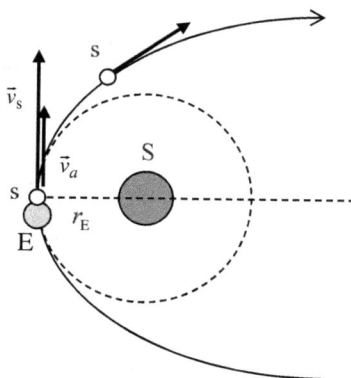

Fig. 2.2

(b) Figure 2.3 shows the coplanar circular orbits of the planets Earth
and Mars in relation to the Sun, as well as the trajectory of the probe,
represented by a sector of an ellipse, from the moment of launch from
Earth (with relative speed $\vec{v}_b$ in the most favorable direction, $v_b < v_a$)
until the moment of intersection with the orbit of Mars, when the
components of the velocity vector of the probe in relation to the
Sun, $\vec{u}$, with respect to the circular orbit of Mars are, respectively,
$\vec{v}_n$ and $\vec{v}_t$:

$$\vec{u} = \vec{v}_n + \vec{v}_t.$$

From position 1 and up to position 2, evolving on an arc of an
ellipse, in whose close focus is the Sun ($v_S < v_E$), the probe's trajectory is in the region of the Sun's gravitational field.

According to the laws of conservation of kinetic momentum and
total mechanical energy, written for the moment of the launch of the
space probe and for the moment when the trajectory of the space
probe intersects the orbit of Mars, it follows that:

$$\vec{L}_i = m_s \vec{r}_E \times \vec{v}_s = m_s \vec{r}_E \times (\vec{v}_b + \vec{v}_E) \, ;$$

$$\vec{L}_f = m_s \vec{r}_M \times \vec{v}_s = m_s \vec{r}_M \times (\vec{v}_n + \vec{v}_t) \, ;$$

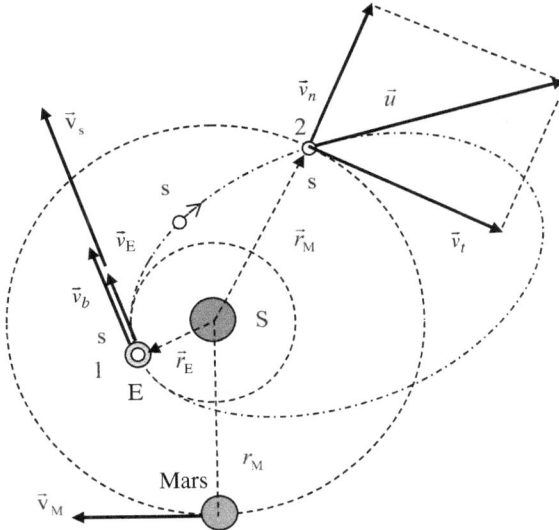

Fig. 2.3

$$\vec{L}_i = \vec{L}_f;$$

$$r_{\mathrm{E}} (v_b + v_{\mathrm{E}}) = r_{\mathrm{M}} v_t; \quad v_t = (v_b + v_{\mathrm{E}}) \frac{r_{\mathrm{E}}}{r_{\mathrm{M}}};$$

$$E_{\mathrm{i}} = -K \frac{m_{\mathrm{s}} M_{\mathrm{S}}}{r_{\mathrm{E}}} + \frac{m_{\mathrm{s}}(v_b + v_{\mathrm{E}})^2}{2};$$

$$E_f = -K \frac{m_{\mathrm{s}} M_{\mathrm{S}}}{r_{\mathrm{M}}} + \frac{m_{\mathrm{s}}(v_n^2 + v_t^2)}{2};$$

$$E_i = E_f;$$

$$-K \frac{m_{\mathrm{s}} M_{\mathrm{s}}}{r_{\mathrm{E}}} + \frac{m_{\mathrm{s}} (v_b + v_{\mathrm{E}})^2}{2} = -K \frac{m_{\mathrm{s}} M_{\mathrm{s}}}{r_{\mathrm{M}}} + \frac{m_{\mathrm{s}} \left(v_n^2 + v_t^2\right)}{2};$$

$$K \frac{M_{\mathrm{s}}}{r_{\mathrm{E}}} = v_{\mathrm{E}}^2; \quad K \frac{M_{\mathrm{s}}}{r_{\mathrm{M}}} = K \frac{M_{\mathrm{s}}}{r_{\mathrm{E}}} \frac{r_{\mathrm{E}}}{r_{\mathrm{M}}} = v_{\mathrm{E}}^2 \frac{r_{\mathrm{E}}}{r_{\mathrm{M}}};$$

$$v_t = (v_b + v_{\mathrm{E}}) \frac{r_{\mathrm{E}}}{r_{\mathrm{M}}};$$

$$-v_E^2 + \frac{(v_b + v_E)^2}{2} = -v_E^2 \frac{r_E}{r_M} + \frac{\left(v_n^2 + v_t^2\right)}{2};$$

$$-v_E^2 + \frac{(v_b + v_E)^2}{2} = -v_E^2 \frac{r_E}{r_M} + \frac{1}{2}v_n^2 + \frac{1}{2}(v_b + v_E)^2 \frac{r_E^2}{r_M^2};$$

$$v_n = \sqrt{(v_b + v_E)^2 \left(1 - \frac{r_E^2}{r_M^2}\right) - 2v_E^2 \left(1 - \frac{r_E}{r_M}\right)}.$$

(c) In variant II, before leaving the solar system, the space probe first leaves the region of the Sun's gravitational field and enters the region of the gravitational field of the planet Mars. Then, leaving the region of the gravitational field of Mars, the space probe leaves the solar system. This means that, in the region of the gravitational field of Mars, the trajectory of the space probe in relation to Mars is a parabola (open trajectory), having the center of the planet Mars in its focus, as indicated in Figure 2.4.

Let $\vec{v}$ be the speed of the probe relative to the planet Mars, corresponding to the moment when the trajectory of the probe intersects the orbit of Mars and enters the region of the gravitational field of Mars. At that time, as noted previously, the speed of the probe relative to the Sun is

$$\vec{u} = \vec{v}_n + \vec{v}_t$$

Using the result from (a), when we demonstrated that, for the probe to exit the region of the Sun's gravitational field, directly leaving the solar system, the speed of the probe relative to Earth upon launching from Earth must be

$$v_a = (\sqrt{2} - 1)v_E,$$

we can write that, for the probe to leave the region of the gravitational field of the planet Mars, where it arrived according to variant II, and then leave the solar system, its speed relative to the planet Mars must be

$$v = (\sqrt{2} - 1)v_M,$$

where v_M is the speed of Mars on its orbit around the Sun;

$$v_M = \sqrt{K \frac{M_S}{r_M}} = v_E \sqrt{\frac{r_E}{r_M}};$$

$$\vec{v} = v_\perp + \vec{v}_{//},$$

where $v_\perp$ and $\vec{v}_{//}$ are the components of the space probe's velocity relative to the planet Mars.

As a result of the collinearity of the vectors $\vec{v}_t$ and $\vec{v}_M$, the interaction with the planet Mars does not change the $\vec{v}_n$ component of the probe's velocity.

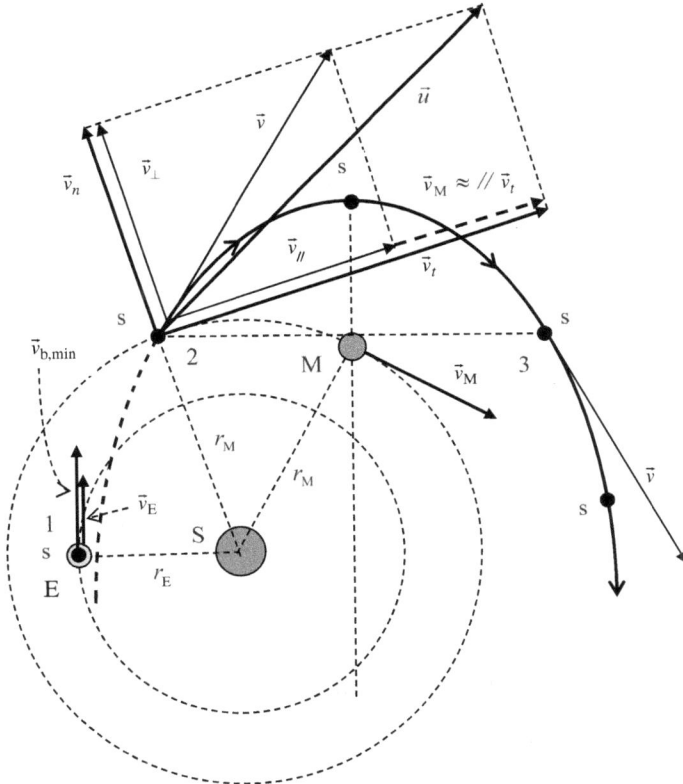

Fig. 2.4

It results that:

$$\vec{v}_\perp = \vec{v}_n; \quad v_\perp = v_n;$$

$$\vec{v}_{//} = \vec{v}_t - \vec{v}_M; \quad v_{//} = v_t - v_M;$$

$$v^2 = v_n^2 + (v_t - v_M)^2;$$

$$v_t = (v_b + v_E)\frac{r_E}{r_M};$$

$$v_n = \sqrt{(v_b + v_E)^2 \left(1 - \frac{r_E^2}{r_M^2}\right) - 2v_E^2 \left(1 - \frac{r_E}{r_M}\right)};$$

$$v = \left(\sqrt{2} - 1\right) v_M; \quad v_M = v_E\sqrt{\frac{r_E}{r_M}};$$

$$\left(\sqrt{2} - 1\right)^2 v_E^2 \frac{r_E}{r_M} = (v_b + v_E)^2 \left(1 - \frac{r_E^2}{r_M^2}\right) - 2v_E^2 \left(1 - \frac{r_E}{r_M}\right)$$
$$+ \left[(v_b + v_E)\frac{r_E}{r_M} - v_E\sqrt{\frac{r_E}{r_M}}\right]^2;$$

$$(v_b + v_E)^2 - 2v_E\frac{r_E}{r_M}\sqrt{\frac{r_E}{r_M}}(v_b + v_E) - 2v_E^2 \left(1 - \sqrt{2}\frac{r_E}{r_M}\right) = 0;$$

$$v_b + v_E = v_E\frac{r_E}{r_M}\sqrt{\frac{r_E}{r_M}} \pm \sqrt{v_E^2\frac{r_E^2}{r_M^2}\frac{r_E}{r_M} + 2v_E^2 \left(1 - \sqrt{2}\frac{r_E}{r_M}\right)};$$

$$v_b = v_E \left[\frac{r_E}{r_M}\sqrt{\frac{r_E}{r_M}} + \sqrt{\frac{r_E^3}{r_M^3} - 2\sqrt{2}\frac{r_E}{r_M} + 2} - 1\right] = v_{b,\min};$$

$$v_{b,\min} \approx 5.5 \text{ km/s}.$$

Problem 3. Oscillations of a Liquid Column

Figure 3.1 shows a U-shaped cylindrical tube with inner diameter d, which is fixed in a vertical position, closed at one end and open at the other. An air cushion is formed due to a column of water with length L. In the equilibrium state of the system, the level difference

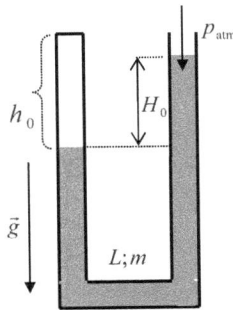

Fig. 3.1

between the two branches of the liquid column is H_o, and the height of the air cushion is h_0.

Due to some disturbance, the water column in the tube starts to oscillate, though the surface of the water in each branch of the tube remains horizontal. Friction and surface tension forces are neglected. It is known that during oscillations, the length of the water column, L, remains constant, and ρ is the density of the water.

(a) *Find* the differential equation of the oscillations of the liquid column, considering that the thermodynamic process evolves isothermally or adiabatically.

Then, *determine* the periods of the small oscillations of the water column, corresponding to the two thermodynamic processes.

The adiabatic exponent of the air (γ), the atmospheric pressure (p_{atm}), the mass of the liquid column in the tube (m), and the gravitational acceleration (g) are known.

(b) *Establish* the time dependence of the air pressure in the tube if the water column oscillations are small and have amplitude A.

Determine the maximum and minimum values of the air pressure in the tube.

(c) When the air pressure in the tube is maximum, z_{max} air molecules pass into the atmosphere through a hole in the wall of the tube per unit time. *Determine* the number of molecules that will pass into the atmosphere per unit time if the hole is opened when the pressure in the tube is minimum, z_{min}. The molecules are considered identical.

Solution

(a)

(1) From diagram (a) in Figure 3.2, which shows the equilibrium state of the liquid column in the tube, corresponding to the initial moment, it follows that:

$$G_0 + p_{atm}S = p_0S; \quad m_0g + p_{atm}S = p_0S; \quad \rho H_0 Sg + p_{atm}S = p_0S;$$

$$p_0 = p_{atm} + \rho g H_0,$$

representing the initial pressure of the air cushion in the tube.

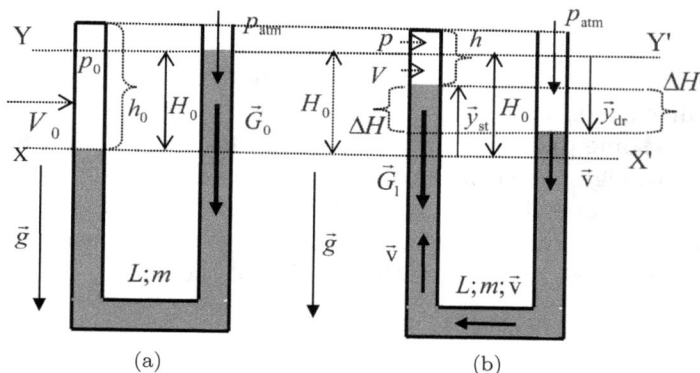

Fig. 3.2

As a result of some disturbance caused by pressure briefly exerted on the surface of the column on the right, the liquid in the tube is uneven, as indicated by diagram (b) in Figure 3.2. Then, the pressure is released. From this moment, the liquid in the tube will return to its initial state. It will pass through the tube, stop, and then start to oscillate.

Diagram (b) of Figure 3.2 shows the column of liquid in the tube at some moment during its oscillations, when the liquid level in the column to the right of the tube drops, having the instantaneous velocity $\vec{v}$ and the instantaneous elongation $\vec{y}_{dr}$ measured relative to its initial level, YY'. In the same diagram, the column of liquid on the left of the tube is also shown, which rises with the instantaneous velocity $\vec{v}$ and the instantaneous elongation $\vec{y}_{st}$ from its initial reference level, XX', where $\vec{y}_{st} = -\vec{y}_{dr}$ and $y_{st} = y_{dr} = y$.

In these conditions, the instantaneous level difference between the free surfaces of the liquid in the two columns, corresponding to the considered moment and highlighted in the same figure, is:

$$\Delta H = y_{dr} - (H_0 - y_{st}) = y_{dr} + y_{st} - H_0; \quad H = 2y - H_0.$$

$$y + h = h_0; \quad y = h_0 - h.$$

Corresponding to the same moment, when the disturbance has disappeared, the result of the forces acting on the entire column of liquid in the tube, opposing the further increase in the level difference of the liquid in the two columns, is

$$\vec{F} = \vec{G}_1 + \vec{F}_p + \vec{F}_{p_\text{atm}},$$

where: $\vec{G}_1$ is the weight of the column of liquid on the left of the tube, having the height ΔH, representing the level difference between the two columns at the considered moment; $\vec{F}_p$ is the pressure force exerted by the air cushion, with pressure p, located in the left-hand column, on the liquid in the tube; and $\vec{F}_{p_\text{atm}}$ is the pressure force exerted by the atmospheric pressure on the liquid in the column on the right of the tube. As a result, this force, relative to the right-hand column, is oriented vertically upwards, so that:

$$F = G_1 + F_p - F_{p_\text{atm}};$$

$$F = m_1 g + pS - p_\text{atm};$$

$$F = \rho \Delta H S g + pS - p_\text{atm} S; \quad \Delta H = 2y - H_0;$$

$$F = \rho(2y - H_0)Sg + (p - p_\text{atm})S.$$

Thus, considering the isothermal evolution of the air cushion, it follows that:

$$pV = p_0 V_0; \quad pSh = p_0 S h_0; \quad ph = p_0 h_0; \quad h = h_0 - y;$$

$$p(h_0 - y) = p_0 h_0; \quad p = \frac{p_0 h_0}{h_0 - y}; \quad p_0 = p_\text{atm} + \rho g H_0;$$

$$p = \frac{(p_\text{atm} + \rho g H_0) h_0}{h_0 - y};$$

$$F = \rho \left(2y - H_0\right) Sg + \left(\frac{\left(p_{\text{atm}} + \rho g H_0\right) h_0}{h_0 - y} - p_{\text{atm}}\right) S;$$

$$F = \rho \left(2y - H_0\right) Sg + \left(\frac{\left(p_{\text{atm}} + \rho g H_0\right) h_0 - p_{\text{atm}} \left(h_0 - y\right)}{h_0 - y}\right) S;$$

$$F = \rho \left(2y - H_0\right) Sg + \left(\frac{p_{\text{atm}} h_0 + \rho g H_0 h_0 - p_{\text{atm}} h_0 + p_{\text{atm}} y}{h_0 - y}\right) S;$$

$$F = \rho \left(2y - H_0\right) Sg + \left(\frac{\rho g H_0 h_0 + p_{\text{atm}} y}{h_0 - y}\right) S;$$

$$F = \frac{\rho \left(2y - H_0\right) \left(h_0 - y\right) g + \rho g H_0 h_0 + p_{\text{atm}} y}{h_0 - y} S;$$

$$F = \frac{2\rho y h_0 g - 2\rho y^2 g - \rho H_0 h_0 g + \rho H_0 y g + \rho g H_0 h_0 + p_{\text{atm}} y}{h_0 - y} S;$$

$$F = \frac{2\rho y h_0 g - 2\rho y^2 g + \rho H_0 y g + p_{\text{atm}} y}{h_0 - y} S;$$

$$F = \frac{2\rho y g \left(h_0 - y\right) + y \left(\rho H_0 g + p_{\text{atm}}\right)}{h_0 - y} S;$$

$$F = 2\rho g S y + \frac{y \left(\rho H_0 g + p_{\text{atm}}\right)}{h_0 - y} S;$$

$$F = 2\rho g S y + \frac{y \left(\rho H_0 g + p_{\text{atm}}\right)}{h_0 \left(1 - \frac{y}{h_0}\right)} S;$$

$$F = 2\rho g S y + \frac{y}{h_0} \frac{\left(\rho H_0 g + p_{\text{atm}}\right)}{\left(1 - \frac{y}{h_0}\right)} S;$$

$$\frac{1}{1 - \frac{y}{h_0}} = \left(1 - \frac{y}{h_0}\right)^{-1}; \quad \frac{y}{h_0} \ll 1; \quad \frac{1}{1 - \frac{y}{h_0}} \approx \left(1 - (-1)\frac{y}{h_0}\right);$$

$$\frac{1}{1 - \frac{y}{h_0}} = \left(1 + \frac{y}{h_0}\right);$$

$$F = 2\rho g S y + \left(\rho H_0 g + p_{\text{atm}}\right) S \frac{y}{h_0} \left(1 + \frac{y}{h_0}\right);$$

$$F = 2\rho g S y + (\rho H_0 g + p_{\text{atm}})\, S \left(\frac{y}{h_0} + \frac{y^2}{h_0^2} \right);$$

$$y \ll h_0; \quad \Rightarrow \quad \frac{y^2}{h_0^2} \ll \frac{y}{h_0}; \quad \Rightarrow \quad \frac{y}{h_0} + \frac{y^2}{h_0^2} \approx \frac{y}{h_0};$$

$$F = 2\rho g S y + (\rho H_0 g + p_{\text{atm}})\, S \frac{y}{h_0};$$

$$F = \left(2\rho g + \frac{\rho H_0 g + p_{\text{atm}}}{h_0} \right) S y;$$

$$k = \left(2\rho g + \frac{\rho H_0 g + p_{\text{atm}}}{h_0} \right) S; \quad S = \frac{\pi d^2}{4}; \quad F = k y; \quad \vec{F} = -k \vec{y}_{dr};$$

$$\vec{F} = -k \vec{y},$$

which proves that, in isothermic conditions, the oscillations of the liquid column in the tube are harmonic;

$$k = m\omega^2 = m \frac{4\pi^2}{T^2}; \quad T = 2\pi \sqrt{\frac{m}{k}};$$

$$T = 2\pi \sqrt{\frac{m}{\left(2\rho g + \frac{\rho H_0 g + p_{\text{atm}}}{h_0} \right) S}} \, T = 2\pi \sqrt{\frac{m h_0}{(2\rho g h_0 + \rho g H_0 + p_{\text{atm}})\, S}},$$

representing the period of the harmonic oscillations of the liquid column in the U-tube in isothermic conditions.

(2) If the process of the air cushion is adiabatic, it results that:

$$p V^\gamma = p_0 V_0^\gamma;$$

$$p = p_0 \left(\frac{V_0}{V} \right)^\gamma = p_0 \left(\frac{h_0}{h_0 - y} \right)^\gamma = p_0 \frac{h_0^\gamma}{h_0^\gamma} \left(\frac{1}{1 - \frac{y}{h_0}} \right)^\gamma = p_0 \left(1 - \frac{y}{h_0} \right)^{-\gamma};$$

$$\left(1 - \frac{y}{h_0} \right)^{-\gamma} \approx 1 - (-\gamma) \frac{y}{h_0} = 1 + \gamma \frac{y}{h_0};$$

$$p = p_0 \left(1 + \gamma \frac{y}{h_0} \right);$$

$$F = \rho\left(2y - H_0\right)Sg + \left(p - p_{\text{atm}}\right)S;$$

$$F = \rho\left(2y - H_0\right)Sg + \left(p_0 + \gamma p_0 \frac{y}{h_0} - p_{\text{atm}}\right)S;$$

$$p_0 = p_{\text{atm}} + \rho g H_0; \quad p_0 - p_{\text{atm}} = \rho g H_0;$$

$$F = \rho\left(2y - H_0\right)Sg + \left(\rho g H_0 + \gamma p_0 \frac{y}{h_0}\right)S;$$

$$F = \rho\left(2y - H_0\right)Sg + \rho g H_0 S + \gamma p_0 \frac{y}{h_0}S;$$

$$F = 2\rho Sgy - \rho H_0 Sg + \rho g H_0 S + \gamma p_0 \frac{S}{h_0}y;$$

$$F = 2\rho Sgy + \gamma p_0 \frac{S}{h_0}y;$$

$$F = \left(2\rho g + \gamma \frac{p_0}{h_0}\right)Sy; \quad p_0 = p_{\text{atm}} + \rho g H_0;$$

$$F = \left(2\rho g + \gamma \frac{p_{\text{atm}} + \rho g H_0}{h_0}\right)Sy;$$

$$\left(2\rho g + \gamma \frac{p_{\text{atm}} + \rho g H_0}{h_0}\right)S = k;$$

$$F = ky; \quad \vec{F} = -k\vec{y},$$

which proves that, in the adiabatic scheme, the oscillations of the liquid column in the U-tube are also harmonic;

$$k = m\omega^2 = m\frac{4\pi^2}{T^2}; \quad T = 2\pi\sqrt{\frac{m}{k}};$$

$$T = 2\pi\sqrt{\frac{m}{\left(2\rho g + \frac{\rho H_0 g + p_{\text{atm}}}{h_0}\right)S}} \quad T = 2\pi\sqrt{\frac{m h_0}{\left(2\rho g h_0 + \rho g H_0 + p_{\text{atm}}\right)S}}.$$

(3) In order to establish the differential equation of the oscillations of the liquid column in the U-tube, we demonstrate that the force under the action of which the liquid column oscillates in adiabatic conditions due to the air cushion is given by the expression:

$$F = \left(2\rho g + \frac{\rho H_0 g + p_{atm}}{h_0}\right) Sy;$$

$$k = \left(2\rho g + \frac{\rho H_0 g + p_{atm}}{h_0}\right) S; \quad S = \frac{\pi d^2}{4}; \quad F = ky; \quad \vec{F} = -k\vec{y}_{dr};$$

$$\vec{F} = -k\vec{y}.$$

Thus, it results that:

$$F = ky; \quad F = ma = m\frac{d}{dt}\left(\frac{dy}{dt}\right) = m\frac{d^2 y}{dt^2} = ky;$$

$$F = ma = m\frac{d}{dt}\left(\frac{dy}{dt}\right) = m\frac{d^2 y}{dt^2} = ky; \quad \frac{k}{m} = \omega^2;$$

$$\frac{d^2 y}{dt^2} = \omega^2 y; \quad \frac{d^2 y}{dt^2} - \omega^2 y = 0,$$

representing the differential equation of the oscillations of the liquid column, a homogeneous differential equation of the second degree, whose general solution is

$$y = A\sin\omega t.$$

(b) If h_0 is the initial height of the air cushion in the tube, and if A is the amplitude of the oscillations of the liquid level in the tube, then the minimum and maximum heights, respectively, of the air cusion in the tube during the oscillations of the liquid column are given by the expressions

$$h_{min} = h_0 - A; \quad h_{max} = h_0 + A.$$

These heights of the air cushion correspond to the air cushion pressures p_{max} and p_{min}, respectively. In the case of the isothermal process, it results that:

$$p_0 V_0 = p_{max} V_{min}; p_{max} = p_0 \frac{V_0}{V_{min}} = p_0 \frac{Sh_0}{Sh_{min}} = p_0 \frac{h_0}{h_0 - A};$$

$$p_{\text{max, isotherm}} = p_0 \frac{h_0}{h_0 - A}; \quad p_0 = p_{\text{atm}} + \rho g H_0;$$

$$p_0 V_0 = p_{\text{min}} V_{\text{max}}; \quad p_{\text{min}} = p_0 \frac{V_0}{V_{\text{max}}} = p_0 \frac{S h_0}{S h_{\text{max}}} = p_0 \frac{h_0}{h_0 + A};$$

$$p_{\text{min, isotherm}} = p_0 \frac{h_0}{h_0 + A}; \quad p_0 = p_{\text{atm}} + \rho g H_0.$$

In the case of the adiabatic process, it results that:

$$p_0 V_0^{\gamma} = p_{\text{max}} V_{\text{min}}^{\gamma};$$

$$p_{\text{max}} = p_0 \left(\frac{V_0}{V_{\text{min}}} \right)^{\gamma} = p_0 \left(\frac{S h_0}{S h_{\text{min}}} \right)^{\gamma} = p_0 \left(\frac{h_0}{h_0 - A} \right)^{\gamma};$$

$$p_{\text{max, adiabatic}} = p_0 \left(\frac{h_0}{h_0 - A} \right)^{\gamma}; \quad p_0 = p_{\text{atm}} + \rho g H_0;$$

$$p_0 V_0^{\gamma} = p_{\text{min}} V_{\text{max}}^{\gamma};$$

$$p_{\text{min}} = p_0 \left(\frac{V_0}{V_{\text{max}}} \right)^{\gamma} = p_0 \left(\frac{S h_0}{S h_{\text{max}}} \right)^{\lambda} = p_0 \left(\frac{h_0}{h_0 + A} \right)^{\gamma};$$

$$p_{\text{min, adiabatic}} = p_0 \left(\frac{h_0}{h_0 + A} \right)^{\gamma}; \quad p_0 = p_{\text{atm}} + \rho g H_0.$$

(c) According to the simplified model of an ideal gas, the gas molecules in a given container move in three mutually perpendicular directions (in both directions in each direction) with the thermal speed v_{thermic}. So, in a sense, only 1/6 of the total number of molecules in the container will move in a certain direction. If n is the concentration of gas molecules in the air cushion, after a time $\Delta t = t_2 - t_1$, through the opening in the wall of the tube where the air bag cushion is located, a sector with the surface area ΔS, the molecules in a cylindrical column with the surface area ΔS and height $v_{\text{thermic}} \cdot \Delta t$.

The following expression gives the number of these molecules:

$$\Delta N = \frac{1}{6} n \cdot \Delta S \cdot v_{\text{thermic}} \cdot \Delta t; \quad v_{\text{thermic}} = \sqrt{\frac{3kT}{m_0}},$$

where $n = \frac{N}{V}$ is the concentration of gas molecules in the air cushion, k is Boltzmann's constant, m_0 is the mass of a gas molecule, and T is the absolute temperature of the gas in the air cushion.

(1) We will first analyze the variation in the oscillations of the liquid column in the tube when the air cushion below the liquid in the left branch of the tube has an adiabatic evolution.

Considering that the gas molecules can only move in three mutually perpendicular directions, with identical speeds equal to the average thermal speeds, it follows that only 1/6 of the gas molecules will move towards the opening in the tube wall with area ΔS.

Consider the number of gas molecules that escape from the tube through the opening in the tube wall with surface area ΔS during the time interval Δt, when, in its adiabatic evolution, the gas pressure is maximum, p_{max}. So, in its adiabatic evolution, when the gas temperature is maximum, T_{max}, and the thermal speeds of the gas molecules are maximum, v_{max}, we have the expression:

$$N_{max} = \frac{1}{6} n_{max} \cdot v_{max} \cdot \Delta t \cdot \Delta S = \frac{1}{6} n_{max} \cdot \sqrt{\frac{3kT_{max}}{m_0}} \cdot \Delta t \cdot \Delta S,$$

where $n_{max} = \frac{N}{V_{min}}$ is the concentration of gas molecules under the specified conditions, N is the total number of gas molecules, and m_0 is the mass of a gas molecule;

$$N_{max} = \frac{1}{6} \frac{N}{V_{min}} \cdot \sqrt{\frac{3kT_{max}}{m_0}} \cdot \Delta t \cdot \Delta S.$$

The number of gas molecules that leave the tube through the opening with the surface area ΔS during the time interval Δt, when, in its adiabatic evolution, the pressure of the gas is minimal, p_{min}; the gas temperature is minimal, T_{min}; and the thermal velocities of the gas molecules are minimal, v_{min}, is given by the following expression:

$$N_{min} = \frac{1}{6} n_{min} \cdot v_{min} \cdot \Delta t \cdot \Delta S = \frac{1}{6} n_{min} \cdot \sqrt{\frac{3kT_{min}}{m_0}} \cdot \Delta t \cdot \Delta S;$$

$$n_{min} = \frac{N}{V_{max}};$$

$$N_{min} = \frac{1}{6} \frac{N}{V_{max}} \cdot \sqrt{\frac{3kT_{min}}{m_0}} \cdot \Delta t \cdot \Delta S.$$

Under these conditions, it results that:

$$\frac{N_{\min}}{N_{\max}} = \frac{\frac{1}{6} \cdot \frac{N}{V_{\max}} \cdot \sqrt{\frac{3kT_{\min}}{m_0}} \cdot \Delta t \cdot \Delta S}{\frac{1}{6} \cdot \frac{N}{V_{\min}} \cdot \sqrt{\frac{3kT_{\max}}{m_0}} \cdot \Delta t \cdot \Delta S};$$

$$\frac{N_{\min}}{N_{\max}} = \frac{V_{\min}}{V_{\max}} \cdot \sqrt{\frac{T_{\min}}{T_{\max}}};$$

$$p_{\max} V_{\min} = vRT_{\max}; \quad p_{\min} V_{\max} = vRT_{\min};$$

$$\frac{p_{\min} V_{\max}}{p_{\max} V_{\min}} = \frac{T_{\min}}{T_{\max}}; \quad V_{\max} = S h_{\max} = S(h_0 + A);$$

$$V_{\min} = S h_{\min} = S(h_0 - A);$$

$$\frac{V_{\min}}{V_{\max}} = \frac{h_0 - A}{h_0 + A};$$

$$\frac{p_{\min} S(h_0 + A)}{p_{\max} S(h_0 - A)} = \frac{T_{\min}}{T_{\max}} = \frac{p_{\min}(h_0 + A)}{p_{\max}(h_0 - A)};$$

$$\frac{N_{\min}}{N_{\max}} = \frac{h_0 - A}{h_0 + A} \cdot \sqrt{\frac{p_{\min}(h_0 + A)}{p_{\max}(h_0 - A)}};$$

$$p_{\min,\,\text{adiabatic}} = p_0 \left(\frac{h_0}{h_0 + A}\right)^{\gamma};$$

$$p_{\max,\,\text{adiabatic}} = p_0 \left(\frac{h_0}{h_0 - A}\right)^{\gamma};$$

$$\frac{N_{\min}}{N_{\max}} = \frac{h_0 - A}{h_0 + A} \cdot \sqrt{\frac{p_0 \left(\frac{h_0}{h_0 + A}\right)^{\gamma}(h_0 + A)}{p_0 \left(\frac{h_0}{h_0 - A}\right)^{\gamma}(h_0 - A)}};$$

$$\frac{N_{\min}}{N_{\max}} = \frac{h_0 - A}{h_0 + A} \cdot \sqrt{\frac{\left(\frac{1}{h_0 + A}\right)^{\gamma - 1}}{\left(\frac{1}{h_0 - A}\right)^{\gamma - 1}}};$$

$$\frac{N_{\min}}{N_{\max}} = \sqrt{\left(\frac{h_0 - A}{h_0 + A}\right)^{\gamma-1}};$$

$$N_{\min} = N_{\max} \cdot \frac{h_0 - A}{h_0 + A} \cdot \sqrt{\left(\frac{h_0 - A}{h_0 + A}\right)^{\gamma-1}};$$

$$N_{\min} = N_{\max} \cdot \sqrt{\left(\frac{h_0 - A}{h_0 + A}\right)^{2}\left(\frac{h_0 - A}{h_0 + A}\right)^{\gamma-1}};$$

$$N_{\min} = N_{\max} \cdot \sqrt{\left(\frac{h_0 - A}{h_0 + A}\right)^{\gamma+1}};$$

$$\frac{N_{\min}}{\Delta t} = \frac{N_{\max}}{\Delta t} \cdot \sqrt{\left(\frac{h_0 - A}{h_0 + A}\right)^{\gamma+1}};$$

$$z_{\min} = z_{\max} \cdot \sqrt{\left(\frac{h_0 - A}{h_0 + A}\right)^{\gamma+1}} < z_{\max}.$$

(2) Let us now analyze the variation in the oscillations of the liquid column in the tube when the air cushion above the liquid located in the left branch of the tube has an isothermal evolution. The number of gas molecules leaving the tube through the opening in the tube wall with the surface area ΔS, during the time interval Δt, when, in its isothermal evolution, the gas pressure is maximum, $p_{\max}$; the temperature of the gas is constant, T; and, therefore, the thermal velocities of the gas molecules are constant, v, is given by the expression

$$N_{\max} = \frac{1}{6} \cdot n_{\max} \cdot v \cdot \Delta t \cdot \Delta S = \frac{1}{6} \cdot n_{\max} \cdot \sqrt{\frac{3kT}{m_0}} \cdot \Delta t \cdot \Delta S,$$

where $n_{\max} = \frac{N}{V_{\min}}$ is the concentration of gas molecules under the specified conditions, N is the total number of gas molecules, and m_0 is the mass of a gas molecule;

$$N_{\max} = \frac{1}{6}\frac{N}{V_{\min}} \cdot \sqrt{\frac{3kT}{m_0}} \cdot \Delta t \cdot \Delta S; \quad p_{\max}V_{\min} = vRT;$$

$$V_{\min} = \frac{vRT}{p_{\max}};$$

$$N_{\max} = \frac{1}{6} \frac{N \cdot p_{\max}}{vRT} \cdot \sqrt{\frac{3kT}{m_0}} \cdot \Delta t \cdot \Delta S.$$

The number of gas molecules that leave the tube through the opening in the tube wall with the surface area ΔS during the time interval Δt, when, in its isothermal evolution, the gas pressure is minimum, $p_{\min}$; the gas temperature is constant, T; and the thermal velocities of the gas molecules are constant, v, is given by the following expression:

$$N_{\min} = \frac{1}{6} n_{\min} \cdot v \cdot \Delta t \cdot \Delta S = \frac{1}{6} n_{\min} \cdot \sqrt{\frac{3kT}{m_0}} \cdot \Delta t \cdot \Delta S; \quad n_{\min} = \frac{N}{V_{\max}};$$

$$N_{\min} = \frac{1}{6} \frac{N}{V_{\max}} \cdot \sqrt{\frac{3kT}{m_0}} \cdot \Delta t \cdot \Delta S; \quad p_{\min} V_{\max} = vRT; \quad V_{\max} = \frac{vRT}{p_{\min}};$$

$$N_{\min} = \frac{1}{6} \frac{N \cdot p_{\min}}{vRT} \cdot \sqrt{\frac{3kT}{m_0}} \cdot \Delta t \cdot \Delta S.$$

Under these conditions, it results that:

$$\frac{N_{\min}}{N_{\max}} = \frac{\frac{1}{6} \cdot \frac{N \cdot p_{\min}}{vRT} \cdot \sqrt{\frac{3kT}{m_0}} \cdot \Delta t \cdot \Delta S}{\frac{1}{6} \cdot \frac{N \cdot p_{\max}}{vRT} \cdot \sqrt{\frac{3kT}{m_0}} \cdot \Delta t \cdot \Delta S}; \quad \frac{N_{\min}}{N_{\max}} = \frac{p_{\min}}{p_{\max}};$$

$$p_{\max} V_{\min} = vRT; \quad p_{\min} V_{\max} = vRT;$$

$$\frac{p_{\min} V_{\max}}{p_{\max} V_{\min}} = 1; \quad \frac{p_{\min}}{p_{\max}} = \frac{V_{\min}}{V_{\max}};$$

$$\frac{N_{\min}}{N_{\max}} = \frac{V_{\min}}{V_{\max}}; \quad V_{\max} = S h_{\max} = S(h_0 + A);$$

$$V_{\min} = S h_{\min} = S(h_0 - A);$$

$$\frac{V_{\min}}{V_{\max}} = \frac{h_0 - A}{h_0 + A}; \quad \frac{N_{\min}}{N_{\max}} = \frac{h_0 - A}{h_0 + A}; \quad N_{\min} = N_{\max} \cdot \frac{h_0 - A}{h_0 + A};$$

$$\frac{N_{\min}}{\Delta t} = \frac{N_{\max}}{\Delta t} \cdot \frac{h_0 - A}{h_0 + A}; \quad z_{\min} = z_{\max} \cdot \frac{h_0 - A}{h_0 + A} < z_{\max}.$$

Problem 4. Liquid in Communicating Vessels

In a system of two cylindrical vessels located on the same horizontal support with cross-sections S_1 and $S_2 < S_1$, which communicate at the base through an orthogonal cylindrical tube with a very small cross-section, having in its middle a closed tap, R, there is a liquid with density ρ. The heights of the liquid columns in the two vessels are h_{01} and $h_{02} < h_{01}$, respectively, as shown in Figure 4.1.

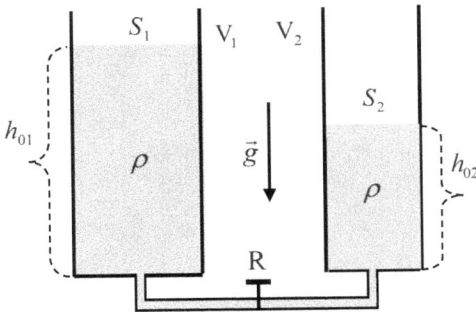

Fig. 4.1

At any moment, t, after opening the valve R, considered as the initial moment, $t = 0$, the mass of the liquid that passes from vessel V_1 to vessel V_2 through the very narrow connecting tube per unit time is directly proportional to the level difference of the liquid in the two vessels corresponding to that moment, t, so:

$$\frac{dm}{dt} = k\,(h_1 - h_2),$$

where k is a constant of proportionality, and $h_1 = h_1(t)$ and $h_2 = h_2(t)$ are the heights of the liquid columns in the two vessels corresponding to the moment t, considered from the moment the tap R is opened.

Determine the time dependences for:

(a) the difference in the heights of the liquid columns in the two vessels, $\Delta h(t) = h_1(t) - h_2(t)$, specifying the result for $t = 0$ and for $t \to \infty$;

(b) the height of the liquid column, $h_1(t)$, in the vessel V_1, specifying the result for $t = 0$ and for $t \to \infty$;

(c) the height of the liquid column, $h_2(t)$, in the vessel V_2, specifying the result for $t = 0$ and for $t \to \infty$.

Solution

(a) From moment t, when the heights of the liquid columns in the two vessels are $h_1(t) < h_{01}$ and $h_2(t) > h_{02}$, respectively, and until the moment $t + dt$, when the heights of the liquid columns in the two vessels are $h_1(t) - dh_1$ and $h_2(t) + dh_2$, respectively, the mass of the liquid that passes from vessel V_1 to vessel V_2 is:

$$dm = \rho S_1 \left[h_1(t) - (h_1(t) - dh_1) \right] = \rho S_2 \left[(h_2(t) + dh_2) - h_2(t) \right];$$

$$dm = -\rho S_1 \cdot dh_1 = \rho S_2 \cdot dh_2.$$

Hence, it results that:

$$-S_1 \cdot dh_1 = S_2 \cdot dh_2; \quad dh_2 = -\frac{S_1}{S_2} \cdot dh_1;$$

$$dh_1 - dh_2 = dh_1 + \frac{S_1}{S_2} \cdot dh_1 = \left(1 + \frac{S_1}{S_2}\right) \cdot dh_1 = \frac{S_1 + S_2}{S_2} \cdot dh_1;$$

$$dh_1 - dh_2 = \frac{S_1 + S_2}{S_1 \cdot S_2} \cdot S_1 \cdot dh_1;$$

$$\frac{dm}{dt} = k(h_1 - h_2); \quad dm = k(h_1 - h_2) \cdot dt;$$

$$dm = -\rho S_1 \cdot dh_1 = \rho S_2 \cdot dh_2;$$

$$k\left(h_1 - h_2\right) \cdot dt = -\rho S_1 \cdot dh_1 = \rho S_2 \cdot dh_2;$$

$$S_1 \cdot dh_1 = -\frac{k}{\rho} \cdot \left[h_1\left(t\right) - h_2\left(t\right)\right] \cdot dt;$$

$$dh_1 - dh_2 = \frac{S_1 + S_2}{S_1 \cdot S_2} \cdot S_1 \cdot dh_1;$$

$$dh_1 - dh_2 = -\frac{k}{\rho}\frac{S_1 + S_2}{S_1 \cdot S_2} \cdot \left[h_1\left(t\right) - h_2\left(t\right)\right] \cdot dt;$$

$$\frac{dh_1 - dh_2}{h_1\left(t\right) - h_2\left(t\right)} = -\frac{k}{\rho}\frac{S_1 + S_2}{S_1 \cdot S_2} \cdot dt;$$

$$\frac{d\left[h_1\left(t\right) - h_2\left(t\right)\right]}{h_1\left(t\right) - h_2\left(t\right)} = -\frac{k}{\rho}\frac{S_1 + S_2}{S_1 \cdot S_2} \cdot dt;$$

$$\int \frac{d\left[h_1\left(t\right) - h_2\left(t\right)\right]}{h_1\left(t\right) - h_2\left(t\right)} = -\frac{k}{\rho} \cdot \frac{S_1 + S_2}{S_1 S_2} \int dt;$$

$$\ln\left[h_1\left(t\right) - h_2\left(t\right)\right] = -\frac{k}{\rho} \cdot \frac{S_1 + S_2}{S_1 S_2} \cdot t + \ln C_1;$$

$$\ln\left[h_1\left(t\right) - h_2\left(t\right)\right] - \ln C_1 = -\frac{k}{\rho} \cdot \frac{S_1 + S_2}{S_1 S_2} \cdot t;$$

$$h_1\left(t\right) - h_2\left(t\right) = C_1 \cdot \exp\left(-\frac{k}{\rho} \cdot \frac{S_1 + S_2}{S_1 S_2} \cdot t\right);$$

$$t = 0; \quad h_1\left(t = 0\right) = h_{01}; \quad h_2\left(t = 0\right) = h_{02};$$

$$C_1 = h_{01} - h_{02};$$

$$h_1\left(t\right) - h_2\left(t\right) = \left(h_{01} - h_{02}\right) \cdot \exp\left(-\frac{k}{\rho} \cdot \frac{S_1 + S_2}{S_1 S_2} \cdot t\right),$$

representing the level difference of the liquid in the two vessels after the time t from the opening of tap R. From this, for $t = 0$, it results that:

$$h_1\left(t = 0\right) - h_2\left(t = 0\right) = \left(h_{01} - h_{02}\right) \cdot \exp\left(-\frac{k}{\rho} \cdot \frac{S_1 + S_2}{S_1 S_2} \cdot 0\right);$$

$$h_{01} - h_{02} = \left(h_{01} - h_{02}\right) \cdot e^0; \quad h_{01} - h_{02} = \left(h_{01} - h_{02}\right).$$

For $t \to \infty$, it results that:

$$h_1(t) - h_2(t) = (h_{01} - h_{02}) \cdot \exp\left(-\frac{k}{\rho} \cdot \frac{S_1 + S_2}{S_1 S_2} \cdot t\right);$$

$$h_1(t) - h_2(t) = (h_{01} - h_{02}) \cdot \frac{1}{\exp\left(\frac{k}{\rho} \cdot \frac{S_1 + S_2}{S_1 S_2} \cdot t\right)};$$

$$h_1(t \to \infty) - h_2(t \to \infty) = (h_{01} - h_{02}) \cdot \frac{1}{\exp\left(\frac{k}{\rho} \cdot \frac{S_1 + S_2}{S_1 S_2} \cdot \infty\right)};$$

$$h_1(t \to \infty) - h_2(t \to \infty) = (h_{01} - h_{02}) \cdot 0;$$

$$h_1(t \to \infty) - h_2(t \to \infty) = 0;$$

$$h_1(t \to \infty) = h_2(t \to \infty).$$

(b)

$$-S_1 \cdot dh_1 = S_2 \cdot dh_2;$$

$$-S_1 \int dh_1 = S_2 \int dh_2; \quad -S_1 \cdot h_1(t) = S_2 \cdot h_2(t) + C_2;$$

$$t = 0; \quad h_1(t = 0) = h_{01}; \quad h_2(t = 0) = h_{02};$$

$$C_2 = -S_1 \cdot h_{01} - S_2 \cdot h_{02};$$

$$-S_1 \cdot h_1(t) - S_2 \cdot h_2(t) = -S_1 \cdot h_{01} - S_2 \cdot h_{02};$$

$$S_1 \cdot h_1(t) + S_2 \cdot h_2(t) = S_1 \cdot h_{01} + S_2 \cdot h_{02};$$

$$h_1(t) - h_2(t) = (h_{01} - h_{02}) \cdot \exp\left(-\frac{k}{\rho} \cdot \frac{S_1 + S_2}{S_1 S_2} \cdot t\right);$$

$$S_2 \cdot h_1(t) - S_2 \cdot h_2(t) = S_2 \cdot (h_{01} - h_{02})$$

$$\cdot \exp\left(-\frac{k}{\rho} \cdot \frac{S_1 + S_2}{S_1 S_2} \cdot t\right);$$

$$S_2 \cdot h_1(t) - S_2 \cdot h_2(t) + S_1 \cdot h_1(t) + S_2 \cdot h_2(t)$$

$$= S_2 \cdot (h_{01} - h_{02}) \cdot \exp\left(-\frac{k}{\rho} \cdot \frac{S_1 + S_2}{S_1 S_2} \cdot t\right)$$

$$+ S_1 \cdot h_{01} + S_2 \cdot h_{02};$$

$$S_2 \cdot h_1(t) + S_1 \cdot h_1(t) = S_2 \cdot (h_{01} - h_{02}) \cdot \exp\left(-\frac{k}{\rho} \cdot \frac{S_1 + S_2}{S_1 S_2} \cdot t\right)$$

$$+ S_1 \cdot h_{01} + S_2 \cdot h_{02};$$

$$h_1(t) \cdot (S_1 + S_2) = S_2 \cdot (h_{01} - h_{02}) \cdot \exp\left(-\frac{k}{\rho} \cdot \frac{S_1 + S_2}{S_1 S_2} \cdot t\right)$$

$$+ S_1 \cdot h_{01} + S_2 \cdot h_{02};$$

$$h_1(t) = \frac{S_2}{S_1 + S_2} \cdot (h_{01} - h_{02}) \cdot \exp\left(-\frac{k}{\rho} \cdot \frac{S_1 + S_2}{S_1 S_2} \cdot t\right)$$

$$+ \frac{S_1 \cdot h_{01}}{S_1 + S_2} + \frac{S_2 \cdot h_{02}}{S_1 + S_2};$$

$$h_1(t) = \frac{S_1 \cdot h_{01}}{S_1 + S_2} + \frac{S_2}{S_1 + S_2}$$

$$\times \left[h_{02} + (h_{01} - h_{02}) \cdot \exp\left(-\frac{k}{\rho} \cdot \frac{S_1 + S_2}{S_1 S_2} \cdot t\right)\right],$$

representing the height of the liquid column in vessel V_1, after the time t. From this, the height of the liquid column at $t = 0$ is:

$$h_1(t = 0) = \frac{S_1 \cdot h_{01}}{S_1 + S_2} + \frac{S_2}{S_1 + S_2}$$

$$\times \left[h_{02} + (h_{01} - h_{02}) \cdot \exp\left(-\frac{k}{\rho} \cdot \frac{S_1 + S_2}{S_1 S_2} \cdot 0\right)\right];$$

$$h_1(t = 0) = \frac{S_1 \cdot h_{01}}{S_1 + S_2} + \frac{S_2}{S_1 + S_2}\left[h_{02} + (h_{01} - h_{02})\right];$$

$$h_1(t = 0) = \frac{S_1 \cdot h_{01}}{S_1 + S_2} + \frac{S_2 \cdot h_{01}}{S_1 + S_2}; \quad h_1(t = 0) = h_{01}.$$

For $t \to \infty$, it results that:

$$h_1(t) = \frac{S_1 \cdot h_{01}}{S_1 + S_2} + \frac{S_2}{S_1 + S_2}$$

$$\times \left[h_{02} + (h_{01} - h_{02}) \cdot \exp\left(-\frac{k}{\rho} \cdot \frac{S_1 + S_2}{S_1 S_2} \cdot t \right) \right];$$

$$h_1(t) = \frac{S_1 \cdot h_{01}}{S_1 + S_2} + \frac{S_2}{S_1 + S_2}$$

$$\times \left[h_{02} + (h_{01} - h_{02}) \cdot \frac{1}{\exp\left(\frac{k}{\rho} \cdot \frac{S_1 + S_2}{S_1 S_2} \cdot t \right)} \right];$$

$$h_1(t \to \infty) = \frac{S_1 \cdot h_{01}}{S_1 + S_2} + \frac{S_2}{S_1 + S_2}$$

$$\times \left[h_{02} + (h_{01} - h_{02}) \cdot \frac{1}{\exp\left(\frac{k}{\rho} \cdot \frac{S_1 + S_2}{S_1 S_2} \cdot \infty \right)} \right];$$

$$h_1(t \to \infty) = \frac{S_1 \cdot h_{01}}{S_1 + S_2} + \frac{S_2}{S_1 + S_2} [h_{02} + (h_{01} - h_{02}) \cdot 0];$$

$$h_1(t \to \infty) = \frac{S_1 \cdot h_{01}}{S_1 + S_2} + \frac{S_2 \cdot h_{02}}{S_1 + S_2};$$

$$h_1(t \to \infty) = \frac{S_1 \cdot h_{01} + S_2 \cdot h_{02}}{S_1 + S_2} < h_{01}.$$

(c)

$$h_1(t) - h_2(t) = (h_{01} - h_{02}) \cdot \exp\left(-\frac{k}{\rho} \cdot \frac{S_1 + S_2}{S_1 S_2} \cdot t \right);$$

$$h_2(t) = h_1(t) - (h_{01} - h_{02}) \cdot \exp\left(-\frac{k}{\rho} \cdot \frac{S_1 + S_2}{S_1 S_2} \cdot t \right);$$

$$h_1\left(t\right) = \frac{S_1 \cdot h_{01}}{S_1 + S_2} + \frac{S_2}{S_1 + S_2}$$

$$\times \left[h_{02} + \left(h_{01} - h_{02}\right) \cdot \exp\left(-\frac{k}{\rho} \cdot \frac{S_1 + S_2}{S_1 S_2} \cdot t\right)\right];$$

$$h_2\left(t\right) = \frac{S_1 \cdot h_{01}}{S_1 + S_2} + \frac{S_2}{S_1 + S_2}$$

$$\times \left[h_{02} + \left(h_{01} - h_{02}\right) \cdot \exp\left(-\frac{k}{\rho} \cdot \frac{S_1 + S_2}{S_1 S_2} \cdot t\right)\right]$$

$$- \left(h_{01} - h_{02}\right) \cdot \exp\left(-\frac{k}{\rho} \cdot \frac{S_1 + S_2}{S_1 S_2} \cdot t\right);$$

$$h_2\left(t\right) = \frac{S_1 \cdot h_{01}}{S_1 + S_2} + \frac{S_2 \cdot h_{02}}{S_1 + S_2}$$

$$+ \left[\frac{S_2}{S_1 + S_2} \cdot \left(h_{01} - h_{02}\right) \cdot \exp\left(-\frac{k}{\rho} \cdot \frac{S_1 + S_2}{S_1 S_2} \cdot t\right)\right]$$

$$- \left(h_{01} - h_{02}\right) \cdot \exp\left(-\frac{k}{\rho} \cdot \frac{S_1 + S_2}{S_1 S_2} \cdot t\right);$$

$$h_2\left(t\right) = \frac{S_1 \cdot h_{01} + S_2 \cdot h_{02}}{S_1 + S_2} + \left(h_{01} - h_{02}\right)$$

$$\cdot \exp\left(-\frac{k}{\rho} \cdot \frac{S_1 + S_2}{S_1 S_2} \cdot t\right) \cdot \left(\frac{S_2}{S_1 + S_2} - 1\right);$$

$$h_2\left(t\right) = \frac{S_1 \cdot h_{01} + S_2 \cdot h_{02}}{S_1 + S_2} + \left(h_{01} - h_{02}\right)$$

$$\cdot \exp\left(-\frac{k}{\rho} \cdot \frac{S_1 + S_2}{S_1 S_2} \cdot t\right) \cdot \left(\frac{S_2 - S_1 - S_2}{S_1 + S_2}\right);$$

$$h_2\left(t\right) = \frac{S_1 \cdot h_{01} + S_2 \cdot h_{02}}{S_1 + S_2} - \left(h_{01} - h_{02}\right) \cdot \frac{S_1}{S_1 + S_2}$$

$$\cdot \exp\left(-\frac{k}{\rho} \cdot \frac{S_1 + S_2}{S_1 S_2} \cdot t\right),$$

representing the height of the liquid column in vessel V_2, after the time t. From this, the height of the liquid column at $t = 0$ is:

$$h_2\,(t = 0) = \frac{S_1 \cdot h_{01} + S_2 \cdot h_{02}}{S_1 + S_2} - (h_{01} - h_{02}) \cdot \frac{S_1}{S_1 + S_2}$$

$$\cdot \exp\left(-\frac{k}{\rho} \cdot \frac{S_1 + S_2}{S_1 S_2} \cdot 0\right);$$

$$h_2\,(t = 0) = \frac{S_1 \cdot h_{01} + S_2 \cdot h_{02}}{S_1 + S_2} - (h_{01} - h_{02}) \cdot \frac{S_1}{S_1 + S_2};$$

$$h_2\,(t = 0) = \frac{S_1 \cdot h_{01} + S_2 \cdot h_{02} - S_1 \cdot h_{01} + S_1 \cdot h_{02}}{S_1 + S_2};$$

$$h_2\,(t = 0) = \frac{S_2 \cdot h_{02} + S_1 \cdot h_{02}}{S_1 + S_2}; \quad h_2\,(t = 0) = \frac{h_{02} \cdot (S_1 + S_2)}{S_1 + S_2};$$

$$h_2\,(t = 0) = h_{02}.$$

For $t \to \infty$, it results that:

$$h_2\,(t) = \frac{S_1 \cdot h_{01} + S_2 \cdot h_{02}}{S_1 + S_2} - (h_{01} - h_{02}) \cdot \frac{S_1}{S_1 + S_2}$$

$$\cdot \exp\left(-\frac{k}{\rho} \cdot \frac{S_1 + S_2}{S_1 S_2} \cdot t\right);$$

$$h_2\,(t) = \frac{S_1 \cdot h_{01} + S_2 \cdot h_{02}}{S_1 + S_2} - (h_{01} - h_{02}) \cdot \frac{S_1}{S_1 + S_2}$$

$$\cdot \frac{1}{\exp\left(\frac{k}{\rho} \cdot \frac{S_1 + S_2}{S_1 \cdot S_2} \cdot t\right)};$$

$$h_2\,(t \to \infty) = \frac{S_1 \cdot h_{01} + S_2 \cdot h_{02}}{S_1 + S_2} - (h_{01} - h_{02}) \cdot \frac{S_1}{S_1 + S_2}$$

$$\cdot \frac{1}{\exp\left(\frac{k}{\rho} \cdot \frac{S_1 + S_2}{S_1 \cdot S_2} \cdot \infty\right)};$$

$$h_2\,(t \to \infty) = \frac{S_1 \cdot h_{01} + S_2 \cdot h_{02}}{S_1 + S_2} - (h_{01} - h_{02}) \cdot \frac{S_1}{S_1 + S_2} \cdot 0;$$

$$h_2 \left(t \to \infty \right) = \frac{S_1 \cdot h_{01} + S_2 \cdot h_{02}}{S_1 + S_2} > h_{02};$$

$$h_1 \left(t \to \infty \right) = \frac{S_1 \cdot h_{01} + S_2 \cdot h_{02}}{S_1 + S_2} < h_{01};$$

$$h_1 \left(t \to \infty \right) = h_2 \left(t \to \infty \right);$$

$$h_{02} < h_2 \left(t \to \infty \right) < h_{01}.$$

Chapter 3

International Pre-Olympic Physics Contest 2003, Cluj-Napoca, Romania

Problem 1. Satellite and Mini-Satellite

A special satellite, S, revolves in an ellipse in the plane of the terrestrial equator, having the Earth in one of its foci. When the satellite is at perigee and the distance between it and the center of the Earth is the minimum, r_{min}, a special mini-satellite S_0 is launched from the satellite to fly on a parabola, having the Earth in its focus.

At some point in its elliptical trajectory, the speed of the satellite is

$$v = \sqrt{KM \left(\frac{2}{r} - \frac{1}{a} \right)}.$$

(a) *Determine* the mass ratio of the two satellites if, after the expulsion of the mini-satellite S_0, the satellite S continues its movement in a circular orbit around the Earth. The semi-major axis of the ellipse, a, is known.

(b) *Determine* the visibility duration of the special satellite S, evolving in a circular orbit, for a terrestrial observer located at the equator, at sea level. *We know*: the angular velocity corresponding to the Earth's rotation, ω_0, and the radius of the Earth, R.

(c) Satellite S must be transferred from its circular orbit to another outer concentric circular orbit with radius R_e.

Determine the speed corrections of satellite S necessary for this transfer, specifying the positions of satellite S when these corrections are made, as well as the duration of the satellite transfer if the gravitational transfer orbit is tangent to the two concentric circular orbits.

Given: the mass of the Earth, M; and the constant of universal attraction, K.

Solution

(a) At some point in its elliptical trajectory, the speed of the special satellite S is

$$v = \sqrt{KM\left(\frac{2}{r} - \frac{1}{a_{env}}\right)}.$$

When the special satellite S is at the minimum distance from the center of the Earth, as indicated in Figure 1.1, its speed is:

$$v_{max} = \sqrt{KM\left(\frac{2}{r_{env;min}} - \frac{1}{a_{env}}\right)};$$

$$v_{max} = \sqrt{\frac{KM(\lambda + 1)(1 - 2\lambda)}{r_0(\lambda - 1)}}.$$

If, after its launch, the mini-satellite S_0 evolves on a parabola, its speed at the time of launch is

$$v_{S_0} = v_0 = \sqrt{\frac{2KM}{r_{env;min}}} = \sqrt{-\frac{2KM}{r_0}(\lambda + 1)},$$

and the speed of the satellite S, evolving continuously on the circle, is

$$v_S = v = \sqrt{\frac{KM}{r_{env;min}}} = \sqrt{-\frac{KM}{r_0}(\lambda + 1)}.$$

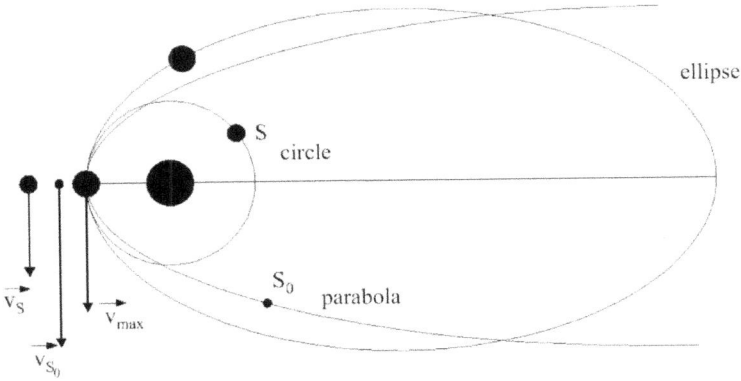

Fig. 1.1

Using the law of conservation of momentum, it follows that:

$$(m + m_0)v_{max} = mv + m_0 v_0;$$

$$\frac{m}{m_0} = \frac{v_0 - v_{max}}{v_{max} - v};$$

$$\frac{m}{m_0} = \frac{\sqrt{2(1 - \lambda)} - \sqrt{1 - 2\lambda}}{\sqrt{1 - 2\lambda} - \sqrt{1 - \lambda}}.$$

(b) For the terrestrial observer, located in position O_r with respect to the center of the Earth, as shown in Figure 1.2, the special satellite S, whose direction in its uniform circular motion we consider to be the same as the direction of the Earth's rotation, appears on the horizon in position R (rise). Due to the Earth's rotation, when the satellite passes below the horizon of the same observation place in position A (set), the terrestrial observer will be in position O compared to the center of the Earth. During the duration t of satellite S's visibility, the observer's vector radius describes the angle at the center α, and the vector radius of the satellite describes the angle at the center β.

If ω_0 and ω, respectively, are the angular velocity of the Earth's rotation and the angular velocity of the uniform circular motion of satellite S, it follows that:

$$\alpha = \omega_0 t;$$

$$\beta = \omega t = \frac{v}{r_{env;min}} t; \quad \beta = \alpha + 2\gamma;$$

$$\cos \gamma = \frac{R}{r_{\text{env;min}}};$$

$$\left(\frac{v}{r_{\text{env;min}}} - \omega_0\right) t = \text{arc cos} \frac{R}{r_{\text{env;min}}};$$

$$t = \frac{2\text{arc cos} \dfrac{R}{r_{\text{env;min}}}}{\dfrac{v}{r_{\text{env;min}}} - \omega_0}.$$

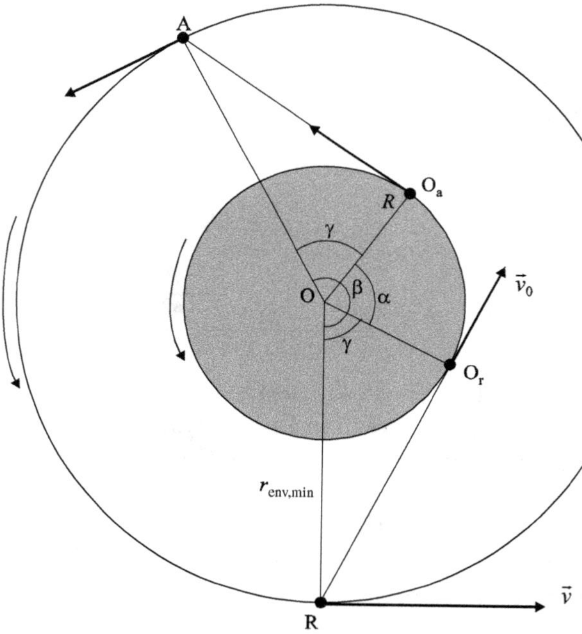

Fig. 1.2

If the direction of the satellite's movement is opposite to the direction of the Earth's rotation, as indicated in Figure 1.3, it results that:

$$\alpha + \beta = 2\gamma; \quad t = \frac{2\text{arc cos} \dfrac{R}{r_{\text{env;min}}}}{\dfrac{v}{r_{\text{env;min}}} + \omega_0}.$$

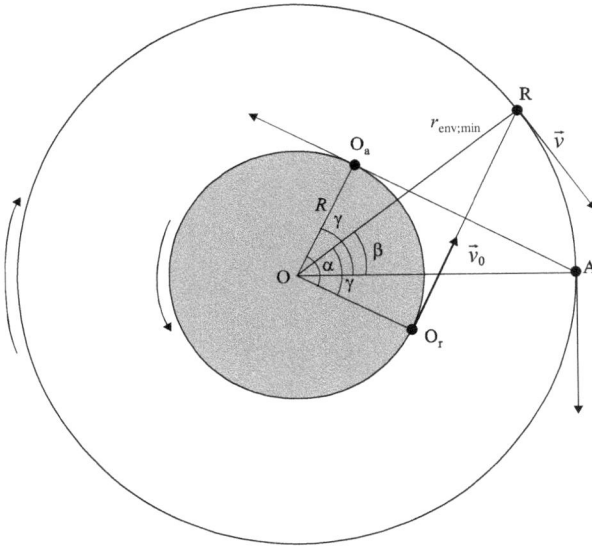

Fig. 1.3

(c) The gravitational transfer of the special satellite S from the circular orbit with radius $r_{\text{env; min}}$ onto the outer concentric circular orbit with radius R_e is achieved as shown in Figure 1.4, creating the conditions for the satellite to revolve on an elliptical orbit tangent to the two circular orbits at the diametrically opposite points P and A, so that the Earth fits in one of the foci of this ellipse.

The elements of the ellipse onto which the special satellite S is transferred are determined as follows:

$$r_{\min} = a(1 - e) = r_{\text{env;min}}; \quad a = \frac{1}{2}(r_{\text{env;min}} + R_e);$$

$$e = 1 - \frac{2r_{\text{env;min}}}{r_{\text{env;min}} + R_e}.$$

At point P on the circular orbit with radius $r_{\text{env; min}}$, an initial correction of the satellite's speed, $\Delta \vec{v}_{\text{in}}$, must be performed so that the resulting speed,

$$\vec{v}_{\text{pg}} = \vec{v} + \Delta \vec{v}_{\text{in}},$$

will be the speed of the satellite at the perigee, P, of the established elliptical orbit:

$$v_{pg} = v_{max} = \sqrt{KM\frac{1+e}{a(1-e)}}.$$

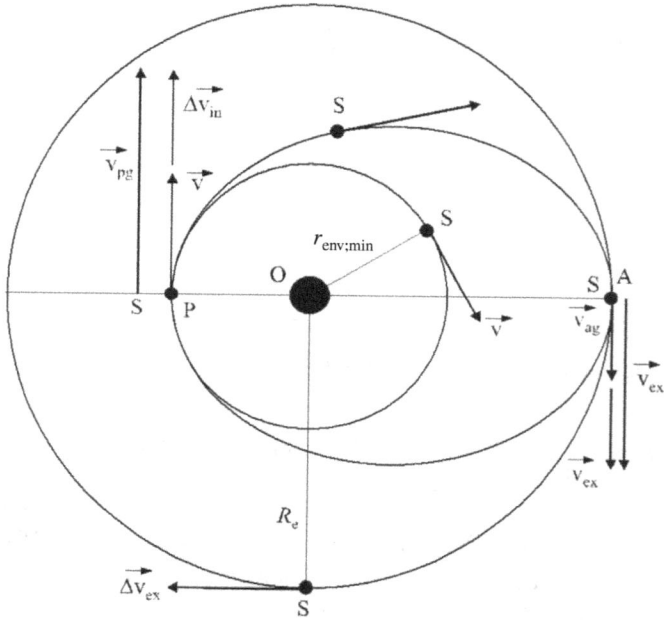

Fig. 1.4

It results that:

$$\Delta v_{in} = v_{max} - v;$$

$$\Delta v_{in} = \sqrt{\frac{KM}{r_{env;min}}}\left(\sqrt{\frac{2R_e}{r_{env;min}+R_e}} - 1\right).$$

When the special satellite S reaches the apogee, A, of its elliptical orbit at the distance $r_{max} = R_e$, its speed is

$$v_{ag} = v_{min} = \sqrt{KM\frac{1-e}{a(1+e)}},$$

or, using the law of conservation of kinetic momentum to find an equivalent solution,

$$v_{\min} = \frac{r_{\min}}{R_e} v_{max}.$$

At point A on the elliptical orbit, a second correction of the satellite's speed, $\Delta \vec{v}_{\text{ex}}$, must be performed so that its resulting speed,

$$\vec{v}_{\text{ex}} = \vec{v}_{\text{ag}} + \Delta \vec{v}_{\text{ex}},$$

is the speed required for the satellite to evolve on a circle with radius R_e:

$$v_{\text{ex}} = \sqrt{\frac{KM}{R_e}}.$$

It results that:

$$\Delta v_{\text{ex}} = v_{\text{ex}} - v_{\min};$$

$$\Delta v_{\text{ex}} = \sqrt{\frac{KM}{R_e}} \left(1 - \sqrt{\frac{2r_{\text{env;min}}}{r_{\text{env;min}} + R_e}} \right).$$

The duration of the transfer of the special satellite S from the inner circular orbit to the outer concentric circular orbit is half of the period of the satellite's movement on the established elliptical orbit:

$$\Delta t = \pi \sqrt{\frac{a^3}{KM}};$$

$$\Delta t = \frac{\pi}{2} (r_{\text{env;min}} + R_e) \sqrt{\frac{r_{\text{env;min}} + R_e}{2KM}}.$$

Both speed corrections increase the local orbital speed values of the satellite transferred from an inner circular orbit to an outer concentric circular orbit.

In the scheme for the transfer of the special satellite S onto an inner concentric circular orbit, as shown in Figure 1.5, the gravitational transfer is performed on an elliptical trajectory, requiring two speed corrections, which are also carried out at diametrically opposite points.

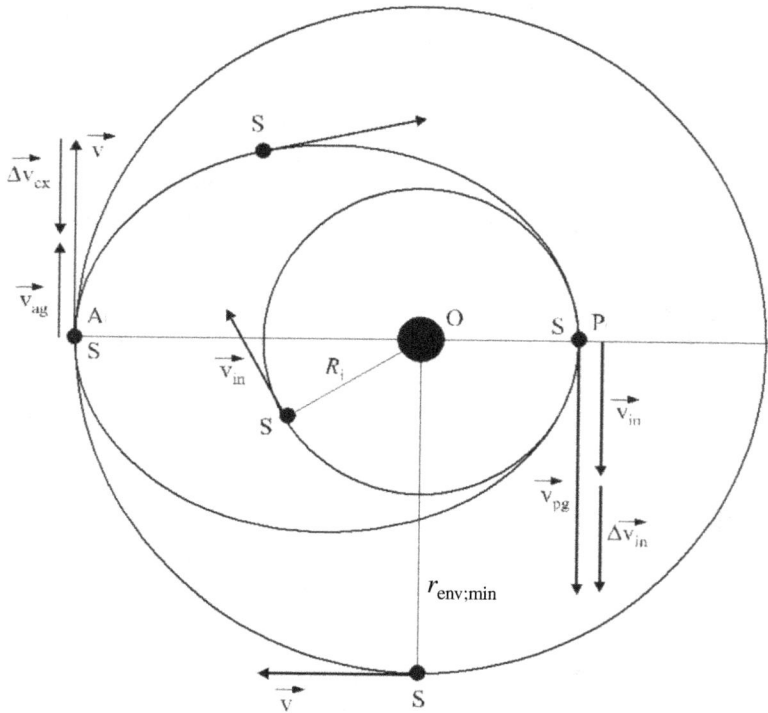

Fig. 1.5

In this case, both speed corrections decrease the orbital speed values of the transferred satellite.

The elements of the ellipse onto which the special satellite S has now been transferred are determined as follows:

$$r_{\max} = a(1 + e) = r_{\text{env; min}}; \quad a = \frac{1}{2}(r_{\text{env; min}} + R_i);$$

$$e = \frac{r_{\text{env;min}} - R_i}{r_{\text{env;min}} + R_i}.$$

At point A on the circular orbit with radius $r_{\text{env; min}}$, an initial correction of the satellite's speed, $\Delta \vec{v}_{\text{ex}}$, must be performed so that the resulting speed,

$$\vec{v}_{\text{ag}} = \vec{v} + \Delta \vec{v}_{\text{ex}},$$

is the speed of the satellite at the apogee, A, of the established elliptical orbit:

$$v_{ag} = v_{min} = \sqrt{KM \frac{1-e}{a(1+e)}}.$$

It results that:

$$\Delta v_{ex} = v - v_{min};$$

$$\Delta v_{ex} = \sqrt{\frac{KM}{r_{env;min}}} \left(1 - \sqrt{\frac{2R_i}{r_{env;min} + R_i}}\right).$$

When the special satellite S reaches the perigee, P, of its elliptical orbit at the distance $r_{min} = R_i$, its speed is

$$v_{pg} = v_{max} = \sqrt{KM \frac{1+e}{a(1-e)}}.$$

At point P on the elliptical orbit, a second correction of the speed of the satellite, $\Delta \vec{v}_{in}$, must be performed in such a way that its resulting speed,

$$\vec{v}_{in} = \vec{v}_{pg} + \Delta \vec{v}_{in},$$

is the necessary speed for the satellite to evolve on a circle with the radius R_i:

$$v_{in} = \sqrt{\frac{KM}{R_i}}.$$

It results that:

$$\Delta v_{in} = v_{max} - v_{in};$$

$$\Delta v_{in} = \sqrt{\frac{KM}{R_i}} \left(\sqrt{\frac{2r_{env;min}}{r_{env;min} + R_i}} - 1\right).$$

The duration of the transfer of the special satellite S from the outer circular orbit to the inner concentric circular orbit is half the time of the satellite's displacement on the established elliptical orbit, meaning that:

$$\Delta t = \pi \sqrt{\frac{a^3}{KM}};$$

$$\Delta t = \frac{\pi}{2}(r_{\text{env; min;}} + R_{\text{i}})\sqrt{\frac{r_{\text{env;min}} + R_{\text{i}}}{2KM}}.$$

Problem 2. The Enveloping of Satellites' Orbits

From a point P_0, located at a distance r_0 from the center O of an idealized spherical Earth, n identical satellites are launched in different directions, all confined to the plane of the Earth's equator. The moduli of their speeds, with respect to the center of the Earth, are identical, v_0, in such a way that

$$r_0 v_0^2 < 2KM,$$

where K is the constant of universal attraction and M is the mass of the Earth.

(a) Formulate the equation and determine the orbital elements of a special satellite S that, under the influence of Earth's gravitational attraction, follows a trajectory that acts as an external envelope to the orbits of the n identical satellites.

  Determine the position of the injection point P_0 with respect to the trajectory (envelope) of the special satellite S.

  Finally, analyze the feasibility of such an enveloping trajectory under the action of Earth's gravity for the following two scenarios:

$$r_0 v_0^2 = 2KM \quad \text{and} \quad r_0 v_0^2 > 2KM.$$

(b) When the special satellite S is at the minimum distance from the center of the Earth, a special mini-satellite S_0 is launched from it. This satellite will revolve on a parabola with the Earth in its focus. As a result, satellite S continues its movement in a circle around the Earth.

  Determine the ratio of the masses of the two special satellites, m/m_0, as well as the visibility duration of the special satellite S, evolving in a circular orbit, for a terrestrial observer located on the Equator at sea level.

We know: the angular speed of the Earth corresponding to its own rotation, ω_0, as well as the radius of the Earth, R. The mass of the fuel burned during the launch maneuvers of the mini-satellite S_0 is negligible.

(c) The satellite S must then be transferred to an outer concentric circular orbit with radius R_e. *Determine* the speed corrections of satellite S necessary for this transfer, specifying its positions when these corrections are made and the duration of the satellite transfer if the transfer orbit is gravitationally tangent to the two concentric circular orbits.

Analyze the case of satellite S's transfer onto an inner concentric circular orbit with the radius R_i, highlighting the differences compared to the previous case.

The durations of the satellite speed corrections, as well as the mass of the fuel burned while performing these corrections, are neglected.

(d) Satellite S's gravitational transfer from the inner circular orbit to the outer circular orbit can be achieved on a parabolic trajectory, with the Earth in its focus.

Determine the speed corrections of satellite S necessary for this transfer, knowing that their moduli are equal. Specify the positions of the satellite when these corrections are made, as well as the duration of the satellite transfer if the gravitational transfer orbit is tangent to the inner circular orbit and intersects the outer circular orbit.

It is known that

$$\int \frac{dx}{\cos^4 x} = \frac{\sin x}{3\cos^3 x} + \frac{2}{3}\operatorname{tg} x.$$

Solution

Suppose that a material point S (for example, a satellite with mass m) moves in the gravitational field of the Earth (a fixed material point, E, with mass M) so that the instantaneous plane of the vectors $\vec{r}$ and $\vec{v}$ is the XY plane shown in Figure 2.1. If $\vec{F}$ is the gravitational force that the planet exerts on the satellite, it results that:

$$m\ddot{\vec{r}} = \vec{F}; \quad m\vec{r} \times \ddot{\vec{r}} = \vec{r} \times \vec{F} = 0;$$

$$m\frac{d}{dt}(\vec{r} \times \dot{\vec{r}}) = 0; \quad \frac{d}{dt}(\vec{r} \times m\vec{v}) = 0;$$

$$\vec{r} \times m\vec{v} = \vec{L}; \quad \frac{d\vec{L}}{dt} = 0;$$

$$\vec{L} = \text{constant}; \quad \vec{L} \perp \vec{r}; \quad \vec{L} \perp \vec{v}.$$

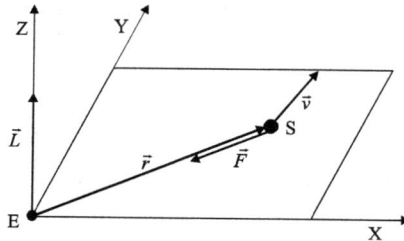

Fig. 2.1

In the following moments, the vectors $\vec{r}$ and $\vec{v}$ will change their orientations (and moduli), but the vector $\vec{L}$ will have to maintain a constant orientation (and constant modulus). As a result, the plane of the $\vec{r}$ and $\vec{v}$ vectors (the XY plane) remains constant, which means that the movement of a material point under the action of the central gravitational force is planar.

To study the movement of the satellite, we use the polar coordinate system represented in Figure 2.2 so that we have:

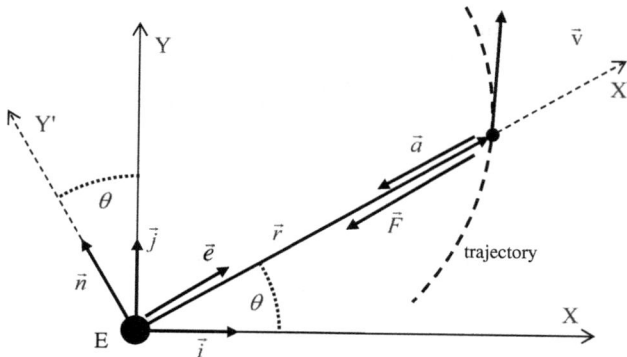

Fig. 2.2

$$\vec{v} = \dot{r}\vec{\rho} + r\dot{\theta}\vec{n};$$

$$\vec{a} = (\ddot{r} - r\dot{\theta}^2)\vec{\rho} + (2\dot{r}\dot{\theta} + r\ddot{\theta})\vec{n};$$

$$m\vec{a} = \vec{F}; \quad \vec{F} = -K\frac{mM}{r^2}\vec{\rho};$$

$$m(\ddot{r} - r\dot{\theta}^2)\vec{\rho} + m(2\dot{r}\dot{\theta} + r\ddot{\theta})\vec{n} = -K\frac{mM}{r^2}\vec{\rho};$$

$$m(\ddot{r} - r\dot{\theta}^2) = -K\frac{mM}{r^2};$$

$$m(2\dot{r}\dot{\theta} + r\ddot{\theta}) = 0;$$

$$\ddot{r} - r\dot{\theta}^2 = -K\frac{M}{r^2};$$

$$2\dot{r}\dot{\theta} + r\ddot{\theta} = 0.$$

From this, multiplying by r, it results that:

$$2r\dot{r}\dot{\theta} + r^2\ddot{\theta} = 0;$$

$$\frac{d}{dt}(r^2\dot{\theta}) = 0; \quad r^2\dot{\theta} = C,$$

where C is the constant of integration;

$$\dot{\theta} = \frac{C}{r^2};$$

$$\dot{r} = \frac{dr}{dt} = \frac{dr}{d\theta}\frac{d\theta}{dt}; \quad \dot{r} = \dot{\theta}\frac{dr}{d\theta} = \frac{C}{r^2}\frac{dr}{d\theta};$$

$$\dot{r} = -C\frac{d}{d\theta}\left(\frac{1}{r}\right);$$

$$\ddot{r} = \frac{d^2r}{dt^2} = \frac{d}{dt}\left(\frac{dr}{dt}\right) = \frac{d}{d\theta}\left(\frac{dr}{dt}\right)\frac{d\theta}{dt} = \dot{\theta}\frac{d}{d\theta}\left(\frac{dr}{dt}\right);$$

$$\ddot{r} = \frac{C}{r^2}\frac{d}{d\theta}\left[-C\frac{d}{d\theta}\left(\frac{1}{r}\right)\right];$$

$$\ddot{r} = -\frac{C^2}{r^2}\frac{d^2}{d\theta^2}\left(\frac{1}{r}\right).$$

$$m(\ddot{r} - r\dot{\theta}^2) = -K\frac{mM}{r^2};$$

$$(\ddot{r} - r\dot{\theta}^2) = -K\frac{M}{r^2};$$

$$-\frac{C^2}{r^2}\frac{d^2}{d\theta^2}\left(\frac{1}{r}\right) - r\frac{C^2}{r^4} = -K\frac{M}{r^2};$$

$$\frac{d^2}{d\theta}\left(\frac{1}{r}\right) + \frac{1}{r} = K\frac{M}{C^2}.$$

This is an inhomogeneous differential equation, for whose integration we first consider the associated homogeneous equation,

$$\frac{d}{d\theta^2}\left(\frac{1}{r}\right) + \frac{1}{r} = 0.$$

Making the substitution $\frac{1}{r} = z_0$, it results that:

$$z_0'' + z_0 = 0; \quad z_0 = e^{k\theta}; \quad k^2 + 1 = 0;$$

$$k_1 = +i; \quad k_2 = -i;$$

$$z_0 = C'e^{i\theta} + C''e^{-i\theta};$$

$$z_0 = C'(\cos\theta + i\sin\theta) + C''(\cos\theta - i\sin\theta);$$

$$z_0 = (C' + C'')\cos\theta + i(C' - C'')\sin\theta;$$

$$C' + C'' = C_1; \quad i(C' - C'') = C_2;$$

$$z_0 = C_1\cos\theta + C_2\sin\theta;$$

$$C_1 = \lambda\cos\theta'; \quad C_2 = \lambda\sin\theta',$$

where θ' is a particular value of θ;

$$z_0 = \lambda\cos(\theta - \theta'),$$

representing the general solution of the homogeneous equation.

From the theory of differential equations, it is known that the general solution of an inhomogeneous equation is equal to the solution of the associated homogeneous equation, with the addition of a particular solution of the inhomogeneous equation,

$$\frac{1}{r} = z_0 + z_p.$$

Since the free term of the inhomogeneous equation is a constant, the particular solution of the inhomogeneous equation is of the form of the free term, so

$$z_p = K\frac{M}{C^2}.$$

It results that:

$$\frac{1}{r} = \lambda\cos(\theta - \theta') + K\frac{M}{C^2};$$

$$\lambda = \frac{e}{p}; \quad K\frac{M}{C^2} = \frac{1}{p};$$

$$\frac{1}{r} = \frac{e}{p}\cos(\theta - \theta') + \frac{1}{p};$$

$$r = \frac{p}{1 + e\cos(\theta - \theta')}.$$

If it is admitted that we have $\theta' = 0$, it follows that the equation of the satellite's trajectory, in relation to the Earth, is of the form

$$r = \frac{p}{1 + e\cos\theta},$$

representing the equation of a conic, where p is the parameter of the conic, and e is the numerical eccentricity of the conic.

Depending on the value of e, the conic can be an ellipse $(0 < e < 1)$, a parabola $(e = 1)$, a hyperbola $(e > 1)$, or a circle $(e = 0)$.

The previous equation,

$$-\frac{C^2}{r^2}\frac{d^2}{d\theta^2}\left(\frac{1}{r}\right) - r\frac{C^2}{r^4} = -K\frac{M}{r^2},$$

can be put into the form

$$-\frac{mC^2}{r^2}\frac{d^2}{d\theta^2}\left(\frac{1}{r}\right) - \frac{mC^2}{r^3} = -K\frac{mM}{r^2} = F,$$

$$F = -\frac{mC^2}{r^2}\left[\frac{d^2}{d\theta^2}\left(\frac{1}{r}\right) + \frac{1}{r}\right],$$

known as Binet's equation, which allows the calculation of the central force when the shape of the trajectory of the material point with mass m subjected to the action of that force is known.

Transported from the ground to altitude h by means of a launch vehicle, a satellite is injected into a trajectory at a point P_0, as indicated in Figure 2.3 (where R is the radius of the Earth). There, it is given the velocity $\vec{v}_0$ in a direction that forms the angle α with the direction OP_0.

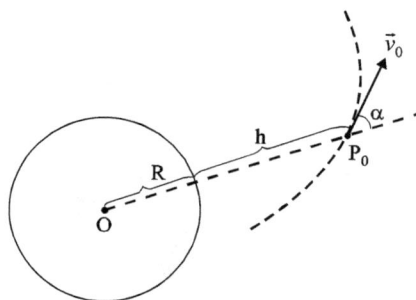

Fig. 2.3

The general study of the movement of a material point in the field of a central gravitational force, carried out using plane polar coordinates, involves the integration of the inhomogeneous differential equation,

$$\frac{d^2}{d\theta^2}\left(\frac{1}{r}\right) + \frac{1}{r} = K\frac{M}{C^2},$$

whose solution is

$$\frac{1}{r} = \frac{1}{p} + \frac{e}{p}\cos(\theta - \theta'),$$

$$r = \frac{p}{1 + e\cos(\theta - \theta')}.$$

This represents the equation of the trajectory of the material point in the field of the central gravitational force, a conic, where p is the parameter of the conic; e is the numeric eccentricity of the conic; θ' is an integration constant; and

$$C = r^2\dot{\theta}, \quad p = \frac{C^2}{KM},$$

where M is the mass of the Earth, K is the constant of universal attraction, and C is the constant of integration.

The solution of the previous differential equation can be obtained in an equivalent form if, according to the theory of differential equations, we proceed as follows:

$$\frac{d^2}{d\theta^2}\left(\frac{1}{r}\right) + \frac{1}{r} = 0;$$

$$\frac{1}{r} = z_0; \quad \frac{d^2 z_0}{d\theta^2} + z_0 = 0;$$

$$z_0'' + z_0 = 0; \quad z_0 = e^{k\theta}; \quad k^2 + 1 = 0;$$

$$k = \pm\sqrt{-1}; \quad k_1 = +i; \quad k_2 = -i;$$

$$z_0 = C' e^{i\theta} + C'' e^{-i\theta};$$

$$z_0 = C'(\cos\theta + i\,\sin\theta) + C''(\cos\theta - i\,\sin\theta);$$

$$z_0 = (C' + C'')\cos\theta + i(C' - C'')\sin\theta;$$

$$C' + C'' = C_1; \quad i(C' - C'') = C_2;$$

$$z_0 = C_1 \cos\theta + C_2 \sin\theta;$$

$$C_1 = \lambda\cos\theta'; \quad C_2 = \lambda\sin\theta';$$

$$z_0 = \lambda\cos(\theta - \theta'),$$

where z_0 is the general solution of homogeneous equations associated with given inhomogeneous equations;

$$\frac{1}{r} = z_0 + z_p,$$

where z_p is a particular solution of an inhomogeneous equation;

$$z_p = K\frac{M}{c^2};$$

$$\frac{1}{r} = K\frac{M}{c^2} + \lambda\cos(\theta - \theta');$$

$$K\frac{M}{c^2} = \frac{1}{p}; \quad \lambda = \frac{e}{p};$$

$$\frac{1}{r} = \frac{1}{p} + \frac{e}{p}\cos(\theta - \theta');$$

$$r = \frac{p}{1 + e\cos(\theta - \theta')}.$$

If the situations, from the initial moment and from some moment t, are those represented in Figure 2.4, we get:

$$\frac{1}{r} = \frac{1 + e\cos(\theta - \theta')}{p};$$

$$\frac{\dot{r}}{r^2} = \frac{e}{p}\dot{\theta}\sin(\theta - \theta');$$

$$\vec{v} = \dot{r}\vec{\rho} + r\dot{\theta}\vec{n};$$

$$t = 0; \quad r = r_0 = R + h; \quad \dot{r} = \dot{r}_0;$$

$$\theta = 0; \quad \dot{\theta} = \dot{\theta}_0;$$

$$v_\rho = v_{\rho 0} = \dot{r}_0 = v_0\cos\alpha;$$

$$v_n = v_{n 0} = r_0\dot{\theta}_0 = v_0\sin\alpha;$$

$$r_0^2\dot{\theta}_0 = r_0 v_0\sin\alpha;$$

$$C = r^2\dot{\theta};$$

$$C = r_0^2\dot{\theta}_0; \quad C = r_0 v_0\sin\alpha;$$

$$p = \frac{C^2}{KM};$$

$$p = \frac{r_0^2 v_0^2\sin^2\alpha}{KM};$$

$$\frac{1}{r} = \frac{1 + e\cos(\theta - \theta')}{p}; \quad \frac{p}{r} = 1 + e\cos(\theta - \theta');$$

$$\theta = 0; \quad \cos(\theta - \theta') = \cos(-\theta') = \cos\theta'; \quad r = r_0;$$

$$\frac{p}{r_0} = 1 + e\cos\theta'; \quad e\cos\theta' = \frac{p}{r_0} - 1;$$

$$e \cos \theta' = \frac{p - r_0}{r_0};$$

$$\frac{\dot{r}}{r^2} = \frac{e}{p} \dot{\theta} \sin(\theta - \theta'); \quad \frac{p\dot{r}}{r^2} = e\dot{\theta} \sin(\theta - \theta');$$

$$\frac{p\dot{r}_0}{r_0^2} = e\dot{\theta}_0 \sin \theta';$$

$$v_\rho = v_{\rho 0} = \dot{r}_0 = v_0 \cos \alpha;$$

$$v_n = v_{n0} = r_0 \dot{\theta}_0 = v_0 \sin \alpha; \quad \dot{\theta}_0 = \frac{v_0}{r_0} \sin \alpha;$$

$$\frac{pv_0 \cos \alpha}{r_0^2} = e\frac{v_0}{r_0} \sin \alpha \sin \theta';$$

$$\frac{p \cos \alpha}{r_0} = e \sin \alpha \sin \theta';$$

$$e \sin \theta' = \frac{p \cos \alpha}{r_0 \sin \alpha}; \quad e \sin \theta' = \frac{p}{r_0} \operatorname{ctg} \alpha;$$

$$\frac{p}{r_0} = 1 + e \cos \theta'; \quad e \cos \theta' = \frac{p}{r_0} - 1; \quad e \cos \theta' = \frac{p - r_0}{r_0};$$

$$\operatorname{tg} \theta' = \frac{p}{p - r_0} \operatorname{ctg} \alpha;$$

$$\operatorname{tg} \theta' = \frac{p}{p - r_0} \operatorname{ctg} \alpha;$$

$$e \sin \theta' = \frac{p}{r_0} \operatorname{ctg} \alpha; \quad e \cos \theta' = \frac{p - r_0}{r_0};$$

$$e^2 \sin^2 \theta' + e^2 \cos^2 \theta' = \frac{p^2}{r_0^2} \frac{\cos^2 \alpha}{\sin^2 \alpha} + \frac{(p - r_0)^2}{r_0^2};$$

$$e^2 = 1 - \frac{2p}{r_0} \left(1 - \frac{p}{2r_0 \sin^2 \alpha}\right);$$

$$p = \frac{r_0^2 v_0^2 \sin^2 \alpha}{KM};$$

$$e^2 = 1 - \frac{2r_0 v_0^2 \sin^2 \alpha}{KM} \left(1 - \frac{r_0 v_0^2}{2KM}\right);$$

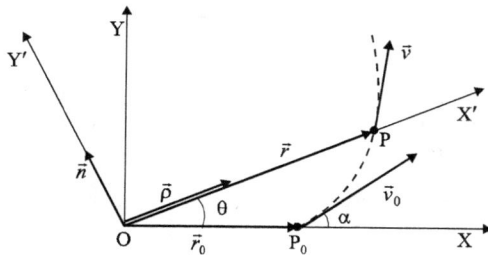

Fig. 2.4

$$r_0 v_0^2 < 2KM, \quad 0 < e < 1, \text{ ellipse};$$

$$r_0 v_0^2 = 2KM, \quad e = 1, \text{ parabola};$$

$$r_0 v_0^2 > 2KM, \quad e > 1, \text{ hyperbola};$$

$$r_0 v_0^2 = KM, \quad \alpha = \frac{\pi}{2}, \ e = 0, \text{ circle}.$$

If, based on Figure 2.4, the launch is made so that $\alpha = 90^0$, it results that:

$$\text{tg } \theta' = \frac{p}{p - r_0} \text{ctg } \alpha; \quad \text{tg } \theta' = 0; \quad \theta' = 0;$$

$$r = \frac{p}{1 + e\cos(\theta - \theta')}; \quad r = \frac{p}{1 + e\cos\theta}.$$

(a) Considering the conditions of the problem, when $r_0 v_0^2 < 2KM$, the orbits of the n satellites form a family of ellipses, as shown in Figure 2.5, having a common focus at the center of the Earth, whose equations in plane polar coordinates are:

$$r = \frac{p}{1 + e\cos(\theta - \theta')}; \quad \frac{1}{r} = \frac{1}{p} + \frac{e}{p}\cos(\theta - \theta');$$

$$p = \frac{c^2}{KM} = \frac{r_0^2 v_0^2 \sin^2 \alpha}{KM};$$

$$\text{tg } \theta' = \frac{p}{p - r_0} \text{ctg } \alpha,$$

where the parameter (angle) α characterizes the individual trajectories of the satellites.

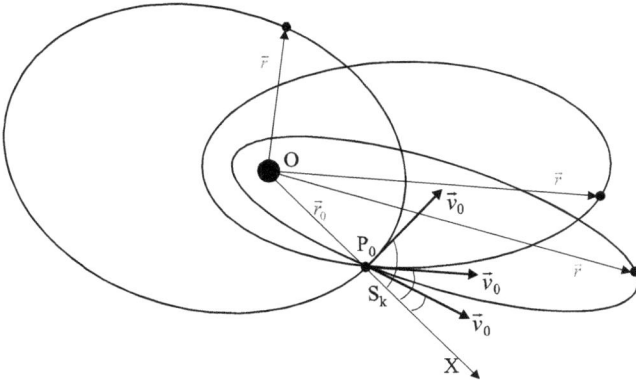

Fig. 2.5

For each element of the family of ellipses, α is a given constant, as indicated by Figure 2.6. In the above form, the equation of the conic is preferable to the original form because it shows that for $\theta = \theta'$, the value of r becomes maximum.

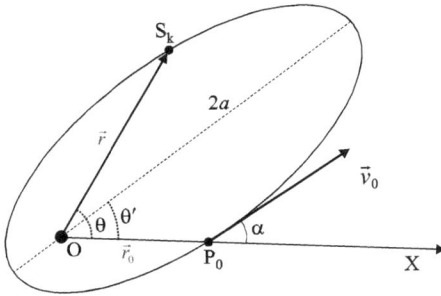

Fig. 2.6

If the total energy of each satellite (S_k)–Earth system is the same for all systems,

$$E = \frac{mv_0^2}{2} - K\frac{mM}{r_0} = -K\frac{mM}{2a},$$

then the semi-major axes of the ellipses described by those n satellites are identical:

$$2a = \frac{r_0}{1 - \frac{r_0 v_0^2}{2KM}}.$$

Finally, the family of trajectories of those n satellites is described by a parametric equation of the form

$$F(r, \theta, \alpha) = \frac{1}{p} + \frac{e}{p}\cos(\theta - \theta') - \frac{1}{r} = 0.$$

The envelope of the trajectories of the n satellites, if it exists, is of course a tangent curve external to all the individual orbits. It satisfies the same equation, $F(r, \theta, \alpha) = 0$, at every point shared with the individual trajectories, because for specific values of r and θ, a point on the envelope coincides with a point on an individual trajectory corresponding to a value of α.

Because the common points between the envelope and the individual trajectories are their tangent points, it follows that the parametric equation of the envelope, $F(r, \theta, \alpha) = 0$, must also fulfill the condition

$$\frac{\partial F}{\partial \alpha} = 0.$$

From $F(r, \theta, \alpha) = 0$ and $\frac{\partial F}{\partial \alpha} = 0$, by eliminating the parameter α, we obtain the equation of the envelope of the family of ellipses, $F(r, \theta) = 0$, representing the trajectory equation of the special satellite S. This satellite evolves under the action of the Earth's gravitational attraction and envelops, through the external tangent, the totality of the orbits of those n satellites launched under the specified conditions.

Under these conditions, from the equation of the ellipse written in plane polar coordinates,

$$\frac{1}{r} = \frac{1}{p} + \frac{e}{p}\cos(\theta - \theta'),$$

by derivation in relation to t, it results that:

$$-\frac{1}{r^2}\frac{dr}{dt} = -\frac{e}{p}\sin(\theta - \theta')\frac{d\theta}{dt};$$

$$\frac{1}{r^2}\dot{r} = \frac{e}{p}\sin(\theta - \theta')\dot{\theta};$$

$$p = \frac{c^2}{KM} = \frac{r_0^2 v_0^2 \sin^2\alpha}{KM};$$

$$r^2 \dot{\theta} = C = r_0 v_0 \sin \alpha;$$

$$\sin^2 \alpha = \frac{1}{1 + \operatorname{ctg}^2 \alpha}; \quad p = \frac{r_0^2 v_0^2}{KM(1 + \operatorname{ctg}^2 \alpha)}.$$

For the initial moment, $t = 0$, when $r = r_0$ and $\theta = \theta_0 = 0$, we have:

$$\dot{r} = \dot{r}_0 = v_0 \cos \alpha; \quad \dot{\theta} = \dot{\theta}_0 = \frac{v_0 \sin \alpha}{r_0};$$

$$\frac{1}{r^2} \dot{r} = \frac{e}{p} \sin(\theta - \theta') \dot{\theta};$$

$$\frac{1}{r_0^2} \dot{r}_0 = \frac{e}{p} \sin(-\theta') \dot{\theta}_0;$$

$$\frac{1}{r_0^2} v_0 \cos \alpha = -\frac{e}{p} \sin \theta' \frac{v_0 \sin \alpha}{r_0};$$

$$\frac{e}{p} \sin \theta' = -\frac{\operatorname{ctg} \alpha}{r_0};$$

$$r = \frac{p}{1 + e \cos(\theta - \theta')};$$

$$\frac{1}{r} = \frac{1}{p} + \frac{e}{p} \cos(\theta - \theta'),$$

$$\theta = 0;$$

$$\frac{1}{r_0} = \frac{1}{p} + \frac{e}{p} \cos(-\theta'); \quad \frac{1}{r_0} = \frac{1}{p} + \frac{e}{p} \cos(\theta');$$

$$\frac{e}{p} \cos \theta' = \frac{1}{r_0} - \frac{1}{p}; \quad \frac{e}{p} \sin \theta' = -\frac{\operatorname{ctg} \alpha}{r_0}.$$

It results that:

$$\frac{1}{r} = \frac{1}{p} + \frac{e}{p} \cos(\theta - \theta'),$$

$$\frac{1}{r} = \frac{1}{p} + \frac{e}{p} \cos \theta \cos \theta' + \frac{e}{p} \sin \theta \sin \theta';$$

$$\frac{e}{p} \cos \theta' = \frac{1}{r_0} - \frac{1}{p}; \quad \frac{e}{p} \sin \theta' = -\frac{\operatorname{ctg} \alpha}{r_0};$$

$$\frac{1}{r} = \frac{1}{p} + \left(\frac{1}{r_0} - \frac{1}{p}\right)\cos\theta - \frac{\text{ctg}\,\alpha}{r_0}\sin\theta;$$

$$\frac{1}{r} = \frac{1}{p}(1 - \cos\theta) + \frac{\cos\theta}{r_0} - \frac{\text{ctg}\,\alpha}{r_0}\sin\theta;$$

$$p = \frac{r_0^2 v_0^2}{KM(1 + \text{ctg}^2\,\alpha)};$$

$$\frac{1}{r} = (1 - \cos\theta)\frac{KM(1 + \text{ctg}^2\,\alpha)}{r_0^2 v_0^2} + \frac{\cos\theta}{r_0} - \frac{\text{ctg}\,\alpha}{r_0}\sin\theta;$$

$$q = \frac{\text{the initial gravitational potential energy of the satellite--Earth system}}{\text{the initial kinetic energy of the system}},$$

$$q < 0;$$

$$q = \frac{-K\frac{mM}{r_0}}{\frac{mv_0^2}{2}} = -\frac{2KM}{r_0 v_0^2} < -1;$$

$$r_0 v_0^2 < 2KM; \quad q = -\frac{2KM}{r_0 v_0^2} < -1;$$

$$\frac{1}{r} = (1 - \cos\theta)\frac{KM(1 + \text{ctg}^2\,\alpha)}{r_0^2 v_0^2} + \frac{\cos\theta}{r_0} - \frac{\text{ctg}\,\alpha}{r_0}\sin\theta;$$

$$q = -\frac{2KM}{r_0 v_0^2}; \quad \frac{q}{r_0} = -\frac{2KM}{r_0^2 v_0^2}; \quad \frac{KM}{r_0^2 v_0^2} = -\frac{q}{2r_0};$$

$$\frac{1}{r} = -(1 - \cos\theta)\frac{q}{2r_0}(1 + \text{ctg}^2\,\alpha) + \frac{\cos\theta}{r_0} - \frac{\text{ctg}\,\alpha}{r_0}\sin\theta;$$

$$\frac{1}{r} = \frac{1}{p}(1 - \cos\theta) + \frac{\cos\theta}{r_0} - \frac{\text{ctg}\,\alpha}{r_0}\sin\theta;$$

$$p = \frac{r_0^2 v_0^2}{KM(1 + \text{ctg}^2\,\alpha)}; \quad q = -\frac{2KM}{r_0 v_0^2};$$

$$F(r, \theta, \alpha) = \frac{1}{p} + \frac{e}{p}\cos(\theta - \theta') - \frac{1}{r} = 0;$$

$$F(r, \theta, \alpha) = -(1 - \cos\theta)\frac{q}{2r_0}(1 + \text{ctg}^2\,\alpha) + \frac{\cos\theta}{r_0}$$

$$- \frac{\text{ctg}\,\alpha}{r_0}\sin\theta - \frac{1}{r} = 0;$$

$$\frac{\partial F}{\partial \alpha} = \frac{\partial F}{\partial(\text{ctg}\,\alpha)}\frac{d(\text{ctg}\,\alpha)}{d\alpha} = -\frac{1}{\sin^2\alpha}\frac{\partial F}{\partial(\text{ctg}\,\alpha)} = 0;$$

$$\alpha \neq 0; \quad \frac{\partial F}{\partial(\text{ctg}\,\alpha)} = 0;$$

$$\frac{\partial F}{\partial(\text{ctg}\,\alpha)} = -(1 - \cos\theta)\frac{q}{r_0}\text{ctg}\,\alpha - \frac{\sin\theta}{r_0} = 0;$$

$$(1 - \cos\theta)\frac{q}{r_0}\text{ctg}\,\alpha = -\frac{\sin\theta}{r_0};$$

$$\text{ctg}\,\alpha = -\frac{\sin\theta}{q(1 - \cos\theta)};$$

$$F(r, \theta, \alpha) = -(1 - \cos\theta)\frac{q}{2r_0}(1 + \text{ctg}^2\,\alpha) + \frac{\cos\theta}{r_0}$$

$$- \frac{\text{ctg}\,\alpha}{r_0}\sin\theta - \frac{1}{r} = 0;$$

$$F(r, \theta, \alpha) = -(1 - \cos\theta)\frac{q}{2r_0}\left(1 + \frac{\sin^2\theta}{q^2(1 - \cos\theta)^2}\right)$$

$$+ \frac{\cos\theta}{r_0} - \frac{\text{ctg}\,\alpha}{r_0}\sin\theta - \frac{1}{r} = 0;$$

$$\frac{\sin\theta}{(1 + \cos\theta)} = \frac{1 - \cos\theta}{\sin\theta}; \quad \sin^2\theta = (1 - \cos\theta)(1 + \cos\theta);$$

$$F(r, \theta, \alpha) = -(1 - \cos\theta)\frac{q}{2r_0}\left(1 + \frac{(1 - \cos\theta)(1 + \cos\theta)}{q^2(1 - \cos\theta)^2}\right)$$

$$+ \frac{\cos\theta}{r_0} - \frac{\text{ctg}\,\alpha}{r_0}\sin\theta - \frac{1}{r} = 0;$$

$$F(r, \theta, \alpha) = -(1 - \cos\theta)\frac{q}{2r_0}\left(1 + \frac{(1 + \cos\theta)}{q^2(1 - \cos\theta)}\right)$$

$$+ \frac{\cos\theta}{r_0} - \frac{\operatorname{ctg}\alpha}{r_0}\sin\theta - \frac{1}{r} = 0;$$

$$\frac{1}{r} = -(1 - \cos\theta)\frac{q}{2r_0}\left(1 + \frac{(1 + \cos\theta)}{q^2(1 - \cos\theta)}\right) + \frac{\cos\theta}{r_0} - \frac{\operatorname{ctg}\alpha}{r_0}\sin\theta;$$

$$\operatorname{ctg}\alpha = -\frac{\sin\theta}{q(1 - \cos\theta)};$$

$$\frac{1}{r} = -(1 - \cos\theta)\frac{q}{2r_0}\left(1 + \frac{(1 + \cos\theta)}{q^2(1 - \cos\theta)}\right) + \frac{\cos\theta}{r_0} + \frac{\sin^2\theta}{r_0 q(1 - \cos\theta)};$$

$$\frac{1}{r} = -(1 - \cos\theta)\frac{q}{2r_0} - \frac{q}{2r_0}\frac{(1 - \cos\theta)(1 + \cos\theta)}{q^2(1 - \cos\theta)}$$

$$+ \frac{\cos\theta}{r_0} + \frac{\sin^2\theta}{r_0 q(1 - \cos\theta)};$$

$$\frac{1}{r} = -(1 - \cos\theta)\frac{q}{2r_0} - \frac{(1 - \cos\theta)(1 + \cos\theta)}{2r_0 q(1 - \cos\theta)}$$

$$+ \frac{\cos\theta}{r_0} + \frac{\sin^2\theta}{r_0 q(1 - \cos\theta)};$$

$$\frac{1}{r} = -(1 - \cos\theta)\frac{q}{2r_0} - \frac{1 - \cos^2\theta}{2r_0 q(1 - \cos\theta)} + \frac{\cos\theta}{r_0} + \frac{\sin^2\theta}{r_0 q(1 - \cos\theta)};$$

$$\frac{1}{r} = -(1 - \cos\theta)\frac{q}{2r_0} - \frac{\sin^2\theta}{2r_0 q(1 - \cos\theta)} + \frac{\cos\theta}{r_0} + \frac{\sin^2\theta}{r_0 q(1 - \cos\theta)};$$

$$\frac{1}{r} = -(1 - \cos\theta)\frac{q}{2r_0} + \frac{\sin^2\theta}{2r_0 q(1 - \cos\theta)} + \frac{\cos\theta}{r_0};$$

$$\frac{1}{r} = \frac{-q^2(1 - \cos\theta)^2 + \sin^2\theta}{2r_0 q(1 - \cos\theta)} + \frac{\cos\theta}{r_0};$$

$$\frac{1}{r} = \frac{-q^2(1 - \cos\theta)^2 + \sin^2\theta + 2q\cos\theta(1 - \cos\theta)}{2r_0 q(1 - \cos\theta)};$$

$$\frac{1}{r} = \frac{-q^2(1 - \cos\theta)^2 + 1 - \cos^2\theta + 2q\cos\theta(1 - \cos\theta)}{2r_0 q(1 - \cos\theta)};$$

$$\frac{1}{r} = \frac{-q^2(1 - \cos\theta)^2 + (1 - \cos\theta)(1 + \cos\theta) + 2q\cos\theta(1 - \cos\theta)}{2r_0 q(1 - \cos\theta)};$$

$$\frac{1}{r} = \frac{-q^2(1 - \cos\theta) + (1 + \cos\theta) + 2q\cos\theta}{2r_0 q};$$

$$\frac{1}{r} = \frac{-q^2 + q^2\cos\theta + 1 + \cos\theta + 2q\cos\theta}{2r_0 q};$$

$$\frac{1}{r} = \frac{1 - q^2 + (1 + 2q + q^2)\cos\theta}{2r_0 q};$$

$$\frac{1}{r} = \frac{(1 - q^2) + (1 + q)^2\cos\theta}{2r_0 q};$$

$$r = \frac{2r_0 q}{(1 - q^2)\left[1 + \frac{(q+1)^2}{1 - q^2}\cos\theta\right]};$$

$$r = \frac{2r_0 q}{(1 - q^2)\left[1 + \frac{(q+1)^2}{(1-q)(1+q)}\cos\theta\right]};$$

$$r = \frac{2r_0 q}{(1 - q^2)\left[1 + \frac{(1+q)}{(1-q)}\cos\theta\right]};$$

$$r = \frac{\frac{2r_0 q}{1 - q^2}}{1 + \frac{1+q}{1-q}\cos\theta};$$

$$q = -\frac{2KM}{r_0 v_0^2} < 0;$$

$$q^2 = \left(-\frac{2KM}{r_0 v_0^2}\right)^2 = \left(\frac{2KM}{r_0 v_0^2}\right)^2;$$

$$r_0 v_0^2 < 2KM; \quad q^2 > 1;$$

$$1 - q^2 < 0; \quad q < 0;$$

$$\frac{2r_0 q}{1 - q^2} > 0;$$

$$\frac{2r_0 q}{1 - q^2} = p_{\text{env}}$$

$$\frac{1 + q}{1 - q} = \frac{1 - |q|}{1 + |q|} < 1;$$

$$\frac{1 + q}{1 - q} = e_{\text{env}}$$

$$r = \frac{p_{\text{env}}}{1 + e \cos \theta_{\text{env}}}.$$

This represents the equation of an ellipse in plane polar coordinates, which proves that the trajectory of the special satellite S that envelops by an external tangent the totality of the orbits of the n satellites is an ellipse, having the Earth in one of its foci. The elements of the orbit (envelope) of the special satellite, S, are determined as follows:

$$r = \frac{\frac{2r_0 q}{1 - q^2}}{1 + \frac{1 + q}{1 - q} \cos \theta};$$

$$r_{\theta=0} = \frac{r_0 q}{q + 1} = r_{\text{env; max}} > 0; \quad r_{\theta=\pi} = -\frac{r_0}{q + 1} = r_{\text{env; min}} > 0;$$

$$r_{\text{env; max}} + r_{\text{env; min}} = 2a_{\text{env}}; \quad a_{\text{env}} = \frac{r_0(q - 1)}{2(q + 1)} > 0;$$

$$q < -1;$$

$$e_{\text{env}} = \sqrt{1 - \frac{b_{\text{env}}^2}{a_{\text{env}}^2}};$$

$$b_{\text{env}} = a_{\text{env}} \sqrt{1 - e_{\text{env}}^2} = \sqrt{a_{\text{env}} p_{\text{env}}} = -\frac{r_0 \sqrt{-q}}{q + 1} > 0;$$

$$b_{\text{env}} = \frac{1}{2} \sqrt{(r_{\text{env;max}} + r_{\text{env; min}})^2 - (r_{\text{env;max}} - r_{\text{env;min}})^2}.$$

The value of $r_{\text{env; max}}$ must correspond to the distance from the center of the Earth to which the satellite is launched from point P_0 at an angle $\alpha = 0$, so that, based on the law of conservation of the total mechanical energy of the system, and using the drawing in Figure 2.7:

$$\frac{mv_0^2}{2} - K\frac{mM}{r_0} = -K\frac{mM}{r_{env;max}}; \quad r_{env;\,max} = \frac{r_0 q}{q+1}; \quad q < -1;$$

$$r_{env;\,max} - r_0 = -\frac{r_0}{1+q} = r_{env;min}.$$

This proves that the injection point of the n satellites (point P$_0$) is one of the foci of the envelope on which the special satellite S revolves.

That the Earth is in one of the foci of the envelope ellipse is proven by Figure 2.7, representing the elliptical orbit of the satellite S$_k$, which was launched at an angle $\alpha = \frac{\pi}{2}$.

Under these conditions, it results that:

$$p = \frac{r_0^2 v_0^2}{KM}; \quad e^2 = \left(1 - \frac{p}{r_0}\right)^2; \quad e = 1 - \frac{p}{r_0} = 1 - \frac{r_0 v_0^2}{KM};$$

$$r_{k;\,min} = \frac{p}{1+e};$$

$$r_{k;\,min} = \frac{r_0}{\frac{2KM}{r_0 v_0^2}\left(1 - \frac{r_0 v_0^2}{2KM}\right)};$$

$$r_{k;\,min} = -\frac{r_0}{1+q} = r_{env;\,min};$$

$$r_{k;\,max} = \frac{p}{1-e} = r_0.$$

If the injections of the satellites at point P$_0$ are done in such a way that the conditions

$$r_0 v_0^2 < 2KM, \quad r_0 v_0^2 = 2KM, \quad r_0 v_0^2 > 2KM,$$

are met, then the orbits of those n satellites form a family of ellipses, parabolas, and hyperbolas, respectively, having the Earth in their focus, for which

$$q = \frac{the\ initial\ gravitational\ potential\ energy\ of\ the\ satellite-}{the\ initial\ kinetic\ energy\ of\ the\ system},$$

according to the variations below:

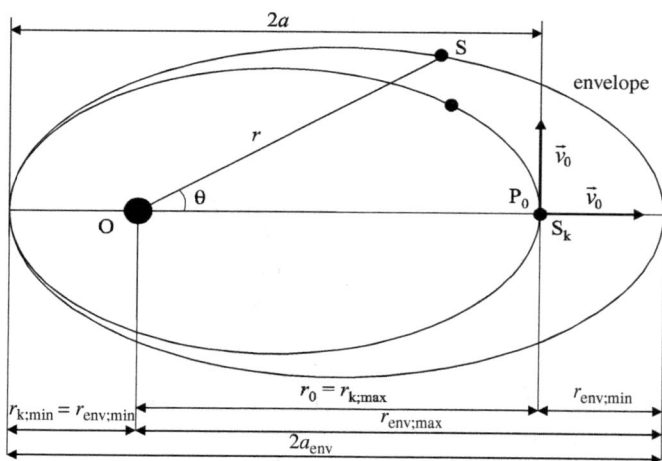

Fig. 2.7

(1)

$$r_0 v_0^2 < 2KM; \quad q = \frac{-K\frac{mM}{r_0}}{\frac{mv_0^2}{2}} = -\frac{2KM}{r_0 v_0^2}; \quad q = -\frac{2KM}{r_0 v_0^2} < 0;$$

$$r = \frac{p_{\text{env}}}{1 + e_{\text{env}} \cos\theta};$$

$$\frac{2r_0 q}{1 - q^2} = p_{\text{env}} > 0; \quad \frac{1+q}{1-q} = e_{\text{env}},$$

which implies the possibility of the special satellite S enveloping the elliptical orbits of the n satellites through the external tangent under the action of the Earth's gravitational attraction.

(2)

$$q = \frac{-K\frac{mM}{r_0}}{\frac{mv_0^2}{2}} = -\frac{2KM}{r_0 v_0^2}; \quad r_0 v_0^2 = 2KM;$$

$$q = -1;$$

$$\frac{1}{r} = \frac{(1 - q^2) + (1 + q)^2 \cos\theta}{2r_0 q} = 0; \quad r \to \infty,$$

which implies the impossibility of the special satellite S enveloping the parabolic orbits of the n satellites through the external tangent under the action of the Earth's gravitational attraction.

(3)

$$q = \frac{-K\frac{mM}{r_0}}{\frac{mv_0^2}{2}} = -\frac{2KM}{r_0 v_0^2}; \quad r_0 v_0^2 > 2KM;$$

$$-1 < q < 0;$$

$$r = \frac{\frac{2r_0 q}{1-q^2}}{1 - \frac{(q+1)^2}{q^2-1}\cos\theta} = \frac{p_{env}}{1 + e_{env}\cos\theta};$$

$$p_{env} < 0,$$

which implies the impossibility of the special satellite S enveloping the hyperbolic orbits of those n satellites through the external tangent under the action of the Earth's gravitational attraction.

(b) At some point on its elliptical trajectory, the speed of the special satellite S is

$$v = \sqrt{KM\left(\frac{2}{r} - \frac{1}{a_{env}}\right)}.$$

When the special satellite S is at the minimum distance from the center of the Earth, as indicated in Figure 2.8, its speed is:

$$v_{max} = \sqrt{KM\left(\frac{2}{r_{env;min}} - \frac{1}{a_{env}}\right)};$$

$$v_{max} = \sqrt{\frac{KM(q+1)(1-2q)}{r_0(q-1)}}.$$

If, after its launch, the mini-satellite S_0 evolves on a parabola, its speed at the moment of launch is

$$v_{S_0} = v_0 = \sqrt{\frac{2KM}{r_{env;min}}} = \sqrt{-\frac{2KM}{r_0}(q+1)},$$

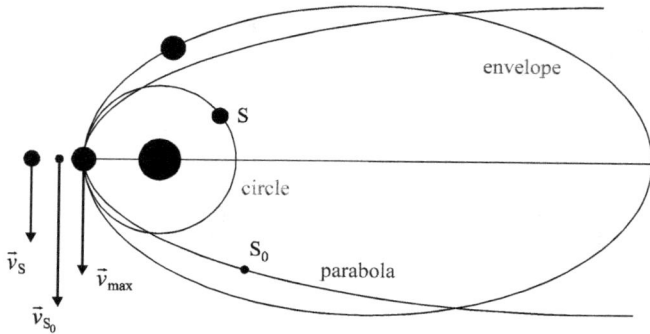

Fig. 2.8

and the speed of satellite S, still revolving in a circle, is

$$v_S = v = \sqrt{\frac{KM}{r_{\mathrm{env;min}}}} = \sqrt{-\frac{KM}{r_0}(q+1)}.$$

Using the law of conservation of momentum, it follows that:

$$(m + m_0)v_{\mathrm{max}} = mv + m_0 v_0;$$

$$\frac{m}{m_0} = \frac{v_0 - v_{\mathrm{max}}}{v_{\mathrm{max}} - v};$$

$$\frac{m}{m_0} = \frac{\sqrt{2(1-q)} - \sqrt{1-2q}}{\sqrt{1-2q} - \sqrt{1-q}}; \quad q < 0; \quad \frac{m}{m_0} > 0.$$

For the terrestrial observer, located at position O_r, facing the center of the Earth (Figure 2.9), the special satellite S, whose direction in its uniform circular motion we consider to be the same as the direction of the Earth's rotation, appears on the horizon in position R (rise). Due to the rotation of the Earth, when the satellite passes below the horizon of the same observation place (set) in position A, the terrestrial observer will be in position O, compared to the center of the Earth.

Within the duration t of the satellite S's visibility, the observer's vector radius describes the angle at center α, and the satellite's vector radius describes the angle at center β.

If ω_0 and ω are the angular velocity of the Earth's rotation and the angular velocity of the uniform circular motion of satellite S, respectively, it follows that:

$$\alpha = \omega_0 t;$$

$$\beta = \omega t = \frac{v}{r_{\text{env; min}}} t;$$

$$\beta = \alpha + 2\gamma;$$

$$\cos\gamma = \frac{R}{r_{\text{env; min}}};$$

$$\left(\frac{v}{r_{\text{env;min}}} - \omega_0 \right) t = \text{arc}\cos\frac{R}{r_{\text{env; min}}};$$

$$t = \frac{2\,\text{arc}\cos\dfrac{R}{r_{\text{env; min}}}}{\dfrac{v}{r_{\text{env;min}}} - \omega_0}.$$

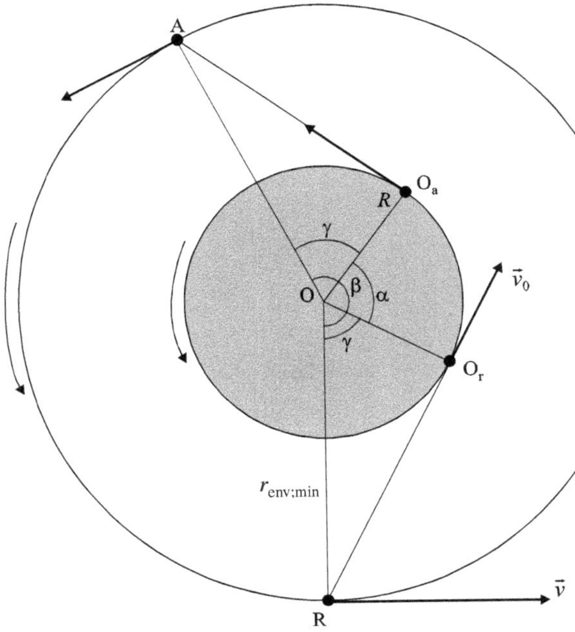

Fig. 2.9

If the direction of the satellite's movement is opposite to the direction of the Earth's rotation, as shown in Figure 2.10, it results that:

$$\alpha + \beta = 2\gamma;$$

$$t = \frac{2\text{arc cos}\dfrac{R}{r_{\text{env;min}}}}{\dfrac{v}{r_{\text{env;min}}} + \omega_0}.$$

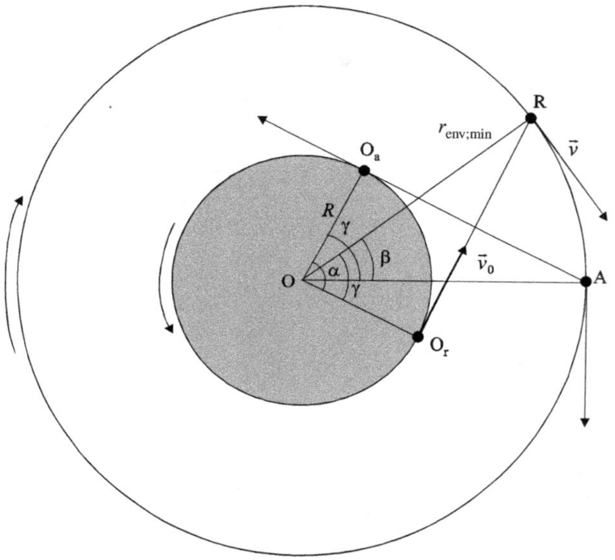

Fig. 2.10

(c) The gravitational transfer of the special satellite S from a circular orbit with radius $r_{\text{env; min}}$ to an outer concentric circular orbit with radius R_e is achieved as shown in Figure 2.11, providing the conditions for the satellite to evolve on an elliptical orbit tangent to the two circular orbits at the diametrically opposite points P and A, so that the Earth is in one of the foci of this ellipse.

The elements of the ellipse onto which the special satellite S is transferred are established as follows:

$$r_{\text{min}} = a(1 - e) = r_{\text{env; min}}; \quad a = \frac{1}{2}(r_{\text{env; min;}} + R_e);$$

$$e = 1 - \frac{2r_{\text{env;min}}}{r_{\text{env;min}} + R_e}.$$

At point P on the circular orbit with radius $r_{\text{env; min}}$, an initial correction of the satellite's speed, $\Delta \vec{v}_{\text{in}}$, must be performed so that the resulting speed,

$$\vec{v}_{\text{pg}} = \vec{v} + \Delta \vec{v}_{\text{in}},$$

is the speed of the satellite at the perigee P of the established elliptical orbit:

$$v_{\text{pg}} = v_{\max} = \sqrt{KM \frac{1+e}{a(1-e)}}.$$

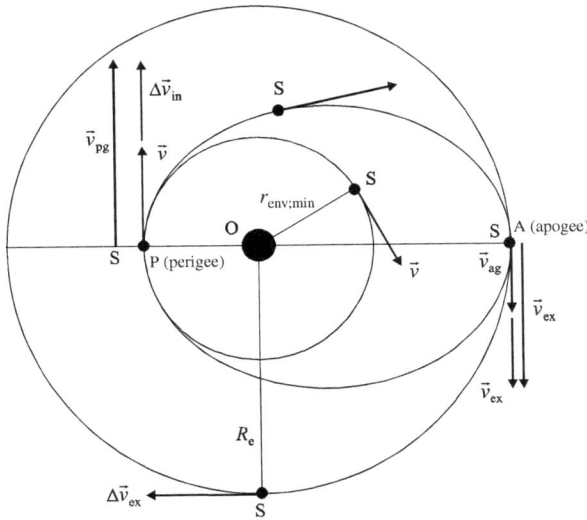

Fig. 2.11

It results that:

$$\Delta v_{\text{in}} = v_{\max} - v;$$

$$\Delta v_{\text{in}} = \sqrt{\frac{KM}{r_{\text{env;min}}}} \left(\sqrt{\frac{2R_e}{r_{\text{env;min}} + R_e}} - 1 \right).$$

When the special satellite S reaches the apogee A of its elliptical orbit, at the distance $r_{\max} = R_e$, its speed is

$$v_{\text{ag}} = v_{\min} = \sqrt{KM \frac{1-e}{a(1+e)}},$$

or, equivalently, using the law of conservation of kinetic momentum,

$$v_{\min} = \frac{r_{\min}}{R_e} v_{\max}.$$

At point A on the elliptical orbit, a second correction of the speed of the satellite, $\Delta \vec{v}_{\text{ex}}$, must be performed so that its resulting speed,

$$\vec{v}_{\text{ex}} = \vec{v}_{\text{ag}} + \Delta \vec{v}_{\text{ex}},$$

is the speed required for the satellite to evolve on a circle with radius R_e:

$$v_{\text{ex}} = \sqrt{\frac{KM}{R_e}}.$$

It results that:

$$\Delta v_{\text{ex}} = v_{\text{ex}} - v_{\min};$$

$$\Delta v_{\text{ex}} = \sqrt{\frac{KM}{R_e}} \left(1 - \sqrt{\frac{2r_{\text{env;min}}}{r_{\text{env;min}} + R_e}} \right).$$

The duration of the transfer of the special satellite S from the inner circular orbit to the outer concentric circular orbit is half the period of the satellite's movement on the established elliptical orbit, that is:

$$\Delta t = \pi \sqrt{\frac{a^3}{KM}};$$

$$\Delta t = \frac{\pi}{2}(r_{\text{env;min}} + R_e)\sqrt{\frac{r_{\text{env;min}} + R_e}{2KM}}.$$

Both speed corrections increase the value of the satellite's local orbital speeds when it is transferred from an inner circular orbit to an outer concentric circular orbit.

In the case of the transfer of the special satellite S to an inner concentric circular orbit, as shown in Figure 2.12, the gravitational transfer is performed on an elliptical trajectory. Two corrections of the velocities are necessary, which are also carried out at diametrically opposite points.

In this case, both velocity corrections cause the local orbital velocity values of the transferred satellite to decrease.

The elements of the ellipse onto which the special satellite S is now transferred are determined as follows:

$$r_{\max} = a(1+e) = r_{\text{env; min}}; \quad a = \frac{1}{2}(r_{\text{env;min}} + R_{\text{i}});$$

$$e = \frac{r_{\text{env;min}} - R_{\text{i}}}{r_{\text{env;min}} + R_{\text{i}}}.$$

At point A on the circular orbit with radius $r_{\text{env; min}}$, an initial correction of the satellite's speed, $\Delta \vec{v}_{\text{ex}}$, must be performed in such a way that the resulting speed,

$$\vec{v}_{\text{ag}} = \vec{v} + \Delta \vec{v}_{\text{ex}},$$

is the speed of the satellite at apogee A of the established elliptical orbit:

$$v_{\text{ag}} = v_{\min} = \sqrt{KM \frac{1-e}{a(1+e)}}.$$

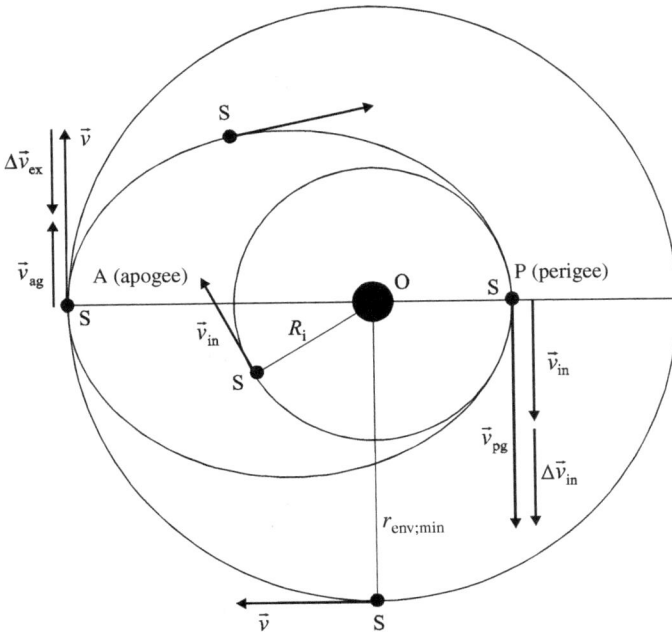

Fig. 2.12

It results that:

$$\Delta v_{ex} = v - v_{min};$$

$$\Delta v_{ex} = \sqrt{\frac{KM}{r_{env;min}}}\left(1 - \sqrt{\frac{2R_i}{r_{env;min} + R_i}}\right).$$

When the special satellite S reaches the perigee P of its elliptical orbit, at the distance $r_{min} = R_i$, its speed is

$$v_{pg} = v_{max} = \sqrt{KM\frac{1+e}{a(1-e)}}.$$

At point P on the elliptical orbit, a second correction of the satellite's speed, $\Delta \vec{v}_{in}$, must be performed in such a way that its resulting speed,

$$\vec{v}_{in} = \vec{v}_{pg} + \Delta \vec{v}_{in},$$

is the speed necessary for the satellite to evolve on a circle of radius R_i:

$$v_{in} = \sqrt{\frac{KM}{R_i}}.$$

It results that:

$$\Delta v_{in} = v_{max} - v_{in};$$

$$\Delta v_{in} = \sqrt{\frac{KM}{R_i}}\left(\sqrt{\frac{2r_{env;min}}{r_{env;min} + R_i}} - 1\right).$$

The duration of the transfer of the special satellite S from the outer circular orbit to the inner concentric circular orbit is half the period of the satellite's displacement on the established elliptical orbit, so:

$$\Delta t = \pi\sqrt{\frac{a^3}{KM}};$$

$$\Delta t = \frac{\pi}{2}(r_{env;\,min;} + R_i)\sqrt{\frac{r_{env;min} + R_i}{2KM}}.$$

(d) The gravitational transfer of the special satellite S from the circular orbit with radius $r_{\text{env; min}}$ to the outer concentric circular orbit with radius R_e is performed as indicated in Figure 2.13. It provides the conditions for the satellite to evololve on a parabolic orbit, with the Earth in its focus, tangent to the inner circular orbit at point A and intersecting the outer circular orbit at point Q.

The first correction of the speed of the satellite S is performed at point A, so that at this point, which becomes the vertex of the parabola, the speed of the satellite is:

$$\vec{v}_A = \vec{v}_{\text{max}} = \vec{v}_{\text{in}} + \Delta\vec{v}_{\text{in}};$$

$$v_{\text{in}} = \sqrt{\frac{KM}{r_{\text{env;min}}}}; \quad v_{\text{max}} = \sqrt{2\frac{KM}{r_{\text{env;min}}}};$$

$$\Delta v_{\text{in}} = v_{\text{max}} - v_{\text{in}} = \sqrt{\frac{KM}{r_{\text{env;min}}}}\left(\sqrt{2} - 1\right).$$

The transfer trajectory of satellite S being a parabola, for which $e = 1$, its equation in plane polar coordinates is

$$r = \frac{p}{1 + \cos\theta} = \frac{p}{2\cos^2\frac{\theta}{2}},$$

where $p = 2r_{\text{env; min}}$.

The intersection of the transfer trajectory with the outer circular orbit is at point Q, for which:

$$\cos\frac{\theta_{\text{max}}}{2} = \frac{p}{2R_e} = \frac{r_{\text{env;min}}}{R_e}; \quad \theta_{\text{max}} = 2\arccos\frac{r_{\text{env;min}}}{R_e};$$

$$v_Q = v_{\text{max}}\cos\frac{\theta_{\text{max}}}{2} = \frac{\sqrt{2KMr_{\text{env;min}}}}{R_e},$$

where the second correction of the satellite speed ($\Delta\vec{v}_{\text{ex}}$; $\Delta v_{\text{ex}} = \Delta v_{\text{in}}$) must be performed, for its arrival on the outer circular orbit with the speed

$$\vec{v}_{\text{ex}} = \vec{v}_Q + \Delta\vec{v}_{\text{ex}},$$

$$v_{\text{ex}} = \sqrt{\frac{KM}{R_e}},$$

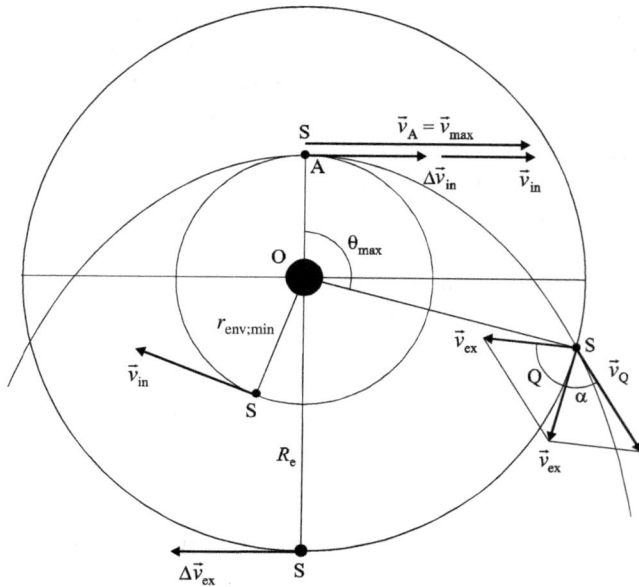

Fig. 2.13

so that we have:

$$v_{\text{ex}}^2 = v_Q^2 + (\Delta v_{\text{ex}})^2 + 2v_Q(\Delta v_{\text{ex}}) \cos \alpha;$$

$$\cos \alpha = \frac{v_{\text{ex}}^2 - v_Q^2 - (\Delta v_{\text{ex}})^2}{2v_Q(\Delta v_{\text{ex}})} = \frac{v_{\text{ex}}^2 - v_Q^2 - (\Delta v_{\text{in}})^2}{2v_Q(\Delta v_{\text{in}})};$$

$$\cos \alpha = \frac{1 - 2\frac{r_{\text{env;min}}}{R_e} - \frac{R_e}{r_{\text{env;min}}}(\sqrt{2} - 1)^2}{2\sqrt{2}(\sqrt{2} - 1)}.$$

We use well-known reasoning to calculate the duration of the gravitational transfer of satellite S between the two concentric circular orbits.

The transfer trajectory of the satellite S being a parabola, for which $e = 1$, its equation in plane polar coordinates is:

$$r = \frac{p}{1 + \cos \theta} = \frac{p}{2 \cos^2 \frac{\theta}{2}},$$

$$C = r^2 \dot{\theta} = r^2 \frac{d\theta}{dt}; \quad p = \frac{C^2}{KM},$$

$$dt = \frac{p^2}{4\sqrt{pKM} \cos^4 \frac{\theta}{2}} d\theta;$$

$$t = \frac{p^2}{4\sqrt{pKM}} \int_0^{\theta_{\max}} \frac{d\theta}{\cos^4 \frac{\theta}{2}};$$

$$t = \frac{p^2}{2\sqrt{pKM}} \left(\frac{2 \sin \frac{\theta}{2}}{3 \cos^3 \frac{\theta}{2}} + \frac{2}{3} \operatorname{tg} \frac{\theta}{2} \right) \Bigg|_0^{\theta_{\max}};$$

$$t = \frac{p^2}{2\sqrt{pKM}} \left(\frac{2 \sin \frac{\theta_{\max}}{2}}{3 \cos^3 \frac{\theta_{\max}}{2}} + \frac{2}{3} \operatorname{tg} \frac{\theta_{\max}}{2} \right).$$

Problem 3. Parallel River and Railway

In a land region in the Northern Hemisphere, with geographic latitude φ, where the gravitational acceleration is g, a river flows from south to north, and, on an adjacent railway, a locomotive moves from south to north along the same meridian, their relative velocities with respect to the Earth being equal and constant, v. The influence of the Earth's rotation on the Earth's gravitational acceleration is neglected.

(a) *Determine* the water level difference between the two banks of the river. *We know*: the width of the river (l) and the angular speed of the Earth's rotation (ω). The speed of the water is the same at any point of the cross-section of the river.

(b) *Determine* the ratio of the vertical normal reactions of the two rails on the locomotive.

It is known that the distance between the two rails is equal to the distance from the center of mass of the locomotive to the plane of the rails.

Solution

(a) The real forces acting on an elementary volume of liquid with mass dm from the surface of the river are pressure, with resultant $d\vec{F}_p$ (from the interaction with the surrounding liquid), and gravity, with $d\vec{G} = \vec{g}dm$.

Apart from these forces, a non-inertial observer participating in the rotating movement of the Earth, located on the bank of the river, must consider that the complementary elementary Coriolis force acts on the elementary volume of liquid, in relative motion on the Earth's surface,

$$d\vec{F}_{\text{cor}} = -2dm\vec{\omega} \times \vec{v},$$

whose orientation is represented in Figure 3.1.

From the fundamental principle of dynamics, relative to the non-inertial reference system X'Y'Z', fixed to the Earth at point O', written in general form, we have

$$dm\vec{a}_{\text{rel}} = d\vec{F} + d\vec{F}_c,$$

where $d\vec{F}$ is the resultant of the real force and $d\vec{F}_c$ is the resultant of the acting complementary forces based on the elementary volume. Considering that its movement is uniform,

$$0 = d\vec{F}_p + d\vec{G} + d\vec{F}_{\text{cor}},$$

where $d\vec{F}_{\text{cor}}$ is the elementary complementary Coriolis force acting on the elementary volume of the liquid:

$$d\vec{F}_{\text{cor}} = -2dm\vec{\omega} \times \vec{v};$$

$$d\vec{R} = d\vec{G} + d\vec{F}_{\text{cor}};$$

$$d\vec{F}_p + d\vec{R} = 0.$$

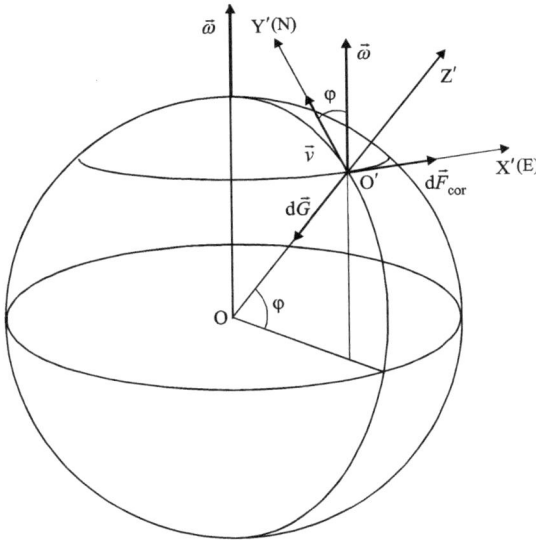

Fig. 3.1

According to the fundamental law of fluid statics, with the free surface of the river water being isobaric ($p = p_{atm} = $ constant), it will have to be a flat surface, perpendicular to the resultant elemental force $d\vec{R}$, inclined at an angle α to the horizontal plane X'O'Y', so that $d\vec{F}_p = -d\vec{R}$, as shown in Figure 3.2. From this, it follows that:

$$\text{tg}\, \alpha = \frac{2\omega v \sin \phi}{g};$$

$$\text{tg}\, \alpha \approx \sin \alpha = \frac{\Delta h}{l};$$

$$\Delta h \approx l \,\text{tg}\, \alpha = \frac{2l\omega v \sin \phi}{g}.$$

(b) In Figure 3.3, in relation to the non-inertial reference system X' Y' Z', fixed to the Earth at point O' on its surface, we represent the complementary Coriolis force in addition to the real forces acting on the locomotive (rail reactions and the weight of the locomotive).

In relation to the reference system considered, the locomotive's movement is a straight line and uniform, so that:

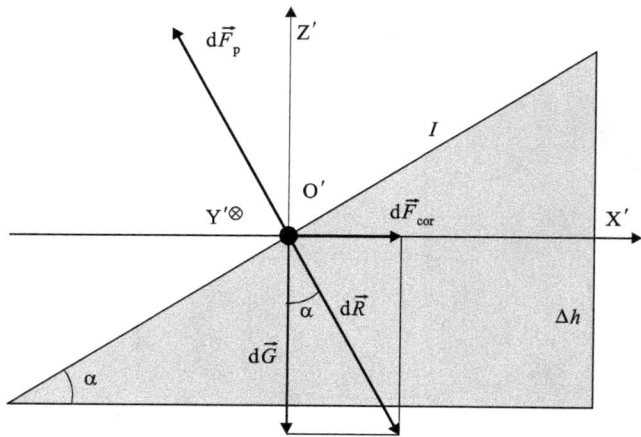

Fig. 3.2

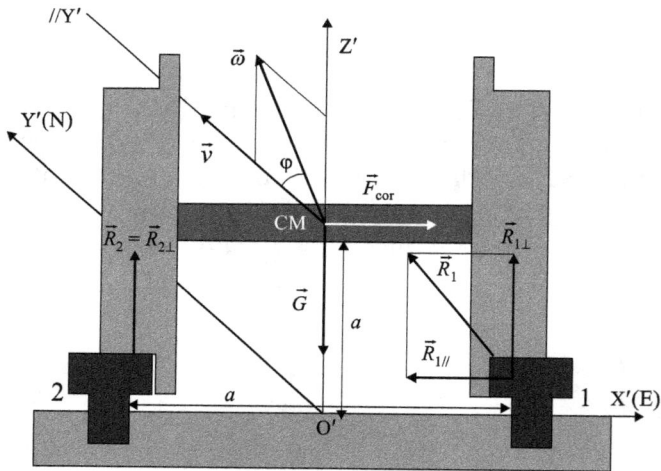

Fig. 3.3

$$\vec{R}_1 + \vec{R}_2 + \vec{G} + \vec{F}_{\text{cor}} = 0;$$

$$\vec{F}_{\text{cor}} = -2m\vec{\omega} \times \vec{v};$$

$$R_{1\perp} + R_{2\perp} = mg;$$

$$R_{1//} = F_{\text{cor}} = 2m\omega v \sin \varphi.$$

The translational movement of the locomotive assumes that, in relation to its center of mass, the resultant of the forces acting on the locomotive is null, so:

$$\vec{M}_{\vec{R}_1} + \vec{M}_{\vec{R}_2} + \vec{M}_{\vec{G}} + \vec{M}_{\vec{F}_{cor}} = 0;$$

$$R_{2\perp} + 2R_{1//} = R_{1\perp}.$$

In these conditions, it results that:

$$R_{1\perp} = \frac{mg}{2} + 2m\omega v \sin\varphi;$$

$$R_{2\perp} = \frac{mg}{2} - 2m\omega v \sin\varphi;$$

$$\frac{R_{1\perp}}{R_{2\perp}} = \left(1 + \frac{4\omega v}{g}\sin\phi\right)\left(1 - \frac{4\omega v}{g}\sin\phi\right)^{-1};$$

$$\frac{4\omega v}{g}\sin\varphi \ll 1;$$

$$\frac{R_{1\perp}}{R_{2\perp}} \approx 1 + \frac{8\omega v}{g}\sin\varphi.$$

Problem 4. Oscillating Rod

Consider a homogeneous horizontal rod suspended by an inextensible wire, which is at rest in the east–west direction and can only rotate around the vertical axis that passes through its center of mass. On this rod, in symmetrical positions, two identical bodies are at rest, each a material point with mass m at distances b from the vertical axis. A special device on the rod simultaneously launches the two bodies towards the axis of rotation with equal speeds; they then stop simultaneously at distances a from the vertical axis, after sliding along a symmetrical path in relation to the rod's center of mass.

(a) *Determine* the angular velocity acquired by the entire system in relation to the Earth. The following are known: I_0 – the moment of inertia of the rod in relation to the vertical axis; ω – the angular speed of rotation of the Earth; φ – the geographical latitude of the place. It will be assumed that the rod remains at rest while the two bodies are moving along the rod.

(b) *Determine* the angular amplitude of the oscillations of the rod, occuring in a horizontal plane, if the elastic suspension wire has the torsion constant C, and write the law of harmonic oscillations of the rod.

Solution

(a) The geometry of the system at the initial moment and at the moment of immobilization of bodies 1 and 2 is represented in Figure 4.1. At an intermediate moment, when the instantaneous velocities of the two bodies are $\vec{v}_1$ and $\vec{v}_2$, respectively, oriented as shown in the drawing, a complementary instantaneous Coriolis force acts on each body:

$$\vec{F}_{\text{cor},1} = -2m\vec{\omega} \times \vec{v}_1;$$

$$\vec{F}_{\text{cor},1} = -2m\vec{\omega}_{y'} \times \vec{v}_1 - 2m\vec{\omega}_{z'} \times \vec{v}_1;$$

$$\vec{F}_{\text{cor},1} = -2m \begin{vmatrix} \vec{i}' & \vec{j}' & \vec{k}' \\ 0 & \omega\cos\phi & \omega\sin\phi \\ -v_1 & 0 & 0 \end{vmatrix}$$

$$= 2m\omega v_1 \sin\varphi\, \vec{j}' - 2m\omega v_1 \cos\varphi \vec{k}';$$

$$v_1 = -\frac{dx'}{dt}; \quad dx' < 0;$$

$$\vec{F}_{\text{cor},1} = -2m\omega\sin\varphi\frac{dx'}{dt}\vec{j}' + 2m\omega\cos\varphi\frac{dx'}{dt}\vec{k}' = \vec{F}_{\text{cor},1,y'} + \vec{F}_{\text{cor},1,z'};$$

$$\vec{F}_{\text{cor},2} = -2m\vec{\omega} \times \vec{v}_2;$$

$$\vec{F}_{\text{cor},2} = -2m\vec{\omega}_{y'} \times \vec{v}_2 - 2m\vec{\omega}_{z'} \times \vec{v}_2;$$

$$\vec{F}_{\text{cor},2} = -2m \begin{vmatrix} \vec{i}' & \vec{j}' & \vec{k}' \\ 0 & \omega\cos\phi & \omega\sin\phi \\ -v_2 & 0 & 0 \end{vmatrix}$$

$$= -2m\omega v_2 \sin\varphi\, \vec{j}' + 2m\omega v_2 \cos\varphi \vec{k}';$$

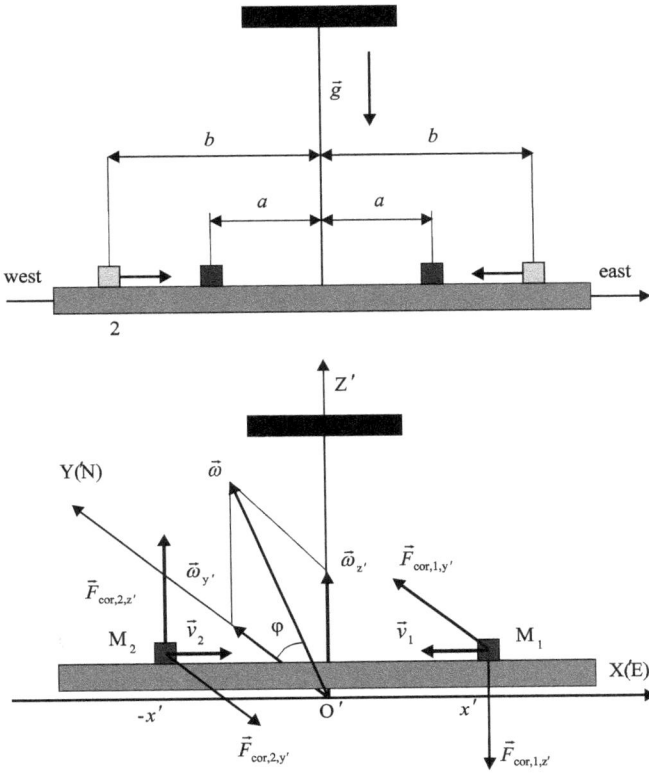

Fig. 4.1

$$v_2 = +\frac{dx'}{dt}; \quad dx' > 0;$$

$$v_2 = v_1; \quad v_2 = -\frac{dx'}{dt}; \quad dx' < 0;$$

$$\vec{F}_{\mathrm{cor},2} = 2m\omega \sin\varphi \frac{dx'}{dt}\vec{j}' - 2m\omega \cos\varphi \frac{dx'}{dt}\vec{k}';$$

$$\vec{F}_{\mathrm{cor},2} = \vec{F}_{\mathrm{cor},2,y'} + \vec{F}_{\mathrm{cor},2,z'}.$$

Considering that the rod can only rotate around the O'Z' axis, the resultant moment of the two Coriolis forces is reduced to the resultant moment of their components parallel to the O'Y axis, so:

$$\vec{M}_{\text{cor},O'Z'} = \overrightarrow{O'M_1} \times \vec{F}_{\text{cor},1,y'} + \overrightarrow{O'M_2} \times \vec{F}_{\text{cor},2,y'};$$

$$M_{\text{cor},O'Z'} = -4m\omega \sin\varphi\, x' \frac{dx'}{dt}.$$

According to the kinetic momentum variation theorem, it follows that:

$$\frac{d\vec{J}}{dt} = \vec{M}_{\text{resultant}} = \vec{M}_{\text{cor},O'Z'};$$

$$dJ = -4m\omega \sin\varphi\, x' dx';$$

$$J - J_0 = -4m\omega \sin\varphi \int_{-b}^{-a} x' dx';$$

$$(I_0 + 2ma^2)\Omega = 2m\omega \sin\varphi (b^2 - a^2);$$

$$\Omega = \frac{2m\omega(b^2 - a^2)\sin\phi}{I_0 + 2ma^2}.$$

(b) From the moment of immobilization of the two bodies, the system is a torsion pendulum, whose movement is described by the equation:

$$(I_0 + 2ma^2)\frac{d^2\theta}{dt^2} + C\theta = 0;$$

$$\frac{d^2\theta}{dt^2} + \omega'^2\theta = 0;$$

$$\omega'^2 = \frac{c}{I_0 + 2ma^2},$$

where ω'' is the pulsation of the oscillations of the torsion pendulum. The general solution of bottom motion equations is

$$\theta = A\cos\omega't + B\sin\omega't.$$

Taking into account the initial conditions, it follows that:

$$t = 0; \quad \theta = 0; \quad \frac{d\theta}{dt} = \Omega;$$

$$A = 0; \quad B = \frac{\Omega}{\omega'};$$

$$\theta = \frac{\Omega}{\omega'} \sin \omega' t = \theta_{\max} \sin \omega' t;$$

$$\theta_{\max} = \frac{\Omega}{\omega'} = \frac{2m\omega(b^2\_a^2) \sin \phi}{\sqrt{C(I_0 + 2ma^2)}}.$$

Chapter 4

International Pre-Olympic Physics Contest 2005, Călimăneşti, Romania

Problem 1. Satellite with a Solar Sail

A special satellite with mass m is evolving around the Sun in the Earth's circular orbit. At a certain time, a "solar sail" (a circular disk with the radius r) opens on the satellite, with one of its sides forming a flat, perfectly reflective mirror, which will be permanently oriented perpendicular to the direction of the Sun.

(a) *Determine* the elements of the pressure force that acts at every moment on the satellite's sail due to the solar radiation with normal incidence on the sail's plane.

(b) *Specify* the type of motion of the satellite after the sail opens.

(c) *Determine* the period of the satellite's rotation around the Sun after the sail opens.

   Given: L, the integral luminosity of the Sun; c, the speed of light in vacuum; R_0, the radius of the Earth's circular orbit around the Sun; M, the mass of the Sun; and K, the constant of the gravitational attraction. The gravitational influence of the Earth on the satellite is neglected.

Consider that

$$m > \frac{Lr^2}{2cKM}.$$

The energy of the total radiation emitted by the Sun per unit time across its entire surface, at all the wavelengths, and in all directions is called the *integral luminosity* of the Sun, L. Dimensionally, luminosity is a power: $L = 3.86 \cdot 10^{26}$ W.

Solution

(a) As Figure 1.1 indicates, let's admit that the Sun is a sphere with surface Σ_0, having the radius R_S.

If Σ is the surface of the circumsolar sphere, whose radius R represents the instantaneous distance between the satellite (the sail of the satellite) and the center of the Sun, then the energy of the solar radiation that crosses the surface Σ per unit time is equal to L.

In these conditions, the energy of the solar radiation that arrives per unit time on the surface of the solar sail with the area πr^2 is

$$x = L\frac{\pi r^2}{4\pi R^2}.$$

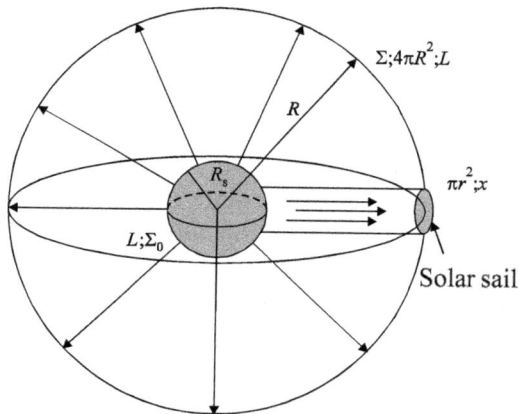

Fig. 1.1

As a result, the illumination of the surface of the sail (the amount of solar radiation energy reaching the sail per unit time per unit area of the sail's surface) is:

$$E = \frac{x}{\pi r^2} = \frac{L}{4\pi R^2}; \quad \langle E \rangle_{SI} = \frac{W}{m^2}.$$

Using Figure 1.2, let us now calculate the variation in the momentum of a photon as a result of the reflection of sunlight on the surface of the sail at a certain time:

$$\Delta \vec{p} = \vec{p}_{\mathrm{r}} - \vec{p}_{\mathrm{i}};$$

$$\Delta p = p_{\mathrm{r}} + p_{\mathrm{i}}; \quad p_{\mathrm{r}} = p_{\mathrm{i}} = p_0 = \frac{h\nu}{c}; \quad \Delta p = 2\frac{h\nu}{c},$$

where h is Planck's constant, ν is the frequency of light, and c is the speed of light in vacuum.

As a result of the principle of reciprocal actions, a force will act on the sail:

$$\vec{f} = -\frac{\Delta \vec{p}}{\Delta t}.$$

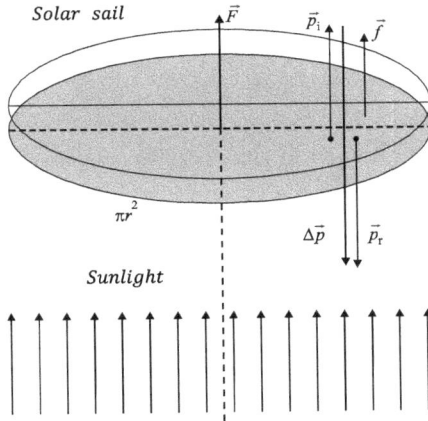

Fig. 1.2

If, during the time interval Δt, a number of photons ΔN_{k} with frequency ν_{k} are reflected on the reflective face of the solar sail, then the force that will act on the sail will be:

$$F_{\mathrm{k}} = f_{\mathrm{k}} \Delta N_{\mathrm{k}} = \Delta N_{\mathrm{k}} \frac{\Delta p_{\mathrm{k}}}{\Delta t} = N_{\mathrm{k}} \pi r^2 \Delta t \frac{2\frac{h\nu_{\mathrm{k}}}{c}}{\Delta t},$$

where N_k is the number of photons with frequency ν_k that arrive per unit area of the sail's surface per unit time;

$$F_k = 2\frac{N_k h\nu_k}{c}\pi r^2;$$

$$\langle N_k h\nu_k \rangle_{SI} = \frac{W}{m^2};$$

$$N_k h\nu_k = E_k,$$

representing the illumination of the sail's surface due to the component of the solar radiation with frequency ν_k;

$$F_k = 2\frac{E_k}{c}\pi r^2;$$

$$P_k = \frac{F_k}{\pi r^2} = 2\frac{E_k}{c},$$

representing the pressure exerted on the sail by the component of the solar radiation with frequency ν_k.

We next calculate the resultant force acting on the sail for all components of the solar radiation $(\nu_1, \nu_2, \ldots, \nu_n)$:

$$F = \sum_{k=1}^{n} F_k = 2\frac{\sum_{k=1}^{n} E_k}{c}\pi r^2;$$

$$\sum_{k=1}^{n} E_k = E,$$

representing the total illumination of the sail's surface due to all components of the solar radiation;

$$F = 2\frac{E}{c}\pi r^2;$$

$$P = \frac{F}{\pi r^2} = 2\frac{E}{c},$$

representing the pressure from the sunlight on the sail;

$$F = 2\frac{L}{c}\frac{\pi r^2}{4\pi R^2} = \frac{Lr^2}{2cR^2};$$

$$\vec{F} = \frac{Lr^2}{2cR^2}\hat{R} = \frac{Lr^2}{2cR^3}\vec{R}.$$

Conclusion: The pressure force acting at any moment on the sail of the satellite due to the solar radiation with normal incidence on the plane of the sail is inversely proportional to the square of the distance between the satellite and the center of the Sun and has the same orientation as the position vector of the satellite relative to the center of the Sun, as shown in Figure 1.3.

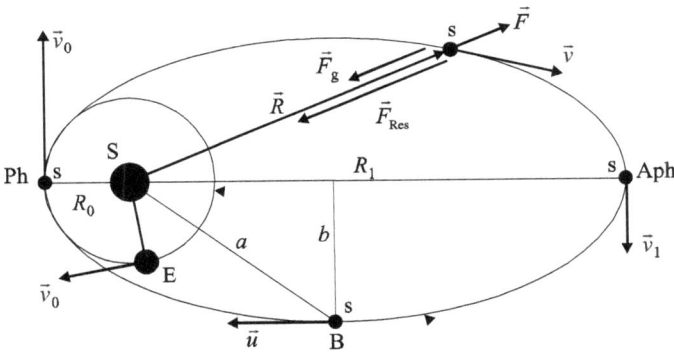

Fig. 1.3

(b) Before the opening of the sail, the movement of the satellite follows the circular orbit of the Earth under the action of the force of gravitational attraction, so that:

$$\frac{mv_0^2}{R_0} = K\frac{mM}{R_0^2} = F_{g0}; \quad v_0 = \sqrt{K\frac{M}{R_0}};$$

$$T_0 = 2\pi\sqrt{\frac{R_0^3}{KM}},$$

where R_0 is the radius of the Earth's orbit.

At a certain moment, after the sail is deployed, the resultant of the forces acting on the satellite, as shown in Figure 1.3, is:

$$\vec{F}_{\text{res}} = \vec{F}_{\text{g}} + \vec{F};$$

$$\vec{F}_{\text{res}} = -\frac{KmM - \frac{Lr^2}{2c}}{R^2}\hat{R}; \quad F_{\text{res}} = \frac{KmM - \frac{Lr^2}{2c}}{R^2}.$$

We identify this as a central force, the effect of which is the movement of the satellite, with the sail open, on an elliptical orbit with the Sun in its focus close to the point where the sail was opened (the perihelion of the elliptical orbit).

According to the laws of conservation of kinetic momentum and total mechanical energy for the *satellite with open sail–Sun* system, it follows that:

$$v_0 R_0 = v_1 R_1;$$

$$\frac{mv_0^2}{2} - \frac{KmM - \frac{Lr^2}{2c}}{R_0} = \frac{mv_1^2}{2} - \frac{KmM - \frac{Lr^2}{2c}}{R_1};$$

$$\alpha = KmM - \frac{mv_0^2}{2} - \frac{KmM - \frac{Lr^2}{2c}}{R_0} = \frac{mv_1^2}{2} - \frac{KmK - \frac{lr^2}{2c}}{R_1};$$

$$\left(mv_0^2 - 2\frac{\alpha}{R_0}\right)R_1^2 + 2\alpha R_1 - mv_0^2 R_0^2 = 0;$$

$$R_1 = \frac{KmM}{KmM - \frac{Lr^2}{c}}R_0; \quad a = \frac{1}{2}(R_0 + R_1),$$

where a is the semi-major axis of the ellipse;

$$a = \frac{2KmM - \frac{Lr^2}{c}}{2KmM - \frac{Lr^2}{2c}}R_0.$$

(c) In accordance with Kepler's third law, if the satellite with the open sail were to evolve around the Sun on a circular orbit with radius r_0, or on an elliptical orbit with semi-major axis a, we could write that:

$$T_{circ}^2 = kr_0^3; \quad T_{ellipse}^2 = ka^3;$$

$$T_{circ} = \frac{2\pi r_0}{v_{circ}}; \quad \frac{mv_{circ}^2}{r_0} = \frac{KmM - \frac{Lr^2}{2c}}{r_0^2};$$

$$\frac{v_{circ}^2}{r_0^2} = \frac{KmM - \frac{Lr^2}{2c}}{mr_0^3}; \quad T_{circ} = 2\pi\sqrt{\frac{mr_0^3}{KmM - \frac{Lr^2}{2c}}};$$

$$\frac{T_{ellipse}^2}{T_{circ}^2} = \frac{a^3}{r_0^3}; \quad T_{ellipse} = 2\pi\sqrt{\frac{ma^3}{KmM - \frac{Lr^2}{2c}}};$$

$$T_{ellipse} = 2\pi R_0 \frac{2KmM - \frac{Lr^2}{c}}{2KmM - \frac{Lr^2}{2c}}\sqrt{\frac{R_0\left(2KmM - \frac{Lr^2}{c}\right)}{\left(2KmM - \frac{Lr^2}{2c}\right)\left(KmM - \frac{Lr^2}{2c}\right)}}.$$

The same result is reached if we admit that the period of rotation of the satellite with the sail open, evolving around the Sun on an ellipse with the semi-major axis a, is equal to the period of rotation of the same satellite if it were to evolve on the confocal circle of the ellipse (a circle with radius a and its center in the center of the Sun, represented in Figure 1.4) with speed u, equal to the speed of the satellite on the elliptical orbit at its minor peak B, the same as the average speed of the satellite on the elliptical orbit.

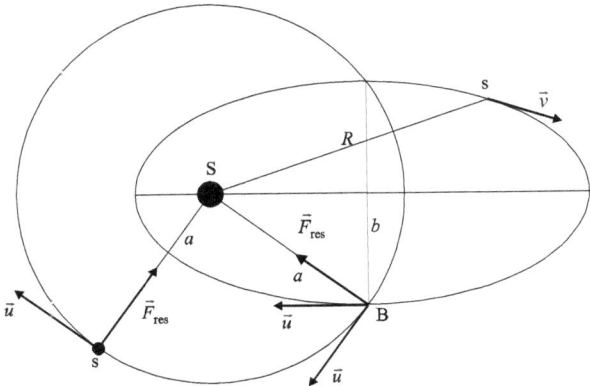

Fig. 1.4

It results that:

$$\frac{mu^2}{a} = F_{\text{res, B}} = \frac{KmM - \frac{Lr^2}{2c}}{a^2}; \quad \frac{a^2}{u^2} = \frac{ma^3}{KmM - \frac{Lr^2}{2c}};$$

$$T = \frac{2\pi a}{u}; \quad T = 2\pi\sqrt{\frac{ma^3}{KmM - \frac{Lr^2}{2c}}}.$$

Problem 2. Bodies Hidden in Identical Cubic Boxes

Inside one of two identical cubic "black boxes", A and B, is a massive metal cylinder; inside the other box is a massive cone.

The masses of the two internal parts, which are homogeneous and made of the same metal, are identical ($m = 280$ g). The two inner bodies are fixed so that their axes of longitudinal symmetry coincide with the axes joining the centers of two opposite lateral faces of each cubic box, and their centers of mass coincide with the centers of mass of the cubic boxes.

(a) *Identify* the opposite lateral faces of each box through whose centers pass the axes of longitudinal symmetry of the objects fixed inside each box.

(b) *Identify* the box inside which the cylinder is located and the box inside which the cone is located.

(c) *Determine* the geometric dimensions of the two bodies inside the two cubic boxes.

Materials available: An empty cubic black box, identical to the two given black boxes; a metal support; a manual electronic timer; a ruler; an inextensible wire; very light and resistant metallic flanges; screws; a wrench for the screws.

We know: the moment of inertia of a homogeneous, massive cylinder relative to its longitudinal axis of symmetry, $I_{0;\text{cylinder}} = \frac{1}{2}mR^2_{\text{cylinder}}$; the moment of inertia of a massive, homogeneous cylinder relative to an axis that passes through its center of mass and is perpendicular to its axis of longitudinal symmetry, $I_{\text{cylinder}} = \frac{1}{12}m(3R^2_{\text{cylinder}} + h^2_{\text{cylinder}})$; the moment of inertia of a homogeneous, massive cone relative to its longitudinal axis of symmetry,

$I_{0;\text{cone}} = \frac{3}{10} m R_{\text{cone}}^2$; the moment of inertia of a massive, homogeneous cone relative to an axis that passes through its center of mass and is perpendicular to its axis of longitudinal symmetry, $I_0 = \frac{3}{20} m \left(R_{\text{cone}}^2 + \frac{1}{4} h_{\text{cone}}^2 \right)$.

Given: the mass of the walls of each cubic box together with the two attached flanges, m_0; the diameter of the shaft on which the wire is wound, d; and the gravitational acceleration, g.

We know that $R_{\text{cylinder}} < \sqrt{\frac{3}{5}} R_{\text{cone}}$.

Solution

(a) With the two flanges mounted on the opposite lateral faces X and X′ of box A, using the thread available, we make a Maxwell pendulum, so that the length of the pendulum is close to the height of the support.

From the higher position, with the two wires wrapped around the horizontal rods of the flanges, we time the duration t_{1A} of the descent of box A after its release until it reaches the lower position. The process is repeated several times, and only three very close values are kept. The arithmetic mean of the three determinations will be accepted as the final result of the descent duration. Knowing the height H from which the box fell and accepting that its center of mass descends in a uniformly accelerated rectilinear motion, regardless of the initial speed, the acceleration of this descent a_{1A} can be determined:

$$a_{1A} = \frac{2H}{t_{1A}^2}.$$

On the other hand, the dynamics of the process, based on Figure 2.1, are as follows:

$$\vec{G}_{\text{total}} + 2\vec{T}_{1A} = M\vec{a}_{1A}; \quad M = m_0 + m;$$

$$Mg - 2T_{1A} = Ma_{1A};$$

$$2T_{1A}r = I_{A;XX'}\varepsilon_{1A}; \quad a_{1A} = \varepsilon_{1A}r;$$

$$Mg - \frac{I_{A;XX'}}{r}\varepsilon_{1A} = Ma_{1A};$$

$$a_{1A} = \frac{Mg}{M + \frac{I_{A;XX'}}{r^2}}; \quad I_{A;XX'} = Mr^2\left(\frac{g}{a_{1A}} - 1\right);$$

$$I_{A;XX'} = I_{box} + I_{object;A;XX'}.$$

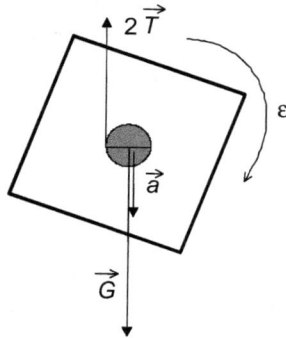

Fig. 2.1

The two flanges are then mounted on the opposite lateral faces Y and Y' and the opposite lateral faces Z and Z', respectively, using the same suspension wire. The box, therefore, descends from the same height H, and the acceleration of the descent is determined each time as follows:

$$a_{2A} = \frac{2H}{t_{2A}^2}; \quad a_{2A} = \frac{Mg}{M + \frac{I_{A;YY'}}{r^2}};$$

$$I_{A;YY'} = Mr^2\left(\frac{g}{a_{2A}} - 1\right);$$

$$I_{A;YY'} = I_{box} + I_{object;A;YY'};$$

$$a_{3A} = \frac{2H}{t_{3A}^2}; \quad a_{3A} = \frac{Mg}{M + \frac{I_{A;ZZ'}}{r^2}};$$

$$I_{A;ZZ'} = Mr^2\left(\frac{g}{a_{3A}} - 1\right);$$

$$I_{A;ZZ'} = I_{box} + I_{object;A;ZZ'}.$$

The results of the determinations made for box A should be presented in the accompanying cumulative table (Table 2.1), from which it follows that:

$$t_{1A} \neq t_{2A}; \quad t_{1A} \neq t_{3A}; \quad t_{2A} \approx t_{3A};$$

$$a_{1A} \neq a_{2A}; \quad a_{1A} \neq a_{3A}; \quad a_{2A} \approx a_{3A};$$

$$I_{A;XX'} \neq I_{A;YY'}; \quad I_{A;XX'} \neq I_{A;ZZ'}; \quad I_{A;YY'} \approx I_{A;ZZ'},$$

which proves that the object in box A has its axis of longitudinal symmetry along the direction of the centers of the opposite lateral faces X and X'.

Table 2.1.

H	t_{1A}	a_{1A}	$I_{A;XX'}$	t_{2A}	a_{2A}	$I_{A;YY'}$	t_{3A}	a_{3A}	$I_{A;ZZ'}$

Proceeding identically but mounting the two flanges on each pair of lateral faces of box B, we establish that:

$$t_{1B} \neq t_{2B}; \quad t_{1B} \neq t_{3B}; \quad t_{2B} \approx t_{3B}; \quad a_{1B} = \frac{Mg}{M + \frac{I_{B;XX'}}{r^2}};$$

$$I_{B;XX'} = Mr^2 \left(\frac{g}{a_{1B}} - 1 \right); \quad I_{B;XX'} = I_{box} + I_{object;B;XX'};$$

$$a_{2B} = \frac{2H}{t_{2B}^2}; \quad a_{2B} = \frac{Mg}{M + \frac{I_{B;YY'}}{r^2}};$$

$$I_{B;YY'} = Mr^2 \left(\frac{g}{a_{2B}} - 1 \right); \quad I_{B;XX'} = I_{box} + I_{object;B;XX'};$$

$$a_{3B} = \frac{2H}{t_{3B}^2}; \quad a_{3B} = \frac{Mg}{M + \frac{I_{B;ZZ'}}{r^2}};$$

$$I_{B;ZZ'} = Mr^2 \left(\frac{g}{a_{3B}} - 1 \right); \quad I_{B;ZZ'} = I_{box} + I_{object;B;ZZ'}.$$

The results of the determinations made for box B should be presented in the accompanying cumulative table (Table 2.2). From this, it follows that:

$$t_{1B} \neq t_{2B}; \quad t_{1B} \neq t_{3B}; \quad t_{2B} \approx t_{3B};$$

$$a_{1B} \neq a_{2B}; \quad a_{1B} \neq a_{3B}; \quad a_{2B} \approx a_{3B};$$

$$I_{B;XX'} \neq I_{B;YY'}; \quad I_{B;XX'} \neq I_{B;ZZ'}; \quad I_{B;YY'} \approx I_{B;ZZ'},$$

which proves that the object in box B has its axis of longitudinal symmetry along the direction of the centers of the opposite lateral faces X and X'.

Table 2.2.

H	t_{1B}	a_{1B}	$I_{B;XX'}$	t_{2B}	a_{2B}	$I_{B;YY'}$	t_{3B}	a_{3B}	$I_{B;ZZ'}$

Conclusion: Each of the two hidden objects has its longitudinal axis of symmetry along the direction of the centers of the opposite sides X and X' of the two boxes.

(b) Let's assume that the cylinder is hidden in box A and the cone is hidden in box B.

As a result, we can write that:

$$I_{A;XX'} = I_{\text{box}} + I_{\text{object};A;XX'} = I_{\text{box}} + I_{0;\text{cylinder}} = I_{\text{box}} + \frac{1}{2}mR_{\text{cylinder}}^2;$$

$$I_{B;XX'} = I_{\text{box}} + I_{\text{object};B;XX'} = I_{\text{box}} + I_{0;\text{cone}} = I_{\text{box}} + \frac{3}{10}mR_{\text{cone}}^2.$$

From the experimental data entered into the two tables, we find that

$$I_{A;XX'} < I_{B;XX'},$$

which can happen only if

$$I_{\text{box}} + \frac{1}{2}mR_{\text{cylinder}}^2 < I_{\text{box}} + \frac{3}{10}mR_{\text{cone}}^2; \quad R_{\text{cylinder}} < \sqrt{\frac{3}{5}}R_{\text{cone}}.$$

These are the exact conditions expressed in the content of the problem.

Conclusion: The cylinder is hidden in box A, and the cone is hidden in box B.

(c) Using the previous measurements, it results that:

$$I_{A;XX'} = I_{box} + I_{0;cylinder} = I_{box} + \frac{1}{2}mR^2_{cylinder};$$

$$R_{cylinder} = \sqrt{\frac{2}{m}(I_{A;XX'} - I_{box})};$$

$$I_{A;YY'} = I_{box} + I_{object;A;YY'} = I_{box} + I_{cylinder};$$

$$I_{cylinder} = I_{A;YY'} - I_{box} = \frac{1}{12}m\left(3R^2_{cylinder} + h^2_{cylinder}\right);$$

$$I_{A;ZZ'} = I_{box} + I_{cylinder} = I_{A;YY'};$$

$$h_{cylinder} = \sqrt{\frac{12}{m}(I_{A;YY'} - I_{box}) - 3R^2_{cylinder}};$$

$$I_{B;XX'} = I_{box} + I_{object;B;XX'} = I_{box} + I_{0;cone} = I_{box} + \frac{3}{10}mR^2_{cone};$$

$$R_{cone} = \sqrt{\frac{10}{3m}(I_{B;XX'} - I_{box})};$$

$$I_{B;YY'} = I_{box} + I_{object;B;YY'}$$

$$= I_{box} + I_{cone} = I_{box} + \frac{3}{20}m\left(R^2_{cone} + \frac{1}{4}h^2_{cone}\right);$$

$$R^2_{cone} + \frac{1}{4}h^2_{cone} = \frac{20}{3m}(I_{B;YY'} - I_{box});$$

$$h_{cone} = 2\sqrt{\frac{20}{3m}(I_{B;YY'} - I_{box}) - R^2_{cone}}.$$

After a similar procedure, using the empty box, it is determined that

$$I_{box} = I_0 = m_0 r^2 \left(\frac{g}{a_0} - 1\right).$$

Conclusion: Figures 2.2 and 2.3 show the geometric dimensions and arrangement of the two bodies inside the two cubic boxes.

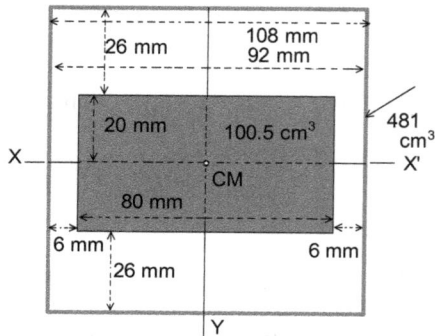

Fig. 2.2

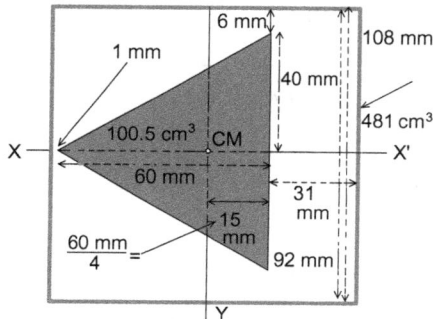

Fig. 2.3

Problem 3. Practical Problem: Lenses, a Plane Mirror, and a Transparent Liquid

Materials Provided (See Figure 3.1)

(1) Support with plug; (2) rods of $\Phi = 10$ mm and $L = 600$ mm; (3) Petri dish cover; (4) plane mirror; (5) biconvex lens; (6) white linear body; (7) graduated ruler; (8) bottle containing water; (9) bottle containing an unkown transparent liquid; (10) plug; (11) funnel.

Determine:

(a) the convergence of the biconvex lens;
(b) the radii of curvature of the lens surfaces;
(c) the refractive index of the unknown liquid.

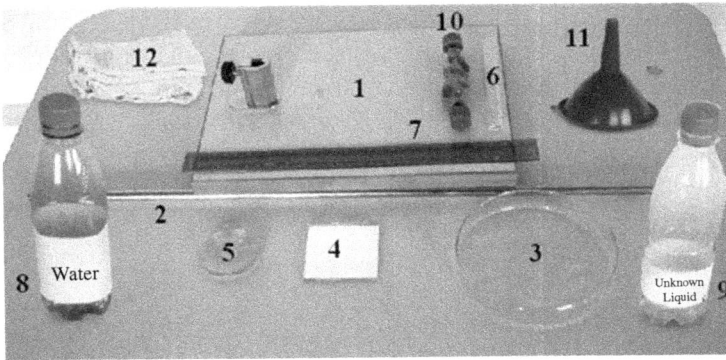

Fig. 3.1

The refractive index of water is known to be $n_{\text{water}} = \frac{4}{3}$.

Solution

(a) Determining the convergence of the lens

To determine the convergence of the biconvex lens, the process shown in Figure 3.2 is carried out, and the position of the object is adjusted, keeping it perpendicular to the main optical axis of the lens so that, looking vertically from top to bottom, from above the object, the object and the image are identical in length and size (width). The distance d, corresponding to this state, is measured.

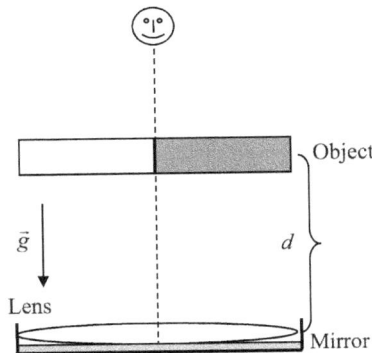

Fig. 3.2

As shown in Figure 3.2, on its way from the object to the eye of the observer, the light traveling from the object to the given optical system passes through the lens, from top to bottom (following the laws of refraction). It is then reflected by the mirror (following the laws of reflection), before passing through the lens again, from bottom to top (according to the laws of refraction), forming the image of the object, in its extension and identical to it, as explained in detail in Figure 3.3.

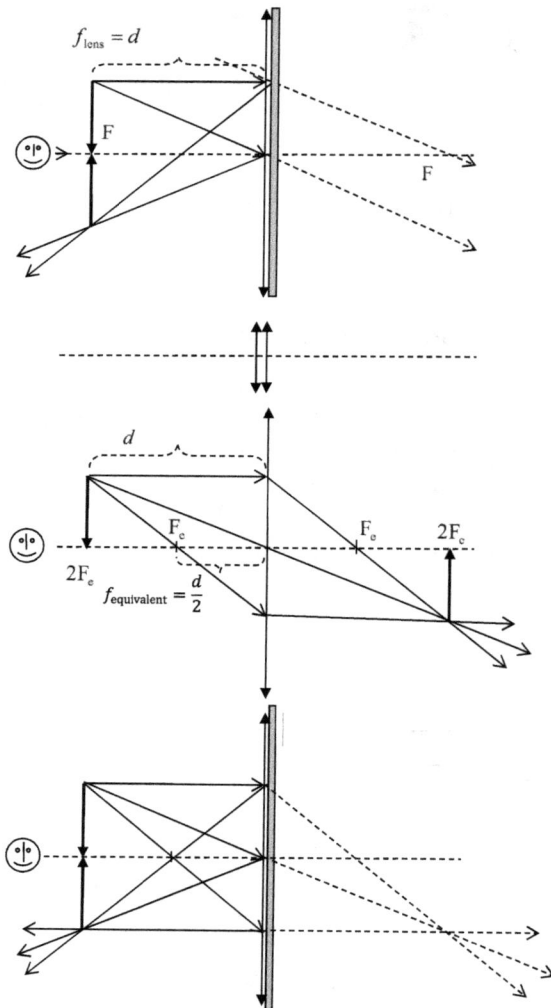

Fig. 3.3

In these conditions, the optical system comprises two identical lenses joined together, equivalent to a single lens with the equivalent focal length given by the following expression:

$$\frac{1}{f_{\text{equivalent}}} = \frac{1}{f_{\text{lens}}} + \frac{1}{f_{\text{lens}}} = \frac{2}{f_{\text{lens}}};$$

$$f_{\text{equivalent}} = \frac{f_{\text{lens}}}{2}.$$

Since "*distance from object*" = "*distance from image*" = d, it results that:

$$\frac{1}{f_{\text{equivalent}}} = \frac{1}{d} + \frac{1}{d} = \frac{2}{d};$$

$$f_{\text{equivalent}} = \frac{d}{2} = \frac{f_{\text{lens}}}{2}; \quad f_{\text{lens}} = d;$$

$$C_{\text{lens}} = \frac{1}{f_{\text{lens}}} = \frac{1}{d}.$$

(b) Determining the radii of curvature of the lens surfaces

(1) To determine the radii of curvature of the two identical biconvex surfaces of the lens, water is placed between the lens and the mirror, as shown in Figure 3.4, forming there a plane divergent concave lens with focal distance f_a.

Knowing that the focal length of a lens is given, in general, by the expression

$$\frac{1}{f} = (n-1)\left(\frac{1}{R_1} + \frac{1}{R_2}\right),$$

it follows that the expression of the focal length of the plane divergent concave lens formed by the water between the lens and the flat mirror surface is

$$\frac{1}{f_{\text{water}}} = -\frac{n_{\text{water}} - 1}{R} < 0,$$

where n_{water} (index of refraction for water) is known.

(2) The experiment is repeated, again adjusting the position of the object, keeping it perpendicular to the main optical axis of the lens,

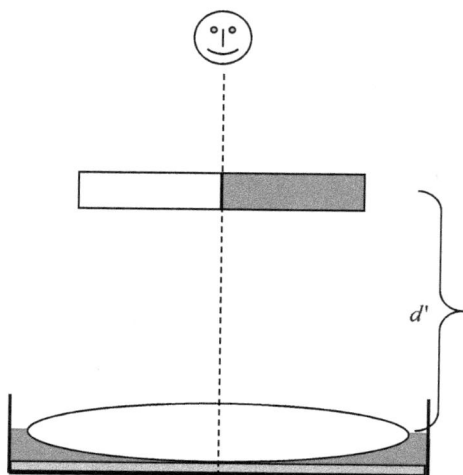

Fig. 3.4

so that, looking vertically from top to bottom, from above the object, the object and its image are elongated and identical in size (width). The distance d' corresponding to this new state is measured.

Since, in this case, the light passes in both directions, through the glass lens and through the water lens, before being reflected by the mirror, the given system comprises four joined lenses, equivalent to a single lens with the equivalent focal length given by the expression

$$\frac{1}{f'_e} = 2\left(\frac{1}{f_{lens}} + \frac{1}{f_{water}}\right) = \frac{2}{d'}.$$

It results that:

$$\frac{1}{f_{lens}} + \frac{1}{f_{water}} = \frac{1}{d'}; \quad \frac{1}{d} - \frac{n_{water} - 1}{R} = \frac{1}{d'};$$

$$R = \frac{dd'(n_{water} - 1)}{d' - d}.$$

(3) In order to repeat the determination of the radii of curvature of the two identical biconvex surfaces of the lens, water is placed in the tub until it covers the lens, as shown in Figure 3.5, forming there two concave plane divergent lenses, each with focal length f_a.

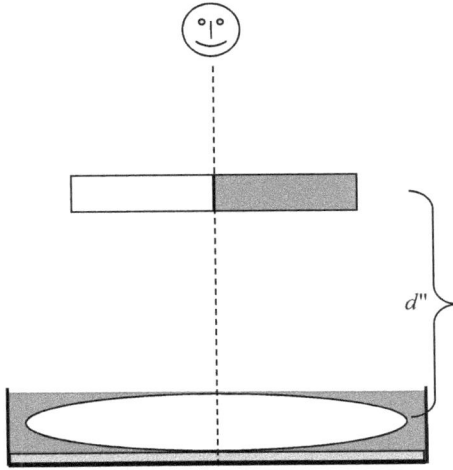

Fig. 3.5

Since, in this case, the light passes in both directions, through the glass lens and the water lens, and is then reflected on the mirror, the given system comprises six lenses joined together, equivalent to a single lens with the equivalent focal length given by the expression

$$\frac{1}{f''_e} = 2\left(\frac{1}{f_{\text{water}}} + \frac{1}{f_{\text{lens}}} + \frac{1}{f_{\text{water}}}\right) = \frac{2}{d''}.$$

It follows that:

$$\frac{1}{f_{\text{lens}}} + \frac{2}{f_{\text{water}}} = \frac{1}{d''}; \quad \frac{1}{d} - 2\frac{n_{\text{water}} - 1}{R} = \frac{1}{d''};$$

$$R = \frac{2dd''(n_{\text{water}} - 1)}{d'' - d}.$$

(c) Determining the refractive index of the unknown liquid

(1) To determine the refractive index of the glass from which the lens is made, we consider the fact that, in general:

$$\frac{1}{f} = (n - 1)\left(\frac{1}{R_1} + \frac{1}{R_2}\right); \quad R_1 = R_2 = R.$$

It follows that:

$$\frac{1}{f_{\text{lens}}} = \frac{2(n-1)}{R} = \frac{1}{d} = C;$$

$$n = 1 + \frac{R}{2d}.$$

(2) To determine the refractive index of the unknown transparent liquid, we repeat the experiment presented in Figure 3.5, replacing the water with the unknown liquid, whose refractive index is n_x, as shown in Figure 3.6.

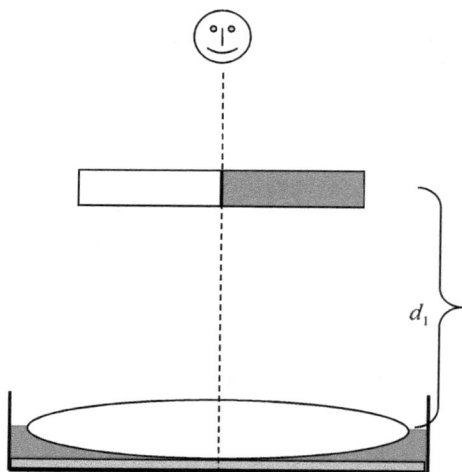

Fig. 3.6

(3) We determine the distance d_1 at which the object and its image appear as a continuous extension of one another. In these conditions, we have:

$$\frac{1}{f_{1,\text{equivalent}}} = 2\left(\frac{1}{f_{\text{lens}}} + \frac{1}{f_x}\right) = \frac{2}{d_1};$$

$$\frac{1}{f_{\text{lens}}} + \frac{1}{f_x} = \frac{1}{d_1};$$

$$\frac{1}{d} - \frac{n_x - 1}{R} = \frac{1}{d_1};$$

$$n_x = 1 + \frac{R(d_1 - d)}{dd_1}.$$

(4) In order to repeat the determination of the refractive index of the unknown transparent liquid, we repeat the experiment presented in Figure 3.5, replacing the water with the unknown liquid, whose refractive index is n_x, as indicated in Figure 3.7.

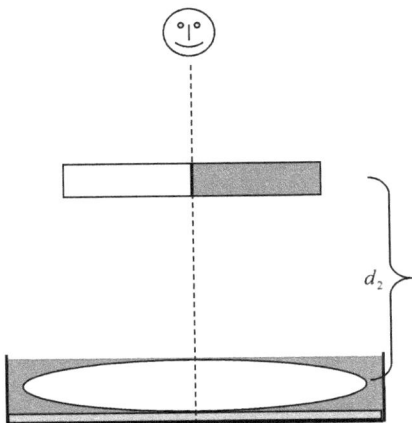

Fig. 3.7

(5) We determine the distance d_2 at which the object and its image appear as a continuous extension of one another.

In these conditions, we have:

$$\frac{1}{f_{2,\text{equivalent}}} = 2\left(\frac{1}{f_x} + \frac{1}{f_{\text{lens}}} + \frac{1}{f_x}\right) = \frac{2}{d_2};$$

$$\frac{1}{f_{\text{lens}}} + \frac{2}{f_x} = \frac{1}{d_2}; \quad \frac{1}{d} - 2\frac{n_x - 1}{R} = \frac{1}{d_2};$$

$$n_x = 1 + \frac{R(d_2 - d)}{2dd_2}.$$

Problem 4. Practical Problem: Ideal Capacitor and Real Coil

Materials Provided (See Figure 4.1)

(1) Coil with unknown characteristics; **(2)** box with two identical ideal capacitors, each with unknown capacity, C, which will be connected in parallel using a switch; **(3)** connecting conductors (10 total); **(4)** ferrite core with handle; **(5)** alternative voltage source with digital display of the voltage value of the terminals; **(6)** digital multimeter (ideal ammeter!); **(7)** switch.

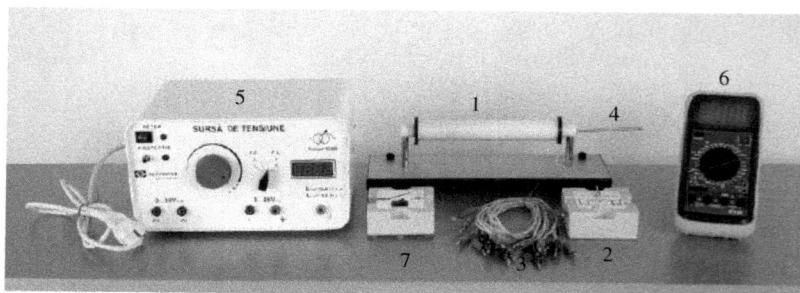

Fig. 4.1

With the given devices, the scheme presented in Figure 4.2 is assembled.

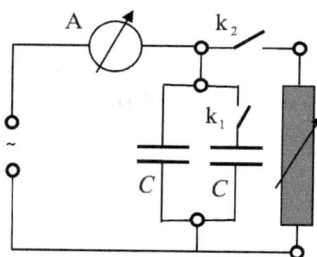

Fig. 4.2

(a) *Determine* the inductance L of the coil for an alternating current, with the ferromagnetic core inside the coil and within the limits of the positives of the core, corresponding to which the indication of the ammeter A is the same regardless of the closed/open state of the switch k_2, while the switch k_1 remains open.

Hint: After each use of the switch k_2 and after each movement of the ferromagnetic core inside the coil, following the same indication of the ammeter, the voltage constant at the generator booms should be adjusted for each of its values, $U = (1, 2, 3, \ldots, 9, 10)$ V.

(b) *Justify* phenomenologically the possibility of the constancy of the inductance L of the coils containing the core, although for each voltage value, $U = (1, 2, 3, \ldots, 9, 10)$ V, the length of the ferromagnetic core inside the coil is different.

(c) Keeping the ferromagnetic core inside the coil within the limits of the previously established positions, *determine* the coil's alternating current equivalent ohmic resistance, R.

    Hint: The voltage constant at the generator terminals should be adjusted for each of its values, $U = (1, 2, 3, \ldots, 9, 10)$ V, after closing the switch k_1, regardless of the position of the switch k_2. The frequency of the alternating voltage used is known, $\nu = 50$ Hz.

Clarifications

The voltage source can only be used for alternating voltages (whose values cannot exceed 12 V to ensure the protection of the capacitors used). Only the proposed scheme should be used, without eliminating any of its elements, by operating the two switches, moving the ferromagnetic core inside the coil, and turning the button of the alternative voltage source. After each switch operation or movement of the ferromagnetic core, restore the voltage to the initial value. Do not change the scale of the ammeter (200 mA a.c.) nor its connecting pins in the circuit (COM—white; mA—red). If you have damaged the ammeter, the source, or the capacitors, replacing them is impossible!

Solution

(a) Determining the inductance L of the coil containing a ferromagnetic core

(1) With the switches k_1 and k_2 open, corresponding to scheme (a) in Figure 4.3, the indication I_1 of the ammeter A is given by the expression

$$I_1 = \frac{U}{X_C} = U\omega C.$$

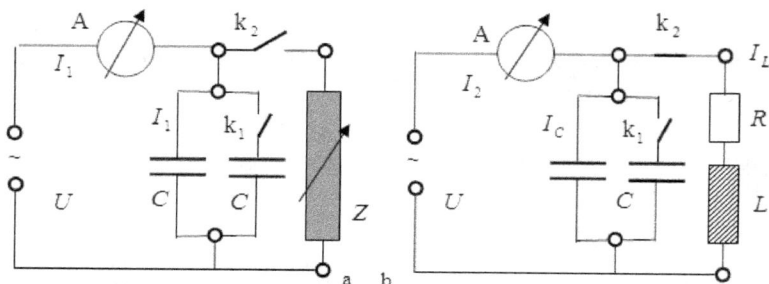

Fig. 4.3

(2) The switch k_1 remains open and the switch k_2 closes, as shown in diagram (b), Figure 4.3. Using the phasor diagrams in Figure 4.4, it follows that the indication of the ammeter, I_2, is given by the expression

$$\bar{I}_2 = \bar{I}_C + \bar{I}_L,$$

where $\bar{I}_2$, $\bar{I}_C$ and $\bar{I}_L$ are phasors associated with the effective values of the current intensities on each side of the network:

$$I_2^2 = I_C^2 + I_L^2 + 2I_C I_L \cos(90^2 + \varphi) = I_C^2 + I_L^2 - 2I_C I_L \sin\varphi;$$

$$I_L = \frac{U}{\sqrt{R^2 + \omega^2 L^2}};$$

$$\sin\varphi = \frac{U_L}{U} = \frac{I_L X_L}{I_L \sqrt{R^2 + \omega^2 L^2}} = \frac{X_L}{\sqrt{R^2 + \omega^2 L^2}}.$$

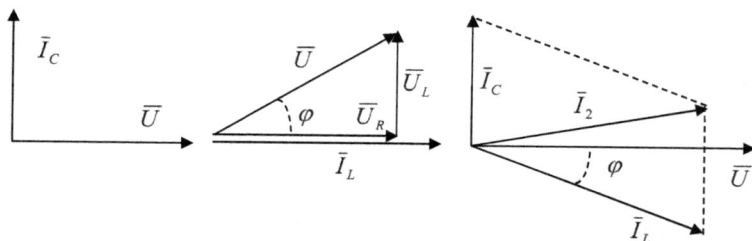

Fig. 4.4

(3) We move the ferromagnetic core inside the coil very slowly until the moment when the indication of the ammeter A is consant, regardless of the closed or open position of the switch k_2. If necessary, we

adjust the source voltage so that its value is equal to the initial one, U. Under these conditions, it results that:

$$I_2 = I_1; \quad I_C = \frac{U}{X_C} = I_1; \quad I_2 = I_1 = I_C;$$

$$I_2^2 = I_1^2 + I_L^2 - 2I_1 I_L \sin \varphi;$$

$$I_1^2 = I_1^2 + I_L^2 - 2I_1 I_L \sin \varphi;$$

$$I_L = 2I_1 \sin \varphi;$$

$$\frac{U}{\sqrt{R^2 + \omega^2 L^2}} = 2\frac{U}{X_C} \frac{X_L}{\sqrt{R^2 + \omega^2 L^2}};$$

$$2X_L = X_C; \quad X_L = \omega L; \quad X_C = \frac{U}{I_1};$$

$$2\omega L = \frac{U}{I_1}; \quad L = \frac{U}{2\omega I_1} = \frac{U}{4\pi\nu I_1}.$$

(4) The experiment is repeated for different values of the voltage U, each time making a very small adjustment to the ferromagnetic core, so that the ammeter indication is constant, regardless of the position of the switch k_2. Complete the attached Table 4.1.

Table 4.1.

No. det.	U (V)	I_1 (mA)	L (H)	L_{average} (H)
1	1.00	3.2	0.4976	0.49407
2	2.00	6.6	0.4825	
3	3.00	9.7	0.4924	
4	4.00	12.8	0.4976	
5	5.00	15.8	0.5039	
6	6.00	19.3	0.4950	
7	7.00	22.6	0.4932	
8	8.00	25.5	0.4995	
9	9.00	28.8	0.4976	
10	10.00	32.4	0.4914	

(b) Phenomenological justification

The intensity of the current through the windings of the coil being variable, additional currents appear in the ferromagnetic core of the coil, called Foucault eddy currents, whose magnetic fields contribute to the coil's magnetic flux, so that, under these conditions, the inductance L of the coil changes, depending not only on the characteristics of the coil. To compensate for the losses in the coil inductance due to Foucault currents, the displacement of the ferromagnetic core inside the coil must be carried out, thus ensuring the constancy of the coil inductance.

(c) Determination of the ohmic resistance of the coil

(1) By closing the switch k_1, the two capacitors are connected in parallel, as shown in Figure 4.5, so that their equivalent capacity is $C' = 2C$. As a result, the indication of the ammeter A (whose internal resistance is negligible) when the switch k_2 remains open will double compared to the previous situation, at the same voltage value, and will be

$$ I'_1 = \frac{U}{X_e} = \frac{U}{\frac{1}{\omega C_e}} = U\omega C_e = 2U\omega C \approx 2I_1. $$

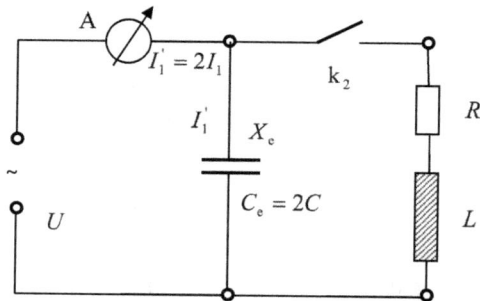

Fig. 4.5

(2) We close the switch k_2, as shown in diagram (a), Figure 4.6, and read indication I'_2 of ammeter A.

If the real impedances of the network elements are associated with complex impedances, according to known rules, using the equivalent

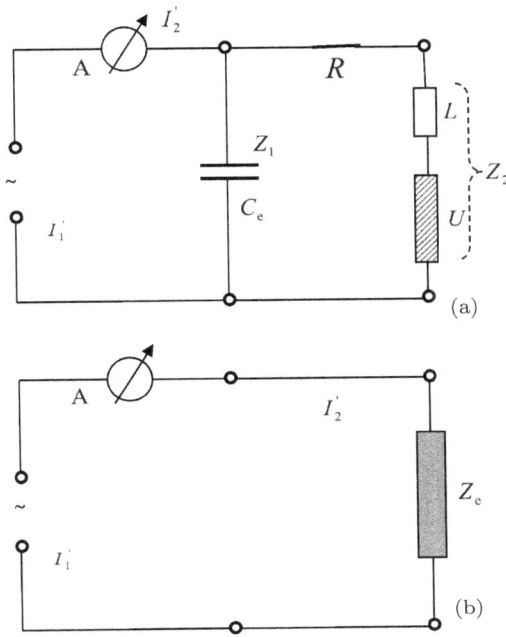

Fig. 4.6

scheme (b) from Figure 4.6, it results that:

$$\bar{Z}_e = \frac{\bar{Z}_1 \bar{Z}_2}{\bar{Z}_1 + \bar{Z}_2};$$

$$\bar{Z}_1 = -jX_e; \quad \bar{Z}_2 = R + jX_L;$$

$$\bar{Z}_e = \frac{-jX_e(R + jX_L)}{-jX_e + R + jX_L} = \frac{X_e X_L - jRX_e}{R + j(X_L - X_e)};$$

$$\bar{Z}_e = \frac{RX_e^2 - jX_e[R^2 + X_L(X_L - X_e)]}{R^2 + (X_L - X_e)^2};$$

$$X_e = \frac{1}{\omega C_e} = \frac{1}{2\omega C} = \frac{X_C}{2} = \frac{2X_L}{2};$$

$$X_e = X_L;$$

$$\bar{Z}_e = \frac{X_L^2}{R} - jX_L;$$

$$Z_e = \frac{U}{I_2'};$$

$$Z_e^2 = \frac{X_L^4}{R^2} + X_L^2 = \frac{U^2}{I_2'^2};$$

$$\frac{X_L^4}{R^2} = \frac{U^2}{I_2'^2} - X_L^2;$$

$$R = \frac{X_L^2}{\sqrt{\left(\frac{U}{I_2'}\right)^2 - X_L^2}};$$

$$R = \frac{\omega^2 L^2}{\sqrt{\left(\frac{U}{I_2'}\right)^2 - \omega^2 L^2}}; \quad 2\omega L = \frac{U}{I_1}; \quad \omega L = \frac{U}{2I_1};$$

$$R = \frac{\left(\frac{U}{2I_1}\right)^2}{\sqrt{\left(\frac{U}{I_2'}\right)^2 - \left(\frac{U}{2I_1}\right)^2}}; \quad R = \frac{\left(\frac{U}{I_1'}\right)^2}{\sqrt{\left(\frac{U}{I_2'}\right)^2 - \left(\frac{U}{I_1'}\right)^2}}.$$

(3) The same experiment is repeated for different values of U. We then complete Tables 4.2 and 4.3.

Table 4.2.

No. det.	U (V)	I_1 (mA)	$I_1' \approx 2I_1$ (mA)	I_2' (mA)	$\left(\frac{U}{I_1'}\right)^2$	$\left(\frac{U}{I_2'}\right)^2$
1	1.00	3.2	6.5	4.5	23668.63	49382.71
2	2.00	6.5	13.4	9.3	22276.67	46248.12
3	3.00	9.6	19.3	13.1	24161.72	52444.49
4	4.00	12.8	25.7	17.3	24224.43	53459.85
5	5.00	16.1	32.4	21.6	23814.96	53583.67
6	6.00	19.2	38.3	25.7	24541.71	54504.98
7	7.00	22.4	44.2	30.4	25081.38	53021.12
8	8.00	26.2	52.4	35.5	23308.66	50783.57
9	9.00	29.2	58.7	39.4	23507.62	52178.61
10	10.00	32.5	65.2	44.4	23523.65	50726.40

Table 4.3.

No. det.	$\sqrt{\left(\dfrac{U}{I_2'}\right)^2 - \left(\dfrac{U}{I_1'}\right)^2}$	$R = \dfrac{\left(\dfrac{U}{I_1'}\right)^2}{\sqrt{\left(\dfrac{U}{I_2'}\right)^2 - \left(\dfrac{U}{I_1'}\right)^2}}$	R_{average} (Ω)
1	160.35 V/A	147.60	142.864
2	154.94	143.77	
3	168.17	143.67	
4	170.98	141.67	
5	172.53	138.03	
6	173.09	141.78	
7	167.15	150.05	
8	165.75	140.62	
9	169.32	138.83	
10	164.93	142.62	

Working Method

1. Generalities

With the help of this source, the following values of voltage and current can be obtained:

- direct current voltages, with continuous adjustment at the output in the range 0–40 V;
- alternating current voltages, with continuous adjustment at the output in the range 0–30 V;
- the maximum current in the loads, 8 A in d.c. and 12 A in a.c.;
- the power charged, max. 160W;
- supply voltage, 220 V, ±20%/50 Hz;
- room temperature, −5, . . . , +40 degrees Celsius;
- relative humidity, max. 65%;
- net mass, 5.5 kg.

2. Description of the Device (See Figure 4.7)

On the front panel are the control and adjustment elements of the source:

(1) the main ON/OFF switch;
(2) LED indicator for the power source;

(3) button for repriming the protection;

(4) LED indicator for the protection status;

(5) terminals for alternating voltage, 0–30V a.c.;

(6) button for voltage adjustment;

(7) terminals for continuous voltage, 0–40V d.c.;

(8) numerical display for voltage and current intensity in d.c. and a.c.;

(9) terminal for grounding or protective neutral;

(10) switch for choosing the value (U or L, d.c. or c.a.) that the display (8) will indicate.

The terminals (7) are not independent of the terminals of the alternating current (5). Thus, the accidental connection of the terminals (5) with the terminals (7) of alternating current produces a short circuit and can cause fuse burning.

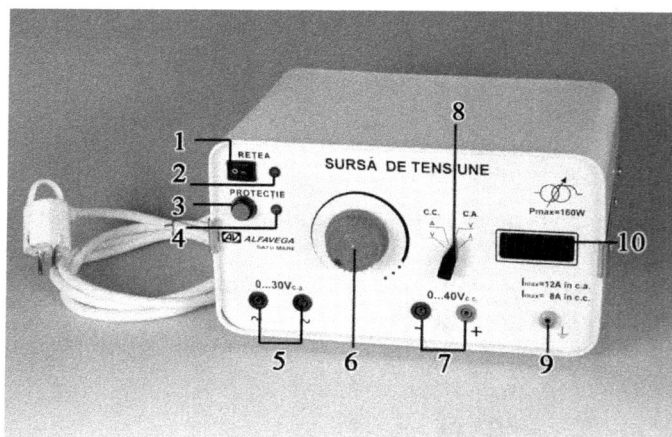

Fig. 4.7

3. Instructions for Use

3.1 *Initiating operation*

We connect the cord ending with a switch and a grounding plug to the network of alternating current at 220 V.

Toggle the ON/OFF switch to the ON position.

The light of the LED (2) indicates the connection of the network.

The button (3) is pressed, and the LED (4) goes out, indicating the protection priming.

3.2 *Working method*

After putting the device into operation, the load is connected with the help of connecting conductors.

Alternating voltage can be collected at the alternating current terminals (5). The voltage value is adjusted using the adjustment button (6) on the side of the device. The same button also adjusts the continuous voltage.

Depending on the position of the switches (10) and (11) on the device display, you can read:

(a) the value of the current through the load corresponding to the voltage applied when the switch (10) is in the A position;
(b) the value of the current or direct voltage from the terminals (7) if the switch (12) is in the d.c. position;
(c) the value of the alternating current or voltage from the booms (5) if the switch (12) is in the a.c. position.

The voltage source produces a continuous (0–40 V) and alternating voltage in the 0–30 V range. It is continuously adjustable, and short-circuit protection is provided.

4. Maintenance and Troubleshooting

The device does not need special maintenance; only cleaning is necessary for the removal of dust.

If there is a lack of voltage at the direct current output, first, safety should be checked.

If necessary, the fuse should only be replaced with a fuse of the same value or 1.6A. If a higher-value fuse is installed, you risk destroying the autotransformer.

Problem 5. Real Capacitor and Ideal Coils

Materials Available

(1) two identical ideal coils with unknown inductances;
(2) a real capacitor, with unknown capacity, C;
(3) connecting conductors—10 pieces;
(4) two ferrite cores with identical handles;
(5) alternating voltage source with digital display of the voltage value at the terminals;
(6) digital multimeter (ideal ammeter!);
(7) circuit breaker.

With the given devices, the scheme presented in Figure 5.1 is assembled.

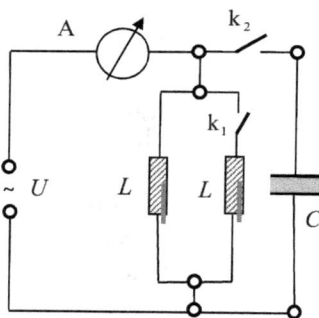

Fig. 5.1

Requirements

(a) *Determine*:

- the ideal elements of the equivalent series circuit for the real coil with the support provided, as well as the relative magnetic permeability of the core of the real coil;
- the ideal elements of the equivalent series circuit for the real supported capacitor.

It is known that the frequency of the alternating voltage from the network is $\nu = 50$ Hz.

Clarification: Ceramic capacitors (ideal capacitors) with identical external appearances can be considered to have identical (unknown) capacities.

The ferromagnetic core is inside the coil, in the position at which the indication of the ammeter A is constant regardless of the closed/open state of the switch k_2, while switch k_1 remains open.

Determine the capacity C of the capacitor. *Hint*: After the actuation of the switch k_2 and each movement of the ferromagnetic core inside the coil, with the indication of the ammeter remaining the same, the voltage constant at the generator terminals should be adjusted to each of its values, $U = (1, 2, 3, \ldots, 9, 10)$ V.

(b) *Justify* phenomenologically the possibility of the constancy of the inductance L of the coil containing the core, although for each voltage value, $U = (1, 2, 3, \ldots, 9, 10)$ V, the length of the ferromagnetic core inside the coil is different.

(c) Keeping the ferromagnetic core inside the coil, within the limits of the positions established previously, *determine* the equivalent ohmic resistance in the alternating current, R, of the capacitor. *Hint*: The voltage constant at the generator terminals should be adjusted to each of its values, $U = (1, 2, 3, \ldots, 9, 10)$ V, after closing the switch k_1, regardless of the position of the switch k_2.

The frequency of the alternating voltage used is known, $\nu = 50$ Hz.

Clarifications

The voltage source can only be used for alternating voltages (whose values cannot exceed 12 V, to ensure the protection of the capacitors used). Only the proposed scheme should be used, without eliminating any elements, by activating the two switches, moving the ferromagnetic core inside the coil, and turning the voltage source's button alternately. After each actuation of a switch or movement of the ferromagnetic core, the voltage should be returned to the initial value. Do not change the scale of the ammeter (200 mA a.c.), nor its connecting pins in the circuit (white—COM; red—mA). If you have damaged the ammeter, source, or capacitor, there is no possibility of replacing them!

Solution

(a) Determining C, the capacity of the capacitor

(1) With switches k_1 and k_2 open, corresponding to scheme (a) in Figure 5.2, the indication I_1 of ammeter A is given by the expression

$$I_1 = \frac{U}{X_L} = \frac{U}{\omega L}.$$

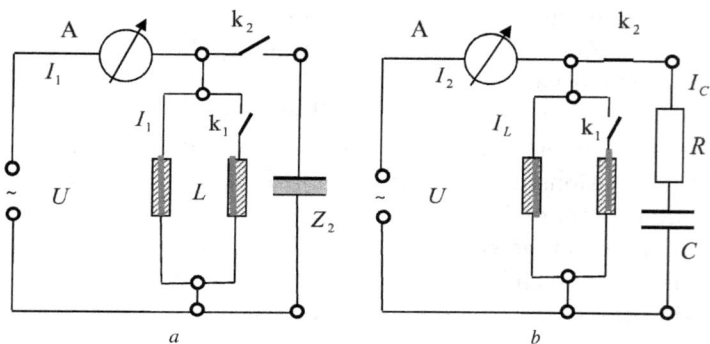

Fig. 5.2

(2) The switch k_1 remains open, and the switch k_2 closes, as shown in Figure 5.2. Using the phasor diagrams in Figure 5.3, it turns out that the indication I_2 of the ammeter is given by the expression

$$\bar{I}_2 = \bar{I}_L + \bar{I}_C,$$

where $\bar{I}_2$, $\bar{I}_L$ and $\bar{I}_C$ are phasors associated with the effective values of the current intensities on each side of the network;

$$I_2^2 = I_L^2 + I_C^2 + 2I_L I_C \cos(90^2 + \varphi);$$

$$I_C = \frac{U}{\sqrt{R^2 + \frac{1}{\omega^2 C^2}}}; \quad \sin\varphi = \frac{U_C}{U} = \frac{I_C X_C}{I_C\sqrt{R^2 + \frac{1}{\omega^2 C^2}}} = \frac{X_C}{\sqrt{R^2 + \frac{1}{\omega^2 C^2}}}.$$

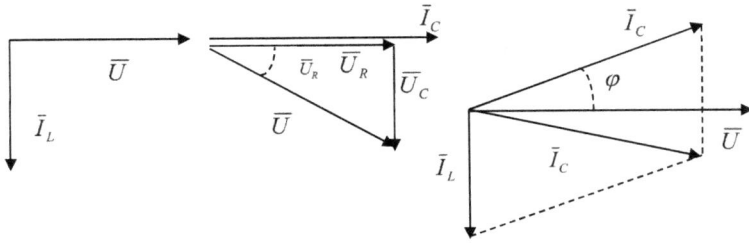

Fig. 5.3

(3) Move the ferromagnetic core inside the coil very slowly until the moment when the indication of the ammeter A is constant, regardless of the closed or open position of the switch k_2. If necessary, adjust the voltage of the source so that its value is equal to the initial one, U. In these conditions, results that:

$$I_2 = I_1; \quad I_L = \frac{U}{X_L} = I_1; \quad I_2 = I_1 = I_L;$$

$$I_2^2 = I_1^2 + I_C^2 - 2I_1 I_C \sin\varphi;$$

$$I_1^2 = I_1^2 + I_C^2 - 2I_1 I_C \sin\varphi;$$

$$I_C = 2I_1 \sin\varphi;$$

$$\frac{U}{\sqrt{R^2 + \frac{1}{\omega^2 C^2}}} = 2\frac{U}{X_L}\frac{X_C}{\sqrt{R^2 + \frac{1}{\omega^2 C^2}}};$$

$$2X_C = X_L;$$

$$2\frac{1}{\omega C} = \frac{U}{I_1}; \quad C = \frac{2I_1}{\omega U} = \frac{I_1}{\pi \nu U}.$$

(4) The experiment is repeated for different values of the voltage U, each time making a very small adjustment to the ferromagnetic core so that the ammeter indication remains the same, regardless of the position of the k_2 switch. Complete the Table 5.1.

(b) Phenomenological justification

As the intensity of the current through the turns of the coil is variable, additional currents appear in the ferromagnetic core of the coil, called Foucault eddy currents, whose magnetic fields contribute to

Table 5.1.

No. det.	U (V)	I_1 (mA)	C (F)	C_{medium} (F)
1				
2				
3				

the coil's magnetic flux, so that, under these conditions, the inductance L of the coil changes, depending not only on the characteristics of the coil. Due to the effect of the Foucault currents, the displacement of the ferromagnetic core inside the coil must be carried out to compensate for the variations in the coil inductance, thus ensuring the constancy of the coil inductance.

(c) Determination of the ohmic resistance R of the capacitor

(1) Closing switch k_1 connects the two coils in parallel so that their equivalent inductance is $L_e = \frac{L}{2}$, as shown in Figure 5.4. As a result, the indication of the ammeter A (whose internal resistance is negligible) when the switch k_2 remains open will double compared to the previous situation at the same voltage value, becoming

$$I_1' = \frac{U}{X_e} = \frac{U}{\omega L_e} = \frac{U}{\omega \frac{L}{2}} = \frac{2U}{\omega L} \approx 2I_1.$$

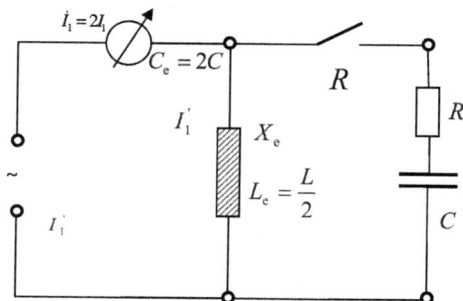

Fig. 5.4

(2) Close the switch k_2, as shown in diagram (a) of Figure 5.5, and read indication I_2 of ammeter A.

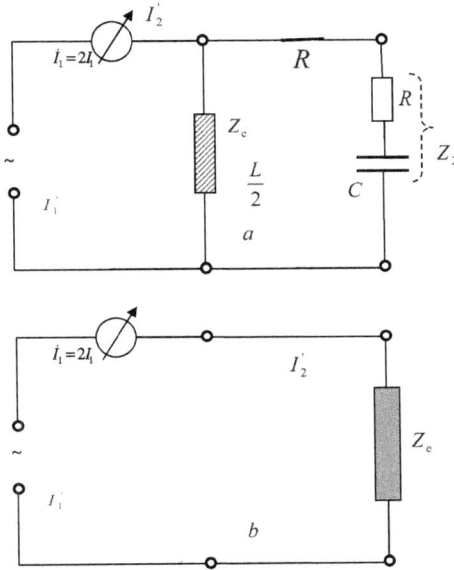

Fig. 5.5

If the real impedances of the relay elements are associated, according to known rules, with complex impedances, using the equivalent scheme (b) in Figure 5.5, it results that:

$$\bar{Z}_e = \frac{\bar{Z}_1 \bar{Z}_2}{\bar{Z}_1 + \bar{Z}_2};$$

$$\bar{Z}_1 = jX_e; \quad \bar{Z}_2 = R - jX_C;$$

$$\bar{Z}_e = \frac{jX_e(R - jX_C)}{jX_e + R - jX_C} = \frac{X_e X_C + jRX_e}{R + j(X_e - X_C)};$$

$$\bar{Z}_e = \frac{RX_e^2 + jX_e[R^2 - X_C(X_e - X_C)]}{R^2 + (X_e - X_C)^2};$$

$$X_e = \omega L_e = \frac{\omega L}{2} = \frac{X_L}{2}; \quad 2X_C = X_L;$$

$$X_e = X_C;$$

$$\bar{Z}_e = \frac{X_C^2}{R} + jX_C;$$

$$Z_e = \frac{U}{I_2'};$$

$$Z_e^2 = \frac{X_C^4}{R^2} - X_C^2 = \frac{U^2}{I_2'^2};$$

$$\frac{X_C^4}{R^2} = \frac{U^2}{I_2'^2} + X_C^2;$$

$$R = \frac{X_C^2}{\sqrt{\left(\frac{U}{I_2'}\right)^2 + X_C^2}};$$

$$R = \frac{\frac{1}{\omega^2 C^2}}{\sqrt{\left(\frac{U}{I_2'}\right)^2 + \frac{1}{\omega^2 C^2}}}; \quad \frac{1}{\omega C} = \frac{U}{2I_1};$$

$$R = \frac{\left(\frac{U}{2I_1}\right)^2}{\sqrt{\left(\frac{U}{I_2'}\right)^2 + \left(\frac{U}{2I_1}\right)^2}}; \quad R = \frac{\left(\frac{U}{I_1'}\right)^2}{\sqrt{\left(\frac{U}{I_2'}\right)^2 + \left(\frac{U}{I_1'}\right)^2}}.$$

(3) The same experiment is repeated for different values of U. Complete Tables 5.2 and 5.3.

Table 5.2.

No. det.	U (V)	I_1 (mA)	$I_1' \approx 2I_1$ (mA)	I_2' (mA)	$\left(\frac{U}{I_1'}\right)^2$	$\left(\frac{U}{I_2'}\right)^2$
1	1.00					
2	2.00					
3	3.00					

Table 5.3.

No. det.	$\sqrt{\left(\frac{U}{I_2'}\right)^2 + \left(\frac{U}{I_1'}\right)^2}$	$R = \dfrac{\left(\frac{U}{I_1'}\right)^2}{\sqrt{\left(\frac{U}{I_2'}\right)^2 + \left(\frac{U}{I_1'}\right)^2}}$	R_{average} (Ω)
1			
2			
3			

Chapter 5

International Pre-Olympic Physics Contest 2007, Călimăneşti, Romania

Problem 1. Near and Far from the Moon!

In two photographs taken using the same camera mounted on an artificial satellite of the Moon, the Moon appears as a circular disk with diameter $d_1 = 46.5\,\text{mm}$ and $d_2 = 40.5\,\text{mm}$, respectively, similar to how it appears from the Earth. The first photo was taken when the satellite was in an elliptical orbit around the Moon and near the periselenium, and the second photo was taken when the satellite was near the aposelenium.

(a) **Eccentricity of the ellipse:** Using all this information, *determine* the numerical eccentricity a of the satellite's orbit. *Estimate* the minimum value of the time interval between the moments of execution of the two photos. It is known that the apparent diameter of the image of the Moon on each photo is directly proportional to the angle at which the Moon is seen from the satellite.

(b) **Stable Lagrange points:** *Locate* the points in the plane of the circular orbit of the Moon around the Earth at which the satellite could be located, such that it would evolve in a circular orbit around the Earth, remaining in the same stable position relative to the Earth and the Moon. *We know*: r_{EM}, the distance between the center of the Earth and the center of the Moon; M_E, the mass of the Earth; and M_M, the mass of the Moon.

(c) **Cosmic impact:** A meteorite, in free fall towards the Moon (along the line that passes through the center of the Moon), collides with an automatic space laboratory, which moves around the Moon in a circular orbit of radius R. After the impact with the meteorite, it remains incorporated into the orbital station, and the system revolves around the Moon in a new orbit. Thus, the minimum distance from the center of the Moon is $r_{\text{min}} = R/2$. *Determine* the speed of the meteorite before the collision with the space station.

Given: the mass of the Moon, M; the constant of universal attraction, K; the mass of the space laboratory, m_1; and the mass of the meteorite, m_1.

(d) **The third cosmic velocity:** *Determine* the approximate minimum value of the escape velocity, with respect to the Earth, that must be given to a body launched from the Earth so that it leaves the Solar System forever (the third cosmic velocity).

Given: $R_{\text{TS}} \approx 1.5 \cdot 10^8$ km; $T_{\text{TS}} = 1$ year; and $v_0 \approx 7.9 \, \frac{\text{km}}{\text{s}}$, the velocity of a terrestrial satellite orbiting the Earth in a very low circular orbit (the first cosmic velocity).

We know that $\frac{M_T}{R_T} \ll \frac{M_S}{R_{\text{TS}}}$. The variation of the body's kinetic energy with respect to the Sun is neglected during the body's evolution from the Earth's surface to the limit of the Earth's gravitational attraction zone.

Solution

(a) The eccentricity of the ellipse

The images of the Moon's hemisphere in the two photos are approximately the same as the image of the Moon's hemisphere that we see from Earth. Considering that the Moon always faces us with the same face, we can conclude that when the two photos were taken, the satellite was located at points approximately between Earth and the Moon. The length of the diameter of the Moon image (d) on each photo is proportional to the angle at which the Moon is seen from the satellite:

$$d \sim \alpha = \frac{D}{r},$$

where D is the real diameter of the Moon, and r is the (variable) distance from the satellite to the center of the Moon for the rest of the month.

The equation of the ellipse, representing the trajectory of the satellite around the Moon, in plane polar coordinates is:

$$r = \frac{p}{1 + e \cos \theta}; \quad p = a(1 - e^2).$$

For the moments corresponding to the passage of the satellite through the aposelenium and the periselenium respectively, it results that:

$$\theta = 0; \quad r_{\text{aposelenium}} = r_{\min} = a(1 - e);$$

$$\theta = \pi; \quad r_{\text{periselenium}} = r_{\max} = a(1 + e).$$

If d_P and d_A are the diameters of the Moon images on the two photos of the Moon taken when the satellite passed exactly through the periselenium and the aposelenium, respectively, it follows that:

$$\frac{d_P}{d_A} = \frac{\alpha_P}{\alpha_A} = \frac{\frac{D}{r_{\min}}}{\frac{D}{r_{\max}}} = \frac{r_{\max}}{r_{\min}} = \frac{1 + e}{1 - e};$$

$$e = \frac{1 - \frac{d_P}{d_A}}{1 + \frac{d_P}{d_A}} = \frac{d_A - d_P}{d_A + d_P}.$$

The photos were taken at moments close to the moments when the satellite passed through the periselenium and aposelenium, respectively, so the lengths of the diameters of the Moon images in the two photographs are:

$$d_1 \approx d_P; \quad d_2 \approx d_A.$$

Under these conditions, it results that:

$$e \approx \frac{d_1 - d_2}{d_1 + d_2}; \quad d_1 = 46.5 \, \text{mm}; \quad d_2 = 40.5 \, \text{mm}; \quad e = 0.07.$$

Considering that the satellite's orbit is more or less stable in space, it will take about half a year to change from "full Moon at periselenium" to "full Moon at aposelenium" (where the Sun is in opposition to the Moon).

(b) Stable Lagrange points

The stable internal Lagrange point, L_{int}, is the point between the centers of the Moon and the Earth where the forces of gravitational attraction exerted by the Earth and the Moon on a material point with a negligible mass ensure its dynamic balance, the material point being permanently in the same position compared to the Earth and the Moon:

$$\omega = \sqrt{\frac{KM_E}{r_{EM}^3}},$$

where M_E is the mass of the Earth, r_{EM} is the distance from the Earth to the Moon, and K is the constant of universal attraction.

If x is the distance from the center of the Moon to the stable internal Lagrange point, L_{int}, then, writing the equilibrium equation of the motion of the material point located in the internally stable Lagrange point, it results that:

$$K\frac{mM_E}{(r_{EM} - x)^2} - K\frac{mM_M}{x^2} = m\omega^2(r_{EM} - x);$$

$$K\frac{M_E}{(r_{EM} - x)^2} - K\frac{M_M}{x^2} = \omega^2(r_{EM} - x);$$

$$K\frac{M_E}{r_{EM}^2\left(1 - \frac{x}{r_{EM}}\right)^2} - K\frac{M_M}{x^2} = \omega^2(r_{EM} - x);$$

$$K\frac{M_E}{r_{EM}^2}\left(1 - \frac{x}{r_{EM}}\right)^{-2} - K\frac{M_M}{x^2} = \omega^2(r_{EM} - x);$$

$$x \ll r_{EM}; \quad \left(1 - \frac{x}{r_{EM}}\right)^{-2} \approx 1 + 2\frac{x}{r_{EM}};$$

$$K\frac{M_E}{r_{EM}^2} + 2K\frac{M_E}{r_{EM}^3}x - K\frac{M_M}{x^2} = K\frac{M_E}{r_{EM}^3}(r_{EM} - x);$$

$$x = r_{EM}\sqrt[3]{\frac{M_M}{3M_E}}; \quad r_{EM} = 384.400\,\text{km};$$

$$\frac{M_M}{M_E} = \frac{1}{81.3}; \quad x = 61.500\,\text{km}.$$

Under similar conditions, it is demonstrated that there is also a stable external Lagrange point, L_{ext}, also located in the direction of the centers of the Moon and the Earth, but on the other side of the Moon, at a distance y from its center, and rotating around the Earth with the same angular velocity as the Moon, for which we have:

$$K\frac{mM_E}{(r_{EM} + y)^2} + K\frac{mM_M}{y^2} = m\omega^2(r_{EM} + y);$$

$$K\frac{M_E}{(r_{EM} + y)^2} + K\frac{M_M}{y^2} = \omega^2(r_{EM} + y);$$

$$K\frac{M_E}{r_{EM}^2 \left(1 + \frac{y}{r_{EM}}\right)^2} + K\frac{M_M}{y^2} = \omega^2(r_{EM} + y);$$

$$K\frac{M_E}{r_{EM}^2}\left(1 + \frac{y}{r_{EM}}\right)^{-2} + K\frac{M_M}{y^2} = \omega^2(r_{EM} + y);$$

$$y \ll r_{EM}; \quad \left(1 + \frac{y}{r_{EM}}\right)^{-2} \approx 1 - 2\frac{y}{r_{EM}};$$

$$K\frac{M_E}{r_{EM}^2} - 2K\frac{M_E}{r_{EM}^3}y + K\frac{M_M}{y^2} = K\frac{M_E}{r_{EM}^3}(r_{EM} + y);$$

$$y = r_{EM}\sqrt[3]{\frac{M_M}{3M_E}} = x.$$

Conclusion: The two stable Lagrange points are symmetrical about the center of the Moon.

(c) Cosmic impact

Notation: $\vec{v}_1$ — velocity of the meteorite before the collision with the orbital station, $v_1 = \sqrt{\frac{KM}{R}}$; $\vec{v}_2$ — velocity of the meteorite before the collision with the orbital station; $\vec{v}$ — speed of the assembly after the collision. According to the law of conservation of momentum, based on Figure 1.1, it results that:

$$m_1\vec{v}_1 + m_2\vec{v}_2 = (m_1 + m_2)\vec{v};$$

$$v = \frac{\sqrt{m_1^2v_1^2 + m_2^2v_2^2}}{m_1 + m_2}.$$

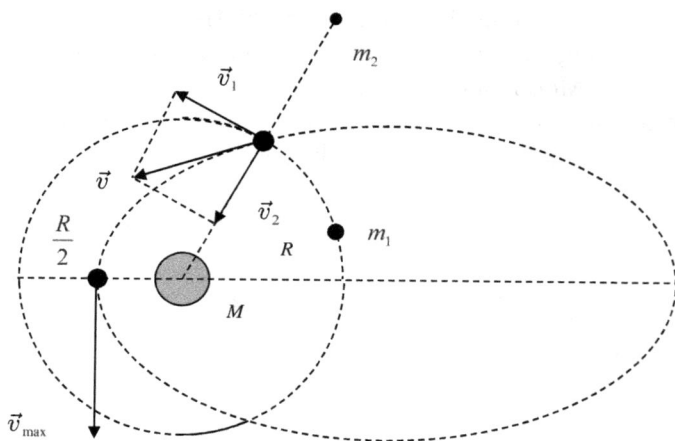

Fig. 1.1

After the collision, the assembly will evolve on an ellipse with the center of the Moon in its focus, so that when the assembly has reached the minimum distance from the center of the Moon, $r_{min} = R/2$, its speed will be $\vec{v}_{max}$. For this, in accordance with the law of conservation of mechanical energy and Kepler's second law (the law of conservation kinetic momentum), it results that:

$$\frac{(m_1 + m_2)v^2}{2} - K\frac{(m_1 + m_2)M}{R}$$

$$= \frac{(m_1 + m_2)v_{max}^2}{2} - K\frac{2(m_1 + m_2)M}{R};$$

$$\frac{v^2}{2} - K\frac{M}{R} = \frac{v_{max}^2}{2} - K\frac{2M}{R};$$

$$v^2 = v_{max}^2 - 2\frac{KM}{R};$$

$$(m_1 + m_2)\vec{v} \times \vec{R} = (m_1 + m_2)\vec{v}_{max} \times \vec{r}_{min};$$

$$v_1 R = v_{max}r_{min} = v_{max}\frac{R}{2}; \quad v_{max} = 2v_1 = 2\sqrt{\frac{KM}{R}};$$

$$v = \sqrt{2\frac{KM}{R}} = \frac{\sqrt{m_1^2 v_1^2 + m_2^2 v_2^2}}{m_1 + m_2};$$

$$\frac{2KM}{R} = \frac{m_1^2 \frac{KM}{R} + m_2^2 v_2^2}{(m_1 + m_2)^2};$$

$$v_2 = \sqrt{\frac{2KM}{R}\left(\frac{m_1}{m_2} + 1\right)^2 - \left(\frac{m_1}{m_2}\right)^2 \frac{KM}{R}}.$$

(d) The third cosmic velocity

(1) Let $\vec{v}_p$ be the velocity, relative to the Sun, of the body C at the time of its launch from the Earth, so that the body reaches the limit of the Sun's gravitational attraction zone and is there at rest in relation to the Sun. Using the details in Figure 1.2, according to the law of conservation of mechanical energy, it results that:

$$R_T \ll R_{TS};$$

$$\frac{mv_p^2}{2} - K\frac{mM_T}{R_T} - K\frac{mM_S}{R_{TS}} = 0;$$

$$\frac{M_T}{R_T} \ll \frac{M_S}{R_{TS}};$$

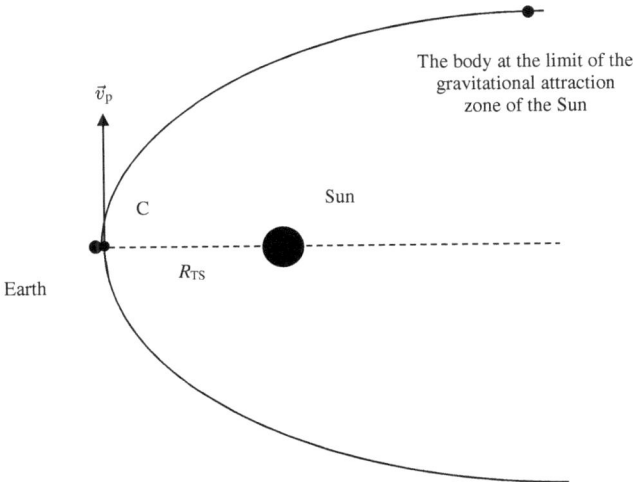

Fig. 1.2

$$\frac{mv_p^2}{2} - K\frac{mM_S}{R_{TS}} = 0;$$

$$v_p = \sqrt{2}\sqrt{K\frac{M_S}{R_{TS}}};$$

$$\sqrt{K\frac{M_S}{R_{TS}}} = v_{TS} = V_{orbital} = V_0,$$

representing the orbital speed of the Earth on its circular path around the Sun;

$$v_p = \sqrt{2}V_0,$$

representing the second cosmic velocity in relation to the Sun (parabolic speed);

$$R_{TS} \approx 1.5 \cdot 10^8 \text{ km}; \quad T_{TS} = 1 \text{ year};$$

$$V_0 = \frac{2\pi R_{TS}}{T_{TS}} \approx 30\ \frac{\text{km}}{\text{s}}; \quad v_p \approx 42.42\ \frac{\text{km}}{\text{s}}.$$

Conclusion: The body launched from Earth follows a parabolic trajectory — relative to the Sun — and reaches the outer boundary of the Sun's gravitational influence, with the Sun located at the focus of the parabola.

(2) Let $\vec{v}$ be the speed, relative to the Earth, of the body C at the time of its release from the Earth, so that the body reaches the limit of the gravitational attraction zone of the Sun and is there at rest in relation to the Sun.

This escape velocity, $\vec{v}$, will have the minimum value when its orientation is the same as the orientation of the vector representing the speed of the Earth in relation to the Sun, $\vec{V_0}$.

As a result, at the limit of the zone of gravitational attraction of the Earth, before reaching the limit of the zone of gravitational attraction of the Sun, the velocity of the body relative to the Earth, $\vec{v_\infty}$, will not be null, $v_\infty \neq 0$. In these conditions, relative to the Earth, in accordance with the law of conservation of mechanical energy, it results that:

$$\frac{mv^2}{2} - K\frac{mM_T}{R_T} = \frac{mv_\infty^2}{2};$$

$$v_\infty^2 = v^2 - 2K\frac{M_T}{R_T};$$

$$\sqrt{K\frac{M_T}{R_T}} = v_{\text{orbital}} = v_0 = 7.9 \ \frac{\text{km}}{\text{s}},$$

representing the orbital speed of the body if it were to evolve in a circle around the Earth at very low altitude (the first cosmic velocity);

$$v_\infty^2 = v^2 - 2v_0^2; \quad v_\infty = \sqrt{v^2 - 2v_0^2},$$

representing the body's speed in relation to the Earth at the limit of the Earth's gravitational attraction zone.

Conclusion: The body launched from Earth follows a hyperbolic trajectory—relative to the Earth—and reaches the boundary of Earth's gravitational influence, with Earth's center located at the focus of the hyperbola.

(3) Let $\vec{v}_{CS\infty}$ be the speed of the body C in relation to the Sun at the limit of the Earth's gravitational attraction zone:

$$\vec{v}_{CS\infty} = \vec{v}_\infty + \vec{v}_{TS} = \vec{v}_\infty + \vec{V_0},$$

where the orientations of the vectors $\vec{v}_\infty$ and $\vec{V_0}$ must be identical;

$$v_{CS\infty} = v_\infty + V_0 = \sqrt{v^2 - 2v_0^2} + V_0.$$

(4) The variation of the body's kinetic energy in relation to the Sun during the body's evolution from the Earth's surface to the limit of the Earth's gravitational attraction zone being negligible, in accordance with the law of conservation of mechanical energy, it results that:

$$v_{CS\infty} \approx v_p = \sqrt{2}V_0;$$

$$\sqrt{2}V_0 = \sqrt{v^2 - 2v_0^2} + V_0;$$

$$v = \sqrt{(\sqrt{2} - 1)^2 V_0^2 + 2v_0^2} \approx 16.7 \ \frac{\text{km}}{\text{s}},$$

representing the speed in relation to the Earth of the body C at the moment of its release from the Earth, so that the body reaches the limit of the gravitational attraction zone of the Sun and is there at rest in relation to the Sun (the third cosmic velocity).

Problem 2. Cylinder in a Vertical Guide

A vertical cylinder, having the plane of its lower base inclined to the vertical with an angle α, can slide without friction in a vertical guide, resting on a horizontal cylinder, with the same mass, m, located on a horizontal support (see Figure 2.1). The sliding friction coefficient between the two cylinders is μ.

(a) *Determine* the accelerations of the two cylinders, the coefficient of friction between the horizontal cylinder and its support, and the acceleration of one of the tangent points between the two cylinders in relation to the vertical cylinder, if the horizontal cylinder does not roll. The gravitational acceleration, g, is known. Initially, the elements of the system are at rest.

(b) After traveling a distance d, the vertical cylinder encounters an obstacle. At that moment, the horizontal cylinder enters the horizontal platform of a cart of mass M at rest, and a rigid, homogeneous, very thin rod engages with the horizontal cylinder along the generator which was in contact with the vertical cylinder.

Determine the mass of the rod and the length of the trolley platform if the cylinder moved without rolling along the entire path and stopped at the opposite end of the platform. The coefficient of friction on the platform is that previously determined. The movement of the trolley on its support occurs without friction.

(c) Having reached the end of the platform, the cylinder is fixed, and the rod on its generator disengages and begins to slide, by translation and without friction, on the surface of the cylinder.

Determine the speed of the rod in relation to the cylinder at the moment of their separation if the vertical displacement of the rod until the moment of separation is Δh. We know that R is the radius of the cylinder.

Particular case: $M + m \gg m_{\text{rod}}$.

Solution

(a) The forces acting on the elements of the fund system are those represented in the drawing in Figure 2.2. It results that:

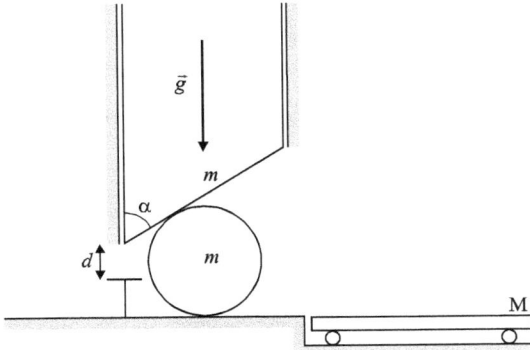

Fig. 2.1

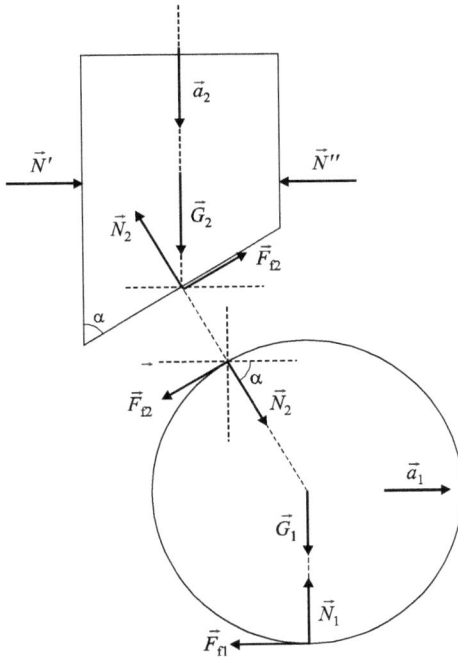

Fig. 2.2

$$m_1 = m; \quad m_2 = m;$$

$$m_2 \vec{a}_2 = \vec{G}_2 + \vec{N}_2 + \vec{F}_{f_2} + \vec{N}' + \vec{N}'';$$

$$ma_1 = N_2(\cos\alpha - \mu\sin\alpha) - \mu' N_1;$$

$$0 = N_2(\sin\alpha + \mu\cos\alpha) + mg - N_1;$$

$$\mu N_2 = \mu' N_1;$$

$$ma_2 = mg - N_2(\sin\alpha + \mu\cos\alpha).$$

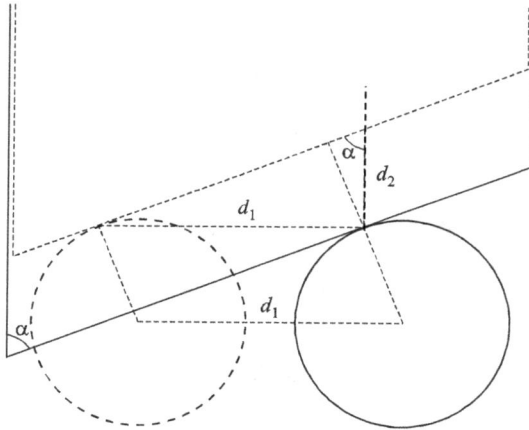

Fig. 2.3

If the relationship between the accelerations a_1 and a_2 deduced from Figure 2.3 is added to these equations, it results that:

$$d_1 = \frac{a_1 t^2}{2}; \quad d_2 = \frac{a_2 t^2}{2};$$

$$a_1 = a_2 \text{tg }\alpha;$$

$$N_2 = \frac{mg\sin\alpha}{1 - \mu\cos\alpha};$$

$$a_2 = \frac{g\cos\alpha(\cos\alpha - \mu - \mu\sin\alpha)}{1 - \mu\cos\alpha};$$

$$a_1 = \frac{g\sin\alpha(\cos\alpha - \mu - \mu\sin\alpha)}{1 - \mu\cos\alpha};$$

$$N_1 = \frac{mg(1 - \mu \cos\alpha + \sin\alpha)}{1 - \mu \cos\alpha};$$

$$\mu' = \frac{\mu \sin\alpha}{1 - \mu \cos\alpha + \sin\alpha}.$$

If the movements of the system elements are uniform, it means that $a_1 = 0$ and $a_2 = 0$. As a result, we have:

$$\mu = \frac{\cos\alpha}{1 + \sin\alpha};$$

$$\mu' = \frac{\sin\alpha \cos\alpha}{1 + \sin\alpha + 2\sin^2\alpha}.$$

For the distance traveled in time t by one of the points of tangency between the two cylinders, we can write the expression

$$d = \frac{at^2}{2},$$

where a is the acceleration of this point relative to the vertical cylinder.

From Figure 2.3, it follows that:

$$a = \frac{a_1}{\sin\alpha} = \frac{a_2}{\cos\alpha};$$

$$a = \frac{g(\cos\alpha - \mu - \mu \sin\alpha)}{1 - \mu \cos\alpha}.$$

(b) Immediately after the separation of the two cylinders, when the forces acting on the horizontal cylinder are those represented in Figure 2.4, it results that:

$$N = (m + m_0)g; \quad F_f = \mu' N;$$

$$m_0 \cos\alpha = \mu'(m_0 + m);$$

$$m_0 = \frac{\mu' m}{\cos\alpha - \mu'};$$

$$m_0 = \frac{\mu m \sin\alpha}{\cos\alpha(1 - \mu \cos\alpha) + \sin\alpha(\cos\alpha - \mu)}.$$

In the drawing, $\vec{a}_0$ is the braking acceleration of the horizontal cylinder immediately after separation, and $\vec{v}_0$ is speed of the horizontal

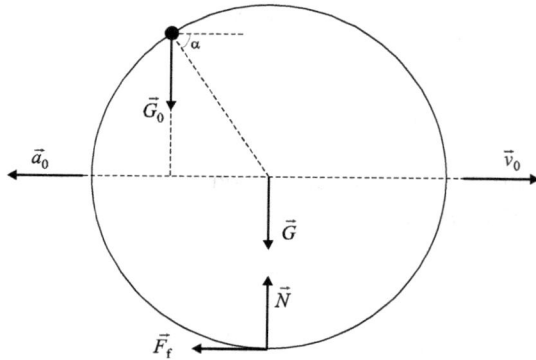

Fig. 2.4

cylinder at the moment of separation. It is easy to prove that

$$v_0 = \sqrt{2a_2 d}\, \mathrm{tg}\,\alpha.$$

In Figure 2.5, $\vec{a}_c$ is the acceleration of the cart relative to the ground, $\vec{a}_0$ is the acceleration of the cylinder relative to the ground, $\vec{a}_r$ is the acceleration of the cylinder relative to the cart, $\vec{v}_0$ is the speed of the cylinder relative to the cart/ground at the moment of entering the trolley, and l is the length of the trolley platform. It results that:

$$F_f = \mu'(m + m_0)g = Ma_c; \quad a_0 = \mu'g;$$

$$a_r = -(a_0 + a_c);$$

$$l = \frac{-v_0^2}{2a_r};$$

$$a_r = -\mu'g\left[1 + \frac{m}{M}\frac{1 + \mu\sin\alpha}{\cos\alpha(1 - \mu\cos\alpha) + \sin\alpha(\cos\alpha - \mu)}\right].$$

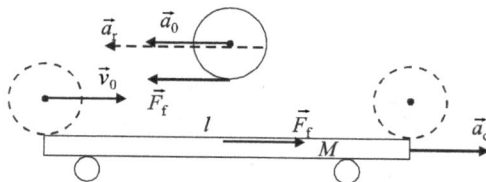

Fig. 2.5

(c) When the horizontal cylinder has reached the opposite end of the platform and is at rest relative to it, the speed of the assembly relative to the ground is:

$$v = a_c t_o,$$

where t_o is the duration of the movement of the cylinder in relation to the platform;

$$t_o = -\frac{v_0}{a_r};$$

$$v = v_0 \frac{a_c}{a_0 + a_c}.$$

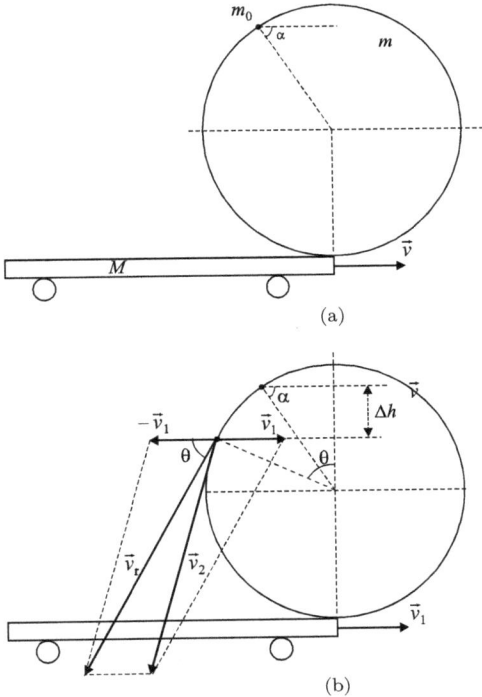

(a)

(b)

Fig. 2.6

In drawing (a) in Figure 2.6, $\vec{v}$ is the speed of the cart–cylinder–rod assembly, in relation to the ground, at the moment of release of the rod from the horizontal cylinder generator. In drawing (b) in the same figure, $\vec{v}_1$ is the speed of the cart–cylinder assembly relative to the ground at the moment of separation of the rod from the cylinder, $\vec{v}_2$ is the speed of the rod in relation to the ground at the moment of separation from the cylinder, and $\vec{v}_r$ is the speed of the rod in relation to the cylinder at the moment of separation.

Using the laws of conservation of mechanical energy and momentum, knowing that the system is isolated only in the horizontal direction, it results that:

$$\frac{(M + m + m_0)v_2}{2} + m_0 g \Delta h = \frac{(M + m)v_1^2}{2} + \frac{m_0 v_2^2}{2},$$

where Δh is the vertical displacement of the rod,

$$\Delta h = R(\sin \alpha - \cos \theta);$$

$$(M + m + m_0)v = (M + m)v_1 - m_0 v_{2x},$$

where $\vec{v}_{2x}$ is the horizontal component of $\vec{v}_2$;

$$\vec{v}_r = \vec{v}_2 - \vec{v}_1;$$

$$\vec{v}_2 = \vec{v}_r + \vec{v}_1,$$

where $\vec{v}_r$ is tangent to the surface of the cylinder;

$$v_{2x} = v_r \cos \theta - v_1;$$

$$(M + m + m_0)v = (M + m + m_0)v_1 - m_0 v_r \cos \theta;$$

$$v_1 = v + \frac{m_0 v_r \cos \theta}{M + m + m_0};$$

$$v_2^2 = v_r^2 + v_1^2 - 2v_r v_1 \cos \theta;$$

$$v_r^2 = \frac{2gR(\sin \alpha - \cos \alpha)}{1 - \frac{m_0 \cos^2 \theta}{M + m + m_0}}.$$

$$v_r^2 = \frac{2g\Delta h}{1 - \frac{m_0}{M + m + m_0} \left(\sin \alpha + \frac{\Delta h}{R} \right)^2}.$$

Particular case: $M + m \gg m_0$; $v_r = \sqrt{2g\Delta h}$.

Problem 3. Rarefied Gases

In two compartments of the same container, with volumes V_1 and V_2, rarefied oxygen is present at the same pressure, with temperatures T_1 and $T_2 > T_1$, respectively.

(a) *Determine* the gas pressures in the two compartments after opening a hole in the dividing wall, if there are Vo moles of gas in the entire container, and the temperatures of the two compartments are kept constant. The universal constant of perfect gases, R, is known.

    Inside a vessel containing a very rarefied gas made up of identical molecules, there is a horizontal, homogeneous plate of mass M whose face has an area of S, supported by four identical, very light springs, each with the elasticity constant k, above another plate, fixed in a horizontal position (Figure 3.1). The length of each spring in the undeformed state is lo. The temperature of the walls of the container and the plate are kept constant, T_1, and the temperature of the bottom plate is kept constant, T_2.

(b) *Determine* the gas pressure in the container if $T_1 < T_2$ and the distance between the plates is d. The gravitational acceleration is known as g.

    It will be considered that the molecules of a rarefied gas have a single speed, represented by the average thermal speed, at a given temperature.

    The balance of some rarefied gases assumes the equality of the molecular flows, in both directions, between one compartment and another compartment.

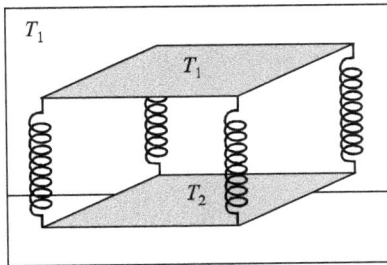

Fig. 3.1

Solution

(a) Gases manifest their kinetic-molecular nature when they are very rarefied; in these conditions, there are no collisions between molecules. Under normal conditions, at appreciable gas densities, the balance means the equality of the pressures in the communicating compartments due to the flow of the gas as a homogeneous medium from the higher-pressure compartment to the lower-pressure compartment.

In conditions of advanced rarefaction, the balance means the equality of the number of molecules that transit from one compartment to the other compartment in a unit of time through the communicating section.

Using the previous result, we know that the number of molecules that cross a unit area of a section per unit time is

$$d = \frac{\Delta Z}{\Delta S \Delta t} = \frac{1}{6} n v_T.$$

Conclusion: The molecular flow rate, d, is directly proportional to n and v_T.

Using Figure 3.2, let's first evaluate the molecular flow rates immediately after opening the hole in the partition wall.

It results that:

$$d_{0;1 \to 2 \sim n_{0_1} v_{T_1}} \sim \frac{p}{kT_1} \sqrt{\frac{3RT_1}{\mu}} \sim \frac{p}{k} \sqrt{\frac{3R}{\mu T_1}};$$

$$d_{0;2 \to 1 \sim n_{02} v_{T_2}} \sim \frac{p}{kT_2} \sqrt{\frac{3RT_2}{\mu}} \sim \frac{p}{k} \sqrt{\frac{3R}{\mu T_2}};$$

$$T_2 > T_1; \quad d_{0;1 \to 2} > d_{0;2 \to 1}.$$

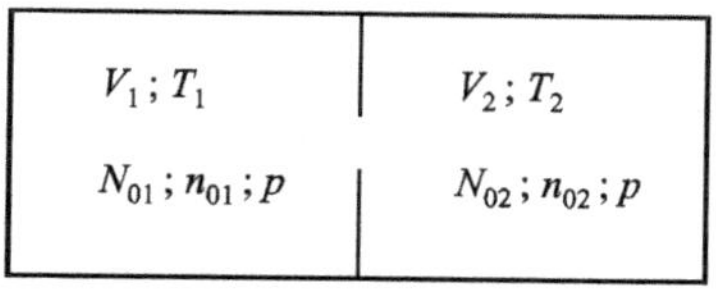

$V_1 ; T_1$	$V_2 ; T_2$
$N_{01} ; n_{01} ; p$	$N_{02} ; n_{02} ; p$

Fig. 3.2

Conclusion: Although the pressures in the two compartments are equal, with the gases being very rarefied, there will be a net transfer of molecules (of gas) from the cold compartment to the hot compartment.

To calculate the pressures in the two compartments at the moment of equilibrium, according to Figure 3.3, it follows that:

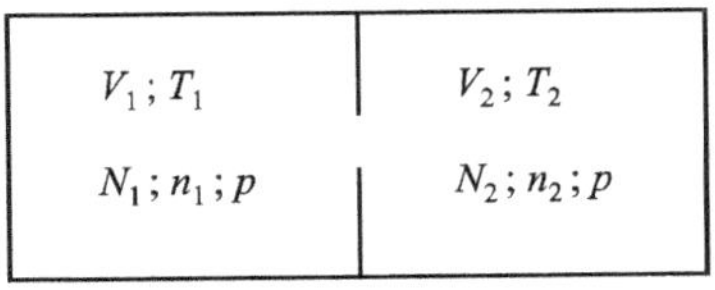

$$
\begin{array}{|c|c|}
\hline
V_1; T_1 & V_2; T_2 \\
N_1; n_1; p & N_2; n_2; p \\
\hline
\end{array}
$$

Fig. 3.3

$$d_{1\to 2} \sim n_1 v_{T_1} \sim \frac{p_1}{kT_1}\sqrt{\frac{3RT_1}{\mu}} \sim \frac{p_1}{k}\sqrt{\frac{3R}{\mu T_1}};$$

$$d_{2\to 1} \sim n_2 v_{T_1} \sim \frac{p_2}{kT_2}\sqrt{\frac{3RT_2}{\mu}} \sim \frac{p_2}{k}\sqrt{\frac{3R}{\mu T_2}};$$

$$d_{1\to 2} = d_{2\to 1};$$

$$\frac{p_1}{\sqrt{T_1}} = \frac{p_2}{\sqrt{T_2}}; \quad T_1 < T_2; \quad p_1 < p_2;$$

$$\frac{n_1 kT_1}{\sqrt{T_1}} = \frac{n_2 kT_2}{\sqrt{T_2}}; \quad n_1\sqrt{T_1} = n_2\sqrt{T_2};$$

$$\frac{N_1}{V_1}\sqrt{T_1} = \frac{N_2}{V_2}\sqrt{T_2};$$

$$N_1 + N_2 = N_{0_1} + N_{0_2} = N_0;$$

$$N_1 = \frac{N_0}{1 + \frac{V_2}{V_1}\sqrt{\frac{T_1}{T_2}}}; \quad N_2 = \frac{N_0}{1 + \frac{V_1}{V_2}\sqrt{\frac{T_2}{T_1}}};$$

$$p_1 = n_1 kT_1 = \frac{N_0 kT_1\sqrt{T_2}}{V_1\sqrt{T_2} + V_2\sqrt{T_1}};$$

$$p_2 = n_2 kT_2 = \frac{N_0 kT_2\sqrt{T_1}}{V_1\sqrt{T_2} + V_2\sqrt{T_1}};$$

$$p_1 = \frac{\nu_0 R T_1 \sqrt{T_2}}{V_1 \sqrt{T_2} + V_2 \sqrt{T_1}};$$

$$p_2 = \frac{\nu_0 R T_2 \sqrt{T_1}}{V_1 \sqrt{T_2} + V_2 \sqrt{T_1}}.$$

(b) The number of molecules that arrive during Δt in a cylindrical, normal beam on a surface with area ΔS is

$$\Delta Z = \frac{1}{6} v_T n \Delta S \Delta t,$$

where n is the concentration of molecules, and v_T is the quadratic velocity of the molecules.

The molecular flux (the number of molecules crossing the unit area per unit time) is

$$\Phi = \frac{\Delta Z}{\Delta S \Delta t} = \frac{1}{6} n v_T.$$

The gas between the plates is a mixture composed of two gases: a component whose molecules have an average thermal velocity v_{T_1}, corresponding to the temperature T_1, and a component whose molecules have an average thermal velocity v_{T_2}, corresponding to the temperature T_2.

It results that:

$$p_i = \frac{1}{3} n_1 m \bar{v}_1^2 + \frac{1}{3} n_1 m \bar{v}_2^2;$$

$$p_i = \frac{1}{3} n_1 m v_{T_1}^2 + \frac{1}{3} n_1 m v_{T_2}^2.$$

From the equality of the flows through the lateral faces of the parallelepiped, as well as between the plates, it follows that:

$$n v_{T_1} = n_1 v_{T_1} + n_2 v_{T_2},$$

$$n v_{T_1} = n_2 v_{T_2},$$

$$n_1 = \frac{1}{2} n;$$

$$n_2 = \frac{1}{2} n \frac{v_{T_1}}{v_{T_2}} = \frac{1}{2} \sqrt{\frac{T_1}{T_2}};$$

$$p_i = \frac{p_e}{2}\left(1 + \sqrt{\frac{T_2}{T_1}}\right) ; \quad T_1 < T_2; \quad p_i > p_e.$$

From the condition of mechanical balance, for the upper plate, it follows that:

$$G = F + 4F_e;$$

$$Mg = (p_{int} - p_{ext})S + 4k(l_0 - d);$$

$$Mg = \left(\frac{p_{ext}}{2}\right)\left(\sqrt{\frac{T_2}{T_1}} - 1\right)S + 4k(l_0 - d);$$

$$p_{ext} = \frac{2[Mg - 4k(l_0 - d)]}{\left(\sqrt{\frac{T_2}{T_1}} - 1\right)S};$$

$$p_{int} = \frac{p_{ext}}{2}\left(1 + \frac{T_2}{T_1}\right);$$

$$p_{int} = \frac{Mg - 4k(l_0 - d)}{\left(\sqrt{\frac{T_2}{T_1}} - 1\right)S}\left(1 + \sqrt{\frac{T_2}{T_1}}\right).$$

Problem 4. Eddy Electric Field

A long cylindrical solenoid, with radius R, having N turns per unit length, is traversed by a current of intensity I. On the same axis as the solenoid is a cylinder with radius $r < R$ and length h, as shown in Figure 4.1(a). The inner cylinder, at rest, carrying on its outer surface the electric charge g, uniformly distributed, is very light, being made of a thin sheet of paper.

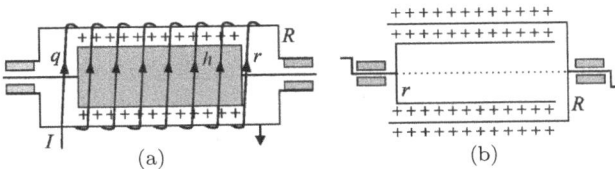

Fig. 4.1

(a) *Justify* the rotational movement acquired by the inner cylinder and *specify* the direction of this rotation in relation to the direction of the current through the solenoid coils when the intensity of the current in the solenoid coils increases by n times. *Determine* the angular velocity of the inner cylinder. *Analyze* the scenario in which the intensity of the current in the solenoid coils decreases by n times. In both scenarios, it will be assumed that $q > 0$ and $q < 0$. Air viscosity is neglected.

(b) *Analyze*, under the same conditions, the scenario $r > R$.

(c) Two very light, coaxial cylinders with radii R and $r < R$, made of a thin sheet of paper, are lying at rest as shown in Figure 4.1(b). They are electrified uniformly, with charges of the same sign and identical surface charge densities. The outer cylinder begins to rotate with uniform acceleration until its angular speed becomes Ω.

Explain the rotational movement acquired by the inner cylinder, specifying the direction of this rotation in relation to the direction of the outer cylinder's rotation, and *determine* the angular velocity of the inner cylinder.

Analyze, under the same conditions, the scenario in which the inner cylinder is rotated. In both scenarios, the following conditions will be considered: the loads of the cylinders are of the same sign; the cylinder loads are of different signs. Air viscosity is neglected.

Hint: A variable magnetic field generates an "eddy" electric field, whose field lines are closed curves, the direction of which is correlated with the direction of the field lines of the variable magnetic field, as indicated in Figure 4.2 (*reverse drill rule*: the direction of the electric field lines is the direction of rotation of the drill, so that it advances in the opposite direction with respect to the direction of $\Delta \vec{B}$).

Solution

(a) When the intensity of the current through the coils of the solenoid increases, the induction of the uniform magnetic field inside the solenoid is variable (increases). This causes the appearance (in the same region) of an electric field whose field lines are closed circles around the magnetic field lines, with their center lying on the solenoid

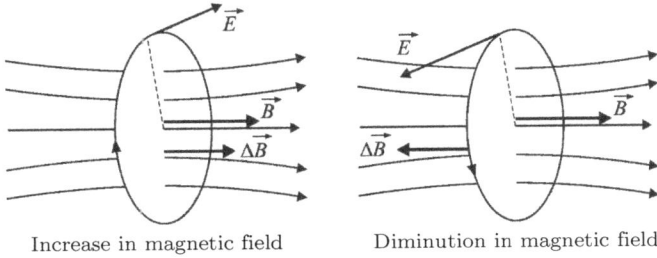

Increase in magnetic field   Diminution in magnetic field

Fig. 4.2

axis. Their direction is illustrated in Figure 4.3, where the electric field line at the cross-section level of the electrified inner cylinder is represented, as well as the electric field intensity vector at different points on the field line $(\vec{E})$.

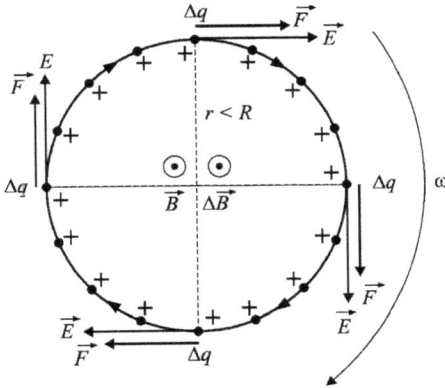

Fig. 4.3

Under these conditions, on each elementary sector on the surface of the paper cylinder, carrying the positive elementary electric charge Δq, will exert one force $\vec{F} = \vec{E}\Delta q$, explaining the direction of rotation of the electrified inner cylinder inside the solenoid when the intensity of the current through its coils increases.

Figure 4.4 illustrates the direction of rotation of the inner cylinder for the proposed scenarios.

The variation of the magnetic induction inside the solenoid, as a result of the variation of the current intensity through the coils of

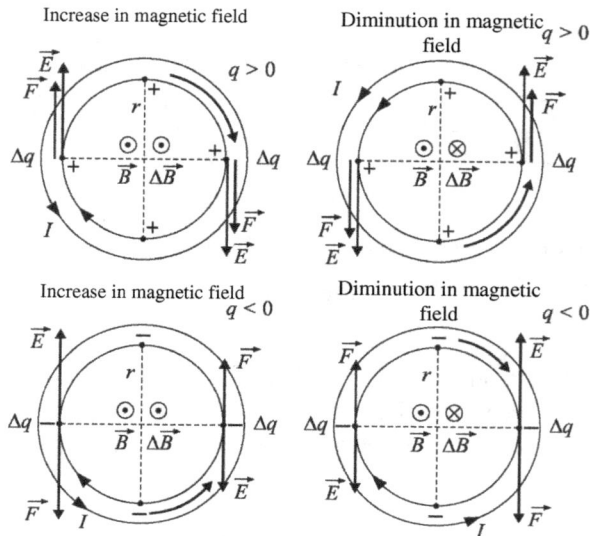

Fig. 4.4

the solenoid, is:

$$\Delta B_{\text{solenoid}} = \mu \frac{N_{\text{total}} \Delta I_{\text{solenoid}}}{l};$$

$$N_{\text{total}} = Nl; \quad \Delta I_{\text{solenoid}} = (n-1)I;$$

$$\Delta B_{\text{solenoid}} = (n-1)\mu NI,$$

so that the variation of the magnetic flux through the inner cylinder, due to the variation of the current intensity through the coils of the solenoid, is

$$\Delta \Phi_{\text{inductor}} = \pi r^2 \Delta B_{\text{solenoid}} = \pi \mu r^2 NI(n-1).$$

From the moment the electrified inner cylinder starts to rotate, it is equivalent to a solenoid, having only one "wide" coil, through which an evenly distributed electric current will pass, whose intensity, when the angular velocity of the cylinder is ω, has the value

$$I_0 = \frac{q}{T} = \frac{q\omega}{2\pi},$$

so that the magnetic induction at a point inside the cylinder is

$$B_{\text{cylinder}} = \mu \frac{I_0}{h} = \mu \frac{q\omega}{2\pi h}.$$

As a result, the variation of the magnetic induction inside the cylinder is

$$\Delta B_{\text{cylinder}} = B_{\text{cylinder}} - 0 = \mu \frac{q\omega}{2\pi h},$$

so that the variation of the magnetic flux through the inner cylinder, due to its rotation, is

$$\Delta \Phi_{\text{induced}} = \pi r^2 \Delta B_{\text{cylinder}} = \mu \frac{r^2 q\omega}{2h}.$$

Since the inner cylinder is very light, the two magnetic fluxes must be equal for energy reasons.

It results that:

$$\Delta \Phi_{\text{inductor}} = \Delta \Phi_{\text{induced}};$$

$$\omega = \frac{2\pi N I h(n-1)}{q}.$$

Similarly, when the intensity of the current in the solenoid's coils decreases by n times, it turns out that

$$\omega = \frac{2\pi(n-1)hNI}{nq}.$$

(b) If $r > R$, as Figure 4.5 indicates, then we can assume that the drawing in Figure 4.3 also depicts the eddy electric field line at the level of the transverse section of the electrified cylinder when it is outside the solenoid ($r > R$), the reasoning remaining the same.

As a result, if the working conditions are maintained (the same direction of the current through the coils of the solenoid, the same variation of the intensity of the current through the solenoid, the same electrical load of the cylinder), the direction of rotation of the electrified cylinder is the same, regardless of the scenario: $r < R$ or $r > R$.

For the four possible scenarios, the direction of rotation of the external electrified cylinder can be established by adapting the

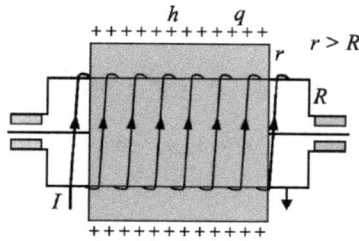

Fig. 4.5

sequences from Figure 4.4 and simply changing the positions of the two circles.

If, from a qualitative point of view, the process is the same in both scenarios ($r < R$; $r > R$), the angular momentum gained by the electrified cylinder cannot be the same in both scenarios.

Indeed, since the magnetic field induction of the solenoid is very small around the solenoid, it follows that, in the scenario $r > R$, we must admit that the magnetic flow varies through the external electrified cylinder due to the varying intensity of the current in the solenoid, which corresponds only to the transverse section of the solenoid, so

$$\Delta\Phi_{\text{inductor}} = \pi R^2 \Delta B_{\text{solenoid}} = \pi\mu R^2 NI(n-1).$$

Well-known reasoning leads us to the variation of the magnetic flux through the outer electrified cylinder:

$$\Delta\Phi_{\text{induced}} = \pi r^2 \Delta B_{\text{cylinder}} = \mu\frac{r^2 q\omega'}{2h}.$$

So, it results that:

$$\Delta\Phi_{\text{inductor}} = \Delta\Phi_{\text{induced}};$$

$$\omega' = \frac{2\pi NIh(n-1)}{q}\frac{R^2}{r^2}; \quad R < r;$$

$$\omega' = \omega\frac{R^2}{r^2}; \quad \omega' < \omega.$$

Similarly, when the current intensity in the coils of the solenoid decreases by n times, it results that:

$$\omega' = \frac{2\pi(n-1)hNI}{nq}\frac{R^2}{r^2} < \omega.$$

(c) If the external electrified cylinder is in rotational motion with angular velocity Ω, then it is equivalent to an electric current having the intensity

$$I = \frac{q}{T} = \frac{\sigma 2\pi R l}{\frac{2\pi}{\Omega}} = \sigma R l \Omega,$$

uniformly distributed in a single "wide" circular coil, equivalent to a solenoid having a large number of coils arranged closely along its entire length so that the intensity of the current through each coil is

$$I_1 = \frac{I}{N_{\text{total}}}.$$

This equivalence is indicated in Figure 4.6, so that the induction of the magnetic field inside the equivalent solenoid (inside the rotating electrified outer cylinder) is

$$B = \mu \frac{N_{\text{total}} I_1}{l} = \frac{\mu I}{l} = \mu \sigma R \Omega.$$

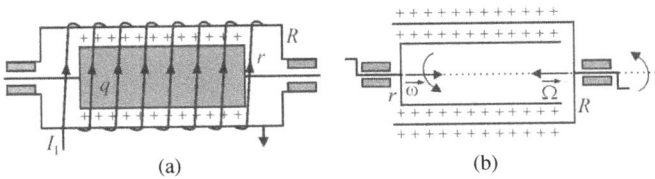

(a)   (b)

Fig. 4.6

In these conditions, the variation of the induction of the magnetic field inside the electrified outer cylinder throughout its accelerated rotation, starting from rest, is

$$\Delta B = B = \mu \sigma R \Omega,$$

where σ is the density of the superficial electric charge, and the variation of the magnetic flux through the inner electrified cylinder, due to the accelerated rotation of the outer electrified cylinder, is

$$\Delta \Phi_{\text{inductor}} = \pi r^2 \Delta B = \pi \sigma r^2 R \Omega \mu.$$

As we already know, the variable (increasing) magnetic field inside the outer cylinder causes the appearance in the same region of an eddy electric current that will rotate the inner electrified cylinder (as indicated in Figure 4.3 for the previously analyzed case). As a result, based on the equivalence of the two systems represented in Figure 4.6, the motion of the inner cylinder, indicated by the first arrow in Figure 4.4, occurs in the opposite direction to the direction of rotation imposed on the outer cylinder (the vectors $\vec{\omega}$ and $\vec{\Omega}$ have opposite directions).

From the moment the electrified inner cylinder starts spinning, it is equivalent to an electric current with the intensity

$$I_0 = \frac{q_0}{T_0} = \frac{\sigma 2\pi r l}{\frac{2\pi}{\omega}} = \sigma r l \omega,$$

uniformly distributed in a single, "wide" circular coil, which is equivalent to a solenoid with a large number of coils closely arranged along its entire length so that the current intensity through each coil is

$$I_0' = \frac{I_0}{N_{\text{total}}},$$

and the self-magnetic induction at a point inside the small cylinder is

$$B' = \mu \frac{N_{\text{total}}' I_0'}{l} = \frac{\mu I_0'}{l} = \mu \sigma r \omega.$$

In these conditions, throughout the accelerated drive of the inner cylinder, the variation of its magnetic induction is

$$\Delta B' = B' - 0 = \mu \sigma r \omega,$$

and the variation of its magnetic flux through the inner cylinder is

$$\Delta \Phi_{\text{induced}} = \pi r^2 \Delta B' = \pi \mu r^3 \sigma \omega.$$

For energy reasons, because the inner cylinder is very light, the two variations of the magnetic fluxes are equal, so it results that

$$\omega = \frac{R}{r}\Omega > \Omega.$$

Let us now assume that the electric charge of the inner cylinder is negative. In this case, shown in the bottom left diagram in Figure 4.4, the inner cylinder rotates in the same direction as the outer cylinder, the angular velocity values remaining in the previously established relationship. Let us now consider the rotation imposed on the inner cylinder with angular velocity $\vec{\Omega}$. Based on its equivalence with the solenoid represented in Figure 4.7, the outer cylinder gains a rotary movement in the opposite direction.

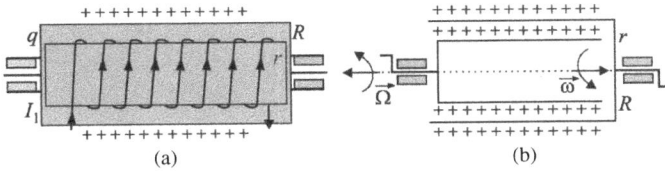

(a)   (b)

Fig. 4.7

Repeating the reasoning presented previously, it follows that

$$\omega = \frac{r^3}{R^3}\Omega < \Omega.$$

In this case, if the outer cylinder's electric charge is negative, it will acquire a rotational movement in the same direction as the inner cylinder.

Problem 5. Astronomical Refraction

The light traveling from a star σ through the interstellar vacuum towards Earth propagates in a straight line until the upper limit of the Earth's atmosphere (Figure 5.1).

Let us admit that the Earth is spherical and that the atmosphere around it is made up of concentric, very thin, homogeneous spherical layers whose density increases uniformly when the altitude decreases, so that the refractive index of the atmospheric air, dependent on its density (and hence dependent on its temperature and pressure) increases from the value 1 (corresponding to the upper limit of the atmosphere) to the average value 1.00029255 (corresponding to the observation point at the base of the lower air layer, at a temperature of 0°C and a pressure of 1 atm).

Under these conditions, respecting the laws of refraction, the light ray entering the Earth's atmosphere will describe a flat curve to reach the eye of the observer, who will see the star in an apparent

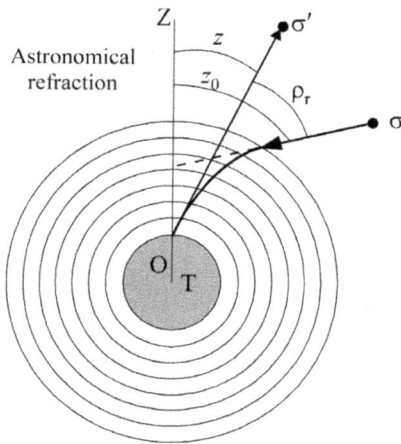

Fig. 5.1

position σ', in the direction tangent to the light ray at the observation point.

The angle ρ_r formed by the initial direction of the light ray (before entering the Earth's atmosphere) with the tangent to the light ray at the point of observation is called *astronomical refraction*.

If z_0 is the true zenithal distance of the star, and z is its observed (apparent) zenithal distance, then $\rho_r = z_0 - z$.

Therefore, the value of the astronomical refraction (the value of the angle ρ_r) fulfills the role of a correction which, applied to the value of the apparent zenith distance (z, determined directly from astronomical observations), allows one to determine the value of the true zenith distance of the star.

The method for determining the value of a star's refraction correction depends on the value (small or large) of its apparent zenithal distance.

(a) *Determine* the correction ρ_r for stars with a small apparent zenithal distance, assuming that the layers of air, with different densities and refractive indices, through which the light beam passes are flat and parallel, as shown in Figure 5.2, which represents two neighboring layers of air with different but very close constant refractive indices, n, and $n + dn$, so that $dn < 0$.

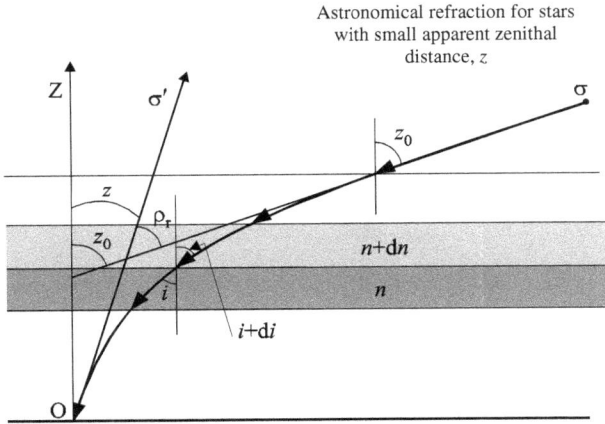

Fig. 5.2

(b) *Determine* the correction ρ_r for stars with a large apparent zenithal distance, assuming that the layers of air surrounding the Earth are spherical, as shown in Figure 5.3, which represents two neighboring spherical layers of air with different but very close constant refractive indices, n and $n + dn$, where $dn < 0$, and with radii r and $r + dr$, respectively, where $dr > 0$.

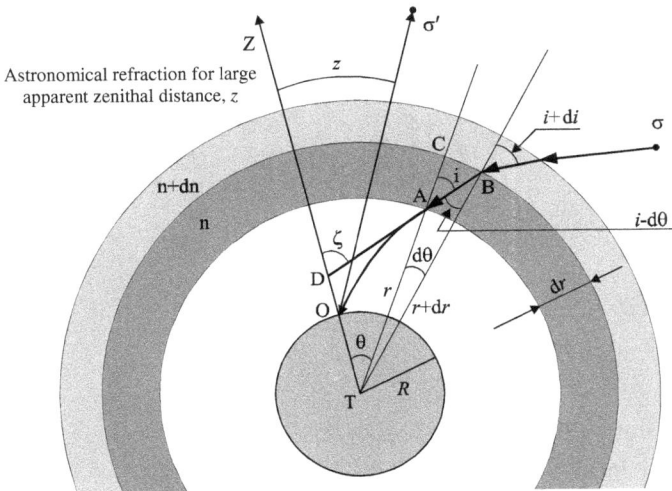

Fig. 5.3

The result of the "refraction integral" is known:

$$\int_1^{n_O} \frac{1}{\sqrt{\left(\frac{nr}{n_O R}\right)^2 - \sin^2 z}} \frac{dn}{n}.$$

Solution

(a) Following a linear path in each of these layers of air, the light ray is refracted at the boundary between them, respecting the laws of refraction:

$$\frac{\sin(i + di)}{\sin i} = \frac{n}{n + dn};$$

$$n \sin i = (n + dn) \sin(i + di).$$

This relation, when applied (written) according to the same procedure for the neighboring layers, starting from the entrance of the light beam into the atmosphere (coming from the cosmic vacuum, for which $n = 1$, under an incidence angle $i = z_0$) and up to the ground level (where $n = n_O$, and $i = z$), leads to:

$$\sin z_0 = n_O \sin z;$$

$$\sin(z + \rho_r) = n_O \sin z;$$

$$\sin z \cos \rho_r + \cos z \sin \rho_r = n_O \sin z;$$

$$\cos \rho_r \approx 1; \quad \sin \rho_r \approx \rho_r,$$

because the angle ρ_r is very small;

$$\rho_r = (n_O - 1) \text{tg} z;$$

$$a_r = n_O = -1;$$

$$\rho_r = a_r \text{tg} z.$$

This relation allows the calculation of the astronomical refraction correction, ρ_r, for a star whose measured apparent zenith distance is z, knowing the astronomical refraction constant of the observation site (a_r).

Under these conditions, the true zenithal distance of the star is

$$z_0 = z + a_r \text{tg } z.$$

(b) Following a linear path in each of these layers of air, the light ray refracts at the limit between them, respecting the laws of refraction:

$$\frac{\sin(i + di)}{\sin(i\_d\theta)} = \frac{n}{n + dn}.$$

In addition, from ΔABT, with the help of the sine theorem, it follows that:

$$\frac{r + dr}{\sin i} = \frac{r}{\sin(i\_d\theta)};$$

$$(n + dn)(r + dr)\sin(i + di) = nr \sin i,$$

a relationship that we interpret as representing an "optical invariant" of the light ray on its way through the atmosphere, so

$$nr \sin i = \text{constant}.$$

The value of this constant is determined by the specifying the invariant at the point of observation, where:

$$n = n_O; \quad r = R; \quad i = z,$$

so that the invariant of the light ray that crosses the Earth's atmosphere, under the specified conditions, is

$$nr \sin i = n_O R \sin z.$$

Logarithmically differentiating (logarithm + differentiation) the previous expression results in

$$\frac{dn}{n} + \frac{dr}{r} + \text{ctg } i \, di = 0.$$

Points A and B, being very close, imply that $d\theta$ and dr are very small, the direction of the segment AB coincides with the direction of the tangent to the light ray at point A, and the sector ACB can be assimilated to a right triangle. From this, it results that

$$\text{tg } i = \frac{(r + dr)d\theta}{dr} \approx \frac{rd\theta}{dr}.$$

If ζ is the angle formed by the tangent to the light ray at point A and the vertical of the place of observation, then, from AADT, it results that

$$\zeta = \theta + i.$$

Thus, using the previous relations, it results that:

$$d\zeta = d\theta + di;$$

$$d\zeta = -\frac{dn}{n}\text{tg}\, i,$$

representing the differential equation of astronomical refraction.

Noting now that if point A coincides with the point where the light ray enters the atmosphere when $\zeta = z_0$, and if point A coincides with the observation point O on the ground when $\zeta = z$, then integrating the previous expression for limits corresponding to the two points results in:

$$\int_z^{z_0} d\zeta = -\int_{n_O}^1 \frac{dn}{n}\text{tg}\, i = z_0 - z = \rho_r;$$

$$\sin i = \frac{n_O R \sin z}{nr};$$

$$\cos i = \sqrt{1 - \frac{n_O^2 R^2 \sin^2 z}{n^2 r^2}};$$

$$\rho_r = \sin z \int_1^{n_O} \frac{1}{\sqrt{\left(\frac{nr}{n_O R}\right)^2 - \sin^2 z}} \frac{dn}{n}.$$

This relationship allows the calculation of the astronomical refraction correction ρ_r for a star whose measured apparent zenith distance is z, knowing the value of the "refraction integral."

Problem 6. A Relativistic Rocket

From the classical study of the dynamics of a material point with variable mass, it is known that, in a field of external forces $\vec{F}_{\text{ext}}$, the equation of motion of a material point with a variable mass (Mescerscki–Levi-Civita equation) is:

$$m\frac{\mathrm{d}\vec{v}}{\mathrm{d}t} = \vec{F}_{\text{ext}} + \vec{v}_{\text{rel}}\frac{\mathrm{d}m}{\mathrm{d}t},$$

where m and $\vec{v}$ are the mass and the velocity, respectively, of the material point at time t, and $\vec{v}_{\text{rel}}$ is the speed, relative to the material point, of the particles captured or expelled by the material point;

$$\vec{v}_{\text{rel}} = \vec{u} - \vec{v},$$

where $\vec{u}$ is the absolute velocity of the particles captured or expelled by the material point.

Let us now assume that the relativistic material point with a variable (decreasing) mass is the rocket represented in Figure 6.1, to which the mobile system O'X'Y'Z' is attached, whose speed, relative to the observer O in the fixed system XYZ at time t, is $\vec{v}$.

Clarification: In relativistic conditions, $\vec{v}_{\text{rel}} = \vec{u} - \vec{v}$ is not identified with the relative velocity $\vec{u}'$.

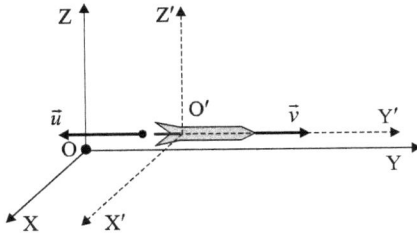

Fig. 6.1

At the initial moment, the absolute speed of the rocket was $\vec{v}_0$, and the initial rest mass of the rocket (in the rocket system) was M'. *Determine* the mass of the rocket (expressed in the rocket system), after part of the fuel has burned, when the absolute speed of the rocket is $\vec{v}$, and $u' = $ constant is the speed of the ejected particles relative to the mobile system. The presence of any external forces will be neglected.

Particular case: $v_0 = 0$; $v \ll c$; $u' = c$.

Solution

If the initial absolute velocity of the rocket was $\vec{v}_0$, and the initial rest mass of the rocket (in the rocket system) was M', then, after some of the fuel has burned and the absolute velocity of the rocket is $\vec{v}$, the rest of the mass of the rocket (in the rocket system) is m' (the next variable), so that, at time t, the mass of the rocket in relation to the observer O is

$$m = \frac{m'}{\sqrt{1 - \frac{v^2}{c^2}}}.$$

If the absolute speed of the ejected particles at the time t is $\vec{u}$, then, in the absence of external forces, the equation of the rocket's motion is:

$$m\frac{dv}{dt} = -(u + v)\frac{dm}{dt};$$

$$\frac{m'}{\sqrt{1 - \frac{v^2}{c^2}}}\frac{dv}{dt} = -(u + v)\frac{d}{dt}\left(\frac{m'}{\sqrt{1 - \frac{v^2}{c^2}}}\right);$$

$$\frac{d}{dt}\left(\frac{m'}{\sqrt{1 - \frac{v^2}{c^2}}}\right) = \frac{1}{\sqrt{1 - \frac{v^2}{c^2}}}\frac{dm'}{dt} + \frac{m'}{c^2}\frac{v}{\left(1 - \frac{v^2}{c^2}\right)^{3/2}}\frac{dv}{dt};$$

$$\frac{m'}{1 - \frac{v^2}{c^2}}\left(1 + \frac{uv}{c^2}\right)\frac{dv}{dt} = -(u + v)\frac{dm'}{dt}.$$

Considering the relativistic velocity composition rule, it follows that:

$$\vec{u}' = \frac{\vec{u} - \vec{v}}{1 - \frac{\vec{u}\vec{v}}{c^2}}; \quad u' = \frac{u + v}{1 + \frac{uv}{c^2}},$$

where u' is the speed of the ejected particles with respect to the mobile system ($u' = \text{constant}$);

$$m'\frac{dv}{dt} = -u'\left(1 - \frac{v^2}{c^2}\right)\frac{dm'}{dt};$$

$$\frac{dm'}{m'} = -\frac{dv}{u'\left(1 - \frac{v^2}{c^2}\right)} < 0,$$

because during the flight of the rocket, its mass is decreasing (meaning that the variation in the mass of the rocket is $dm' < 0$);

$$\frac{1}{1 - \frac{v^2}{c^2}} = \frac{1}{2}\frac{1}{1 - \frac{v}{c}} + \frac{1}{2}\frac{1}{1 + \frac{v}{c}};$$

$$\frac{dm'}{m'} = -\frac{1}{2u'}\frac{dv}{1 - \frac{v}{c}} - \frac{1}{2u'}\frac{dv}{1 + \frac{v}{c}}.$$

From this, by integration, it results that:

$$\int_{M'}^{m'} \frac{dm'}{m'} = -\frac{1}{2u'}\left[\int_{v_0}^{v} \frac{dv}{1 + \frac{v}{c}} + \int_{v_0}^{v} \frac{dv}{1 - \frac{v}{c}}\right];$$

$$\ln\frac{m'}{M'} = -\frac{c}{2u'}\left[\ln\left(1 + \frac{v}{c}\right)\bigg|_{v_0}^{v} - \ln\left(1 - \frac{v}{c}\right)\bigg|_{v_0}^{v}\right];$$

$$\ln\frac{m'}{M'} = -\frac{c}{2u'}\left[\ln\frac{1 + \frac{v}{c}}{1 - \frac{v}{c}} - \ln\frac{1 + \frac{v_0}{c}}{1 - \frac{v_0}{c}}\right];$$

$$\ln\frac{m'}{M'} = -\frac{c}{2u'}\ln\frac{\left(1 + \frac{v}{c}\right)\left(1 - \frac{v_0}{c}\right)}{\left(1 - \frac{v}{c}\right)\left(1 + \frac{v_0}{c}\right)};$$

$$m' = M'\left[\frac{\left(1 + \frac{v}{c}\right)\left(1 - \frac{v_0}{c}\right)}{\left(1 - \frac{v}{c}\right)\left(1 + \frac{v_0}{c}\right)}\right]^{-c/2u'}.$$

In particular, if the acceleration of the rocket starts from rest ($v_0 = 0$), it results that

$$m' = M'\left(\frac{1 + \frac{v}{c}}{1 - \frac{v}{c}}\right)^{-c/2u'},$$

and for low speeds ($v \ll c$), we have:

$$\frac{1 + \frac{v}{c}}{1 - \frac{v}{c}} = \left(1 + \frac{v}{c}\right)\left(1 - \frac{v}{c}\right)^{-1} \approx 1 + 2\frac{v}{c};$$

$$m' = M'\left[\left(1 + \frac{2v}{c}\right)^{\frac{c}{2v}}\right]^{-v/u'};$$

$$\lim_{n\to\infty}\left(1+\frac{1}{n}\right)^n = e;$$

$$m' = M'e^{-v/u'}.$$

This relation is known from the study of non-relativistic rockets.

Note: If the relativistic rocket is photonic, for which $u' = c$, it results that:

$$m' = M'\left(\frac{1+\frac{v}{c}}{1-\frac{v}{c}}\right)^{-1/2}.$$

Problem 7. Relativistic Collision

Two atoms, with rest masses m_{01} and m_{02}, moving rectilinearly and uniformly with the relativistic velocities $\vec{v}_1 = v_{1x}\vec{i}$ and $\vec{v}_2 = v_{2x}\vec{i} + v_{2y}\vec{j}$, respectively, where $\vec{i}$ and $\vec{j}$ are the vertices of the perpendicular axes OX and OY, respectively, collide at a certain moment and form a molecule, considered a material point.

Determine the speed and rest mass of the molecule.

Particular case: $\vec{v}_2 = -v_{2x}\vec{i}$ and $m_1\vec{v}_{1x} = -m_2\vec{v}_{2x}$, where m_1 and m_2 are the moving masses of the two atoms, and c is known.

Solution

Figure 7.1 represents the two atoms before the collision, as well as the molecule resulting from their collision.

The interaction of the two atoms occurs in compliance with the laws of conservation of momentum and total energy, in the relativistic form.

From the momentum conservation law, in projections on the two axes, it follows that:

$$m_1 v_{1x} + m_2 v_{2x} = m v_x;$$

$$\frac{m_{01}v_{1x}}{\sqrt{1-\frac{v_{1x}^2}{c^2}}} + \frac{m_{02}v_{2x}}{\sqrt{1-\frac{v_{2x}^2+v_{2y}^2}{c^2}}} = \frac{m_0 v_x}{\sqrt{1-\frac{v_x^2+v_y^2}{c^2}}},$$

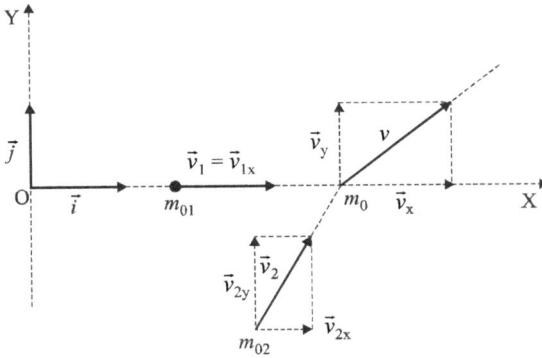

Fig. 7.1

where m_0 is the rest mass of the resulting molecule;

$$m_2 v_{2y} = m v_y;$$

$$\frac{m_{02} v_{2y}}{\sqrt{1 - \frac{v_{2x}^2 + v_{2y}^2}{c^2}}} = \frac{m_0 v_y}{\sqrt{1 - \frac{v_x^2 + v_y^2}{c^2}}}.$$

From the law of conservation of total energy, it follows that:

$$m_1 c^2 + m_2 c^2 = m c^2;$$

$$m_1 + m_2 = m;$$

$$\frac{m_{01}}{\sqrt{1 - \frac{v_{1x}^2}{c^2}}} + \frac{m_{02}}{\sqrt{1 - \frac{v_{2x}^2 + v_{2y}^2}{c^2}}} = \frac{m_0}{\sqrt{1 - \frac{v_x^2 + v_y^2}{c^2}}}.$$

Solving the system formed by the previous three equations, it results that:

$$v_x = \frac{\dfrac{m_{01} v_{1x}}{\sqrt{1 - \frac{v_{1x}^2}{c^2}}} + \dfrac{m_{02} v_{2x}}{\sqrt{1 - \frac{v_{2x}^2 + v_{2y}^2}{c^2}}}}{\dfrac{m_{01}}{\sqrt{1 - \frac{v_{1x}^2}{c^2}}} + \dfrac{m_{02}}{\sqrt{1 - \frac{v_{2x}^2 + v_{2y}^2}{c^2}}}};$$

$$v_y = \frac{\dfrac{m_{02} v_{2y}}{\sqrt{1 - \frac{v_{2x}^2 + v_{2y}^2}{c^2}}}}{\dfrac{m_{01}}{\sqrt{1 - \frac{v_{1x}^2}{c^2}}} + \dfrac{m_{02}}{\sqrt{1 - \frac{v_{2x}^2 + v_{2y}^2}{c^2}}}};$$

$$m_0 = \left[\frac{m_{01}}{\sqrt{1-\frac{v_{1x}^2}{c^2}}} + \frac{m_{02}}{\sqrt{1-\frac{v_{2x}^2+v_{2y}^2}{c^2}}}\right]$$

$$\times \left[1 - \frac{1}{c^2}\frac{\dfrac{m_{01}^2 v_{1x}^2}{1-\frac{v_{1x}^2}{c^2}} + \dfrac{m_{02}^2(v_{2x}^2+v_{2y}^2)}{1-\frac{v_{2x}^2+v_{2y}^2}{c^2}} + \dfrac{2m_{01}m_{02}v_{1x}v_{2x}}{\sqrt{1-\frac{v_{1x}^2}{c^2}}\sqrt{1-\frac{v_{2x}^2+v_{2y}^2}{c^2}}}}{\dfrac{m_{01}^2}{1-\frac{v_{1x}^2}{c^2}} + \dfrac{m_{02}^2}{1-\frac{v_{2x}^2+v_{2y}^2}{c^2}} + \dfrac{2m_{01}m_{02}}{\sqrt{1-\frac{v_{1x}^2}{c^2}}\sqrt{1-\frac{v_{2x}^2+v_{2y}^2}{c^2}}}}\right].$$

If the two atoms move along the OX axis in opposite directions, with equal magnitudes of momentum, the following specifications apply to the previous formulas:

$$v_{2y} = 0;$$

$$\frac{m_{01}v_{1x}}{\sqrt{1-\frac{v_{1x}^2}{c^2}}} = -\frac{m_{02}v_{2x}}{\sqrt{1-\frac{v_{2x}^2}{c^2}}}.$$

It results that:

$$v_x = 0; \quad v_y = 0;$$

$$m_0 = \frac{m_{01}}{\sqrt{1-\frac{v_{1x}^2}{c^2}}} + \frac{m_{02}}{\sqrt{1-\frac{v_{2x}^2}{c^2}}},$$

that is, the molecule resulting from the collision of the two atoms remains at rest, but with a rest mass $m_0 > m_{01} + m_{02}$.

Problem 8. Electromagnetic Submarine

On the two sides of a special submarine made of insulating plastic, two longitudinal metal strips are mounted, as shown in Figure 8.1, connected to the terminals of a continuous voltage generator. An electromagnet is mounted in a vertical position inside the submarine, between the two metal strips.

    Analyze and justify the possibility of moving the submarine through the seawater.

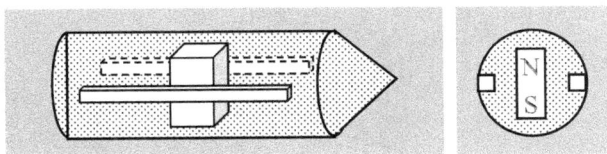

Fig. 8.1

Solution

Due to the dissociation of NaCl molecules, there are Na^+ ions and Cl^- ions in the seawater, with Cl in the free state. The electric fields of the two metal strips will act on the two types of ions, moving them in opposite directions inside the submarine. As a result, ionic currents will be formed around the submarine, through both its upper and lower parts, whose direction is from the $(+)$ band to the $(-)$ band. The two types of ions move in opposite directions, but their movements are equivalent to two electric currents of the same direction.

Moving in the magnetic field of the electromagnet, the Lorentz force acts on these ions, $\vec{F}_L = q\vec{v} \times \vec{B}$. For either of the two types of ions, wherever they are within the vicinity of the submarine, the Lorentz forces are oriented in the same direction along the submarine, as shown in Figure 8.2.

The effect of these Lorentz forces will be to push the apsis along the submarine. Following the principle of reciprocal actions, the submarine will be propelled in the opposite direction.

The submarine's forward direction can be changed by changing the electrical polarities of the two strips or the polarity of the electromagnet.

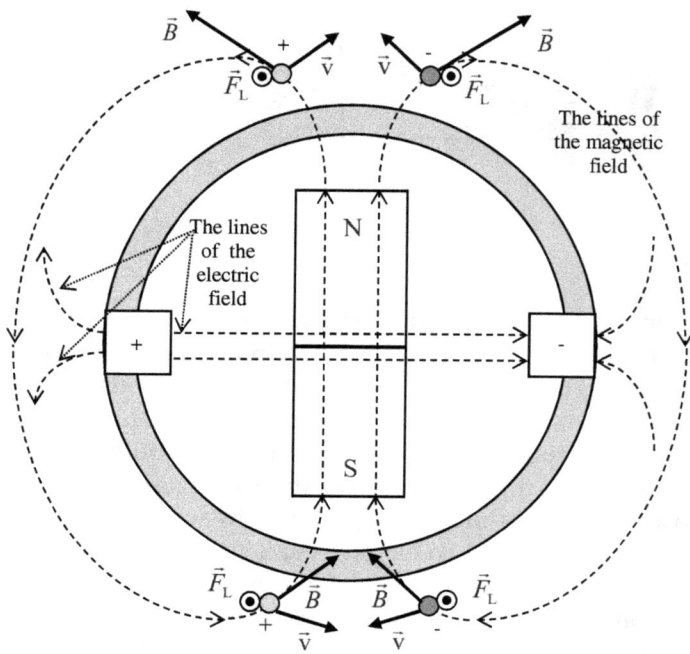

Fig. 8.2

Chapter 6

International Pre-Olympic Physics Contest 2009, Călimăneşti, Romania

Problem 1. Gravitational Equatorial Ring on the Surface of the Sun

This problem was proposed by Andrei Constantin, Romanian Olympian in Physics, participant and medalist at IOPH.

Due to the Sun's rotation, its equatorial radius is slightly larger than its polar radius, so that the distribution of solar matter around the axis of rotation no longer has a spherical symmetry. This is why the precession of the orbits of the planets in our solar system occurs (the slip/deflection/rotation of the perihelion of each planet's orbit). In a simple model, we can consider that the rotation of the Sun has the effect of forming a gravitational equatorial ring on the surface of the solar sphere, with negligible thickness and mass $M << M_S$, where M_S is the mass of the Sun (which includes the mass of the ring).

To prove the existence of this effect, we will consider that the solar mass is distributed uniformly and symmetrically with respect to the axis perpendicular to the plane of the planet's orbit, and the gravitational potential energy of the Sun–planet system, corresponding to the proposed model, when the planet with mass m is at a distance

$r \gg R$ relative to the center of the Sun, is given by the expression

$$E_p = -\frac{k}{r} \left[1 + \frac{1}{2} J_2 \left(\frac{R}{r} \right)^2 - \frac{3}{8} J_4 \left(\frac{R}{r} \right)^4 + \frac{5}{16} J_6 \left(\frac{R}{r} \right)^6 + \cdots \cdots \right],$$

where R is the radius of the Sun. The constants $J_2, J_4, J_6, \ldots$ depend on the exact distribution of the solar matter around the axis of rotation of the Sun.

(a) Using arguments of a qualitative nature, *determine* the constant k from the previous expression of the gravitational potential energy, E_p. The constant of universal attraction, K, is known.

   Determine the constant J_2 from the previous expression of gravitational potential energy, E_p, using the proposed model.

   It will be admitted that: $(1 + \alpha)^n \cong 1 + n\alpha + \frac{1}{2}n(n-1)\alpha^2$; $\alpha \ll 1$.

(b) For elliptical orbits with very small eccentricity, the very small deviation of a planet from the equatorial circular orbit can be explained by admitting that the planet performs radial oscillations with a very small amplitude, around a distance $r = r_0$. These oscillations overlap with the circular motion of the planet around the Sun. (If this process is analyzed under the conditions of observing the law of conservation of kinetic momentum, then it is demonstrated that the radial oscillations of the planet must be accompanied by angular oscillations with the same period. In this problem, this effect is neglected.)

   Determine the period of the radial oscillations of the planet, considering that $J_2 = 0$. The revolution period of the planet around the Sun, T_{rev}, is known.

   Considering $J_2 = 0$, *write* the expression of the planet's kinetic energy depending on: the projection of the planet's momentum in the radial direction, p_{rad}; the kinetic moment of the planet, L; the mass of the planet, m; and the distance from the planet to the center of the Sun, r.

(c) Given that the kinetic moment of the planet $L = $ constant, *identify*, as the *effective gravitational potential* of the point where the planet is located, $V_{ef} = V_{ef}(r)$, all the terms in the expression of the total energy of the Sun–planet system that do not depend on the radial momentum of the planet. The introduction of the concept of effective gravitational potential reduces the planar movement of the planet to a radial movement.

Determine the distance r_0 for which the *effective gravitational force*, $F_{ef} = \frac{dV_{ef}}{dr}$, is null.

(d) Using the approximation proposed in the problem, *determine* the approximate expression of the total energy of the Sun–planet system for $r = r_0 + \Delta r$, where $\Delta r \ll r_0$, keeping most of the terms in R^2. Then, comparing this expression with the expression of the total energy of a harmonic oscillator, *determine* the period of the planet's radial oscillations, T_{osc}, in the case of $J_2 \neq 0$. *Particular case:* $J_2 = 0$.

Solution

(a) From the expression given for the gravitational potential energy of the Sun–planet system, true in the case of deviations from the spherical distribution,

$$E_p = -\frac{k}{r}\left[1 + \frac{1}{2}J_2\left(\frac{R}{r}\right)^2 - \frac{3}{8}J_4\left(\frac{R}{r}\right)^4 + \frac{5}{16}J_6\left(\frac{R}{r}\right)^6 + \cdots\cdots\right],$$

where the additional terms appear only in the case of deviations from the spherical distribution, the known expression of the potential energy of the Sun–planet system when the distribution of the Sun's mass is spherical must be

$$E_p = -K\frac{mM_S}{r}.$$

In these conditions, by neglecting the terms due to deviations from the spherical distribution, it results that

$$k = KmM_S,$$

where m is the mass of the planet, M_S is the mass of the Sun, and K is the constant of universal attraction.

Figure 1.1 shows an elementary sector, with mass dM, on the outline of the circular ring with mass M. The elementary gravitational potential energy of the system formed by the planet with mass m and the elementary sector with mass dM on the outline of the circular ring is:

$$dE_\mathrm{p} = -K \frac{m\,dM}{\sqrt{r^2 + R^2 - 2rR\cos\theta}};$$

$$M = 2\pi R\gamma,$$

where γ is the linear density of the ring;

$$dM = \gamma\,ds = \gamma R\,d\theta; \quad dM = \frac{M}{2\pi}\,d\theta;$$

$$dE_\mathrm{p} = -K \frac{mM\,d\theta}{2\pi\sqrt{r^2 + R^2 - 2rR\cos\theta}}.$$

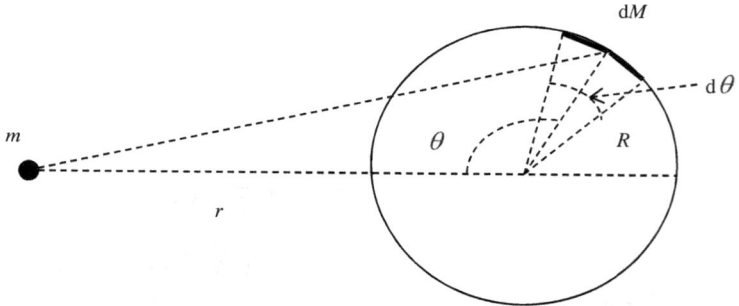

Fig. 1.1

The gravitational potential energy of the system, consisting of the planet with mass m and the entire circular ring with mass M, is:

$$E_\mathrm{p,ring-planet} = -K \frac{mM}{2\pi} \int_0^{2\pi} \frac{d\theta}{\sqrt{r^2 + R^2 - 2rR\cos\theta}};$$

$$E_\mathrm{p,ring-planet} = -K \frac{mM}{2\pi} \int_0^{2\pi} \frac{d\theta}{r\sqrt{1 + \frac{R^2}{r^2} - 2\frac{R}{r}\cos\theta}};$$

$$E_\mathrm{p,ring-planet} = -K \frac{mM}{2\pi} \int_0^{2\pi} \frac{1}{r} \left[1 + \left(\frac{R^2}{r^2} - 2\frac{R}{r}\cos\theta\right)\right]^{-\frac{1}{2}} d\theta;$$

$$\frac{R^2}{r^2} - 2\frac{R}{r}\cos\theta = \alpha << 1;$$

$$(1+\alpha)^n \cong 1 + n\alpha + \frac{1}{2}n(n-1)\alpha^2; \quad \alpha << 1;$$

$$\left[1 + \left(\frac{R^2}{r^2} - 2\frac{R}{r}\cos\theta\right)\right]^{-\frac{1}{2}} \cong 1 - \frac{1}{2}\left(\frac{R^2}{r^2} - 2\frac{R}{r}\cos\theta\right)$$
$$+ \frac{3}{8}\left(\frac{R^2}{r^2} - 2\frac{R}{r}\cos\theta\right)^2;$$

$$\left[1 + \left(\frac{R^2}{r^2} - 2\frac{R}{r}\cos\theta\right)\right]^{-\frac{1}{2}} \cong 1 - \frac{1}{2}\left(\frac{R^2}{r^2} - 2\frac{R}{r}\cos\theta\right)$$
$$+ \frac{3}{8}\left(\frac{R^4}{r^4} - 4\frac{R^3}{r^3}\cos\theta + 4\frac{R^2}{r^2}\cos^2\theta\right);$$

$$\frac{R^4}{r^4} \to 0; \quad \frac{R^3}{r^3} \to 0;$$

$$\left[1 + \left(\frac{R^2}{r^2} - 2\frac{R}{r}\cos\theta\right)\right]^{-\frac{1}{2}} \cong 1 - \frac{1}{2}\left(\frac{R^2}{r^2} - 2\frac{R}{r}\cos\theta\right)$$
$$+ \frac{3}{2}\left(\frac{R^2}{r^2}\cos^2\theta\right);$$

$$\left[1 + \left(\frac{R^2}{r^2} - 2\frac{R}{r}\cos\theta\right)\right]^{-\frac{1}{2}} \cong 1 + \frac{R}{r}\cos\theta + \frac{R^2}{r^2}\frac{3\cos^2\theta - 1}{2};$$

$$E_{\text{p,ring-planet}} = -K\frac{mM}{2\pi}\int_0^{2\pi}\frac{1}{r}\left(1 + \frac{R}{r}\cos\theta + \frac{R^2}{r^2}\frac{3\cos^2\theta - 1}{2}\right)d\theta.$$

The gravitational potential energy of the system consisting of the planet with mass m and the sun in the proposed model is:

$$E_{\text{p,Sun-planet model}} = E_{\text{p,Sun-planet}} + E_{\text{p,ring-planet}};$$

$$E_{p,\text{Sun-planet model}} = -K\frac{m\,(M_S - M)}{r}$$

$$- K\frac{mM}{2\pi} \int_0^{2\pi} \frac{1}{r}\left(1 + \frac{R}{r}\cos\theta + \frac{R^2}{r^2}\frac{3\cos^2\theta - 1}{2}\right)\mathrm{d}\theta;$$

$$\int_0^{2\pi} \frac{1}{r}\left(1 + \frac{R}{r}\cos\theta + \frac{R^2}{r^2}\frac{3\cos^2\theta - 1}{2}\right)\mathrm{d}\theta$$

$$= \frac{1}{r}\int_0^{2\pi}\mathrm{d}\theta + \frac{R}{r^2}\int_0^{2\pi}\cos\theta\mathrm{d}\theta + \frac{3R^2}{2r^3}\int_0^{2\pi}\cos^2\mathrm{d}\theta - \frac{R^2}{2r^3}\int_0^{2\pi}\mathrm{d}\theta;$$

$$\int \cos\theta\mathrm{d}\theta = -\sin\theta; \quad \int \cos^2\theta\mathrm{d}\theta = \frac{1}{2}\left(\theta + \frac{\sin 2\theta}{2}\right);$$

$$\int_0^{2\pi} \frac{1}{r}\left(1 + \frac{R}{r}\cos\theta + \frac{R^2}{r^2}\frac{3\cos^2\theta - 1}{2}\right)\mathrm{d}\theta$$

$$= \frac{1}{r}\,\theta\big|_0^{2\pi} - \frac{R}{r^2}\sin\theta\big|_0^{2\pi} + \frac{3R^2}{2r^3}\frac{1}{2}\left(\theta + \frac{\sin 2\theta}{2}\right)\bigg|_0^{2\pi} - \frac{R^2}{2r^3}\,\theta\big|_0^{2\pi};$$

$$\int_0^{2\pi} \frac{1}{r}\left(1 + \frac{R}{r}\cos\theta + \frac{R^2}{r^2}\frac{3\cos^2\theta - 1}{2}\right)\mathrm{d}\theta$$

$$= \frac{1}{r}2\pi + \frac{3R^2}{2r^3}\frac{1}{2}2\pi - \frac{R^2}{2r^3}2\pi;$$

$$\int_0^{2\pi} \frac{1}{r}\left(1 + \frac{R}{r}\cos\theta + \frac{R^2}{r^2}\frac{3\cos^2\theta - 1}{2}\right)\mathrm{d}\theta$$

$$= \frac{1}{r}2\pi + \frac{3R^2}{2r^3}\frac{1}{2}2\pi - \frac{R^2}{2r^3}2\pi;$$

$$\int_0^{2\pi} \frac{1}{r}\left(1 + \frac{R}{r}\cos\theta + \frac{R^2}{r^2}\frac{3\cos^2\theta - 1}{2}\right)\mathrm{d}\theta = \frac{2\pi}{r}\left(1 + \frac{R^2}{4r^2}\right);$$

$$E_{p,\text{Sun-planet model}} = -K\frac{m\,(M_S - M)}{r} - \frac{KmM}{r}\left(1 + \frac{R^2}{4r^2}\right);$$

$$E_{p,\text{Sun-planet model}} = -K\frac{mM_S}{r} - K\frac{mM}{4r}\frac{R^2}{r^2}.$$

Comparing with the expression

$$E_{\mathrm{p}} = -\frac{k}{r}\left[1 + \frac{1}{2}J_2\left(\frac{R}{r}\right)^2 - \frac{3}{8}J_4\left(\frac{R}{r}\right)^4 + \frac{5}{16}J_6\left(\frac{R}{r}\right)^6 + \cdots\right],$$

from which we retain only the first two terms, it follows that:

$$-K\frac{mM_{\mathrm{S}}}{r} - K\frac{mM}{4r}\frac{R^2}{r^2} = -\frac{k}{r}\left[1 + \frac{1}{2}J_2\left(\frac{R}{r}\right)^2\right]$$

$$= -K\frac{mM_{\mathrm{S}}}{r}\left[1 + \frac{1}{2}J_2\frac{R^2}{r^2}\right];$$

$$J_2 = \frac{1}{2}\frac{M}{M_{\mathrm{S}}}.$$

(b) If $J_2 = 0$, it means that the planet's trajectory around the Sun is an ellipse. In this case, the orbit of the planet being closed, during one complete revolution, representing the revolution period of the planet, T_{rev}, the distance between the planet and the Sun has an exact minimum value (when the planet is at perihelion) and maximum value (when the planet is at aphelion). As a result, the period of small radial oscillations of the planet, T_{osc}, is strictly equal to the period of revolution of the planet around the Sun, $T_{\mathrm{osc}} = T_{\mathrm{rev}}$.

From Figure 1.2, where we decomposed the tangential velocity vector of the planet into two components, it follows that:

$$\vec{v} = \vec{v}_{/\!/} + \vec{v}_\perp; \quad v^2 = v_{/\!/}^2 + v_\perp^2;$$

$$E_{\mathrm{c}} = \frac{mv^2}{2} = \frac{1}{2}mv_{/\!/}^2 + \frac{1}{2}mv_\perp^2;$$

$$\vec{L} = \vec{r} \times m\vec{v}; \quad L = rmv\sin\alpha = rmv_\perp; \quad v_\perp = \frac{L}{mr};$$

$$E_{\mathrm{c}} = \frac{1}{2}mv_{/\!/}^2 + \frac{L^2}{2mr^2}; \quad E_{\mathrm{c}} = \frac{1}{2m}p_{\mathrm{rad}}^2 + \frac{L^2}{2mr^2}.$$

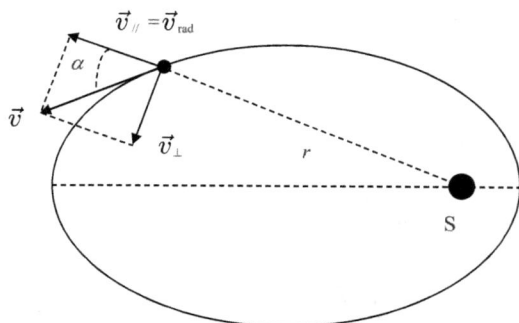

Fig. 1.2

(c) Using the general expression of gravitational potential energy of the Sun–planet system,

$$E_{\mathrm{p}} = -\frac{k}{r}\left[1 + \frac{1}{2}J_2\left(\frac{R}{r}\right)^2 - \frac{3}{8}J_4\left(\frac{R}{r}\right)^4 + \frac{5}{16}J_6\left(\frac{R}{r}\right)^6 + \cdots\right],$$

written for the elliptical orbit of the planet,

$$E_{\mathrm{p}} = -\frac{k}{r}\left[1 + \frac{1}{2}J_2\left(\frac{R}{r}\right)^2\right],$$

as well as the general expression of the kinetic energy, deduced above,

$$E_{\mathrm{c}} = \frac{1}{2m}p_{\mathrm{rad}}^2 + \frac{L^2}{2mr^2},$$

it results that the total energy of the Sun–planet system is

$$E = E_{\mathrm{c}} + E_{\mathrm{p}} = \frac{1}{2m}p_{\mathrm{rad}}^2 + \frac{L^2}{2mr^2} - \frac{k}{r}\left[1 + \frac{1}{2}J_2\left(\frac{R}{r}\right)^2\right].$$

The effective gravitational potential of the point where the planet is located (V_{ef}), being represented by the sum of all the terms in the expression of the total energy of the Sun–planet system, which do not depend on the radial momentum of the planet, it results that:

$$V_{\mathrm{ef}} = \frac{L^2}{2mr^2} - K\frac{mM_{\mathrm{S}}}{r}\left[1 + \frac{1}{2}J_2\frac{R^2}{r^2}\right];$$

$$V_{\text{ef}} = \frac{L^2}{2mr^2} - K\frac{mM_S}{r} - K\frac{mM_S J_2 R^2}{2r^3};$$

$$V_{\text{ef}}\,(r) = \frac{L^2}{2mr^2} - \frac{k}{r} - \frac{kJ_2 R^2}{2r^3}.$$

The distance r_0, relative to the Sun, where the effective force cancels out, is obtained as the solution of the equation

$$F_{\text{ef}} = \frac{\mathrm{d}V_{\text{ef}}}{\mathrm{d}r} = 0.$$

It results that:

$$2kmr^2 - 2L^2 r + 3kmJ_2 R^2 = 0;$$

$$r_{1,2} = \frac{L^2}{2km}\left[1 \pm \sqrt{1 - 6J_2 R^2 \left(\frac{km}{L^2}\right)^2}\right].$$

If $J_2 \to 0$, the solution with "minus" $\to 0$, and the corresponding effective potential $V_{\text{ef}} \to +\infty$. To ensure that the effective potential is a finite value, we keep only the solution with "plus", so we have

$$r_0 = \frac{L^2}{2km}\left[1 + \sqrt{1 - 6J_2 R^2 \left(\frac{km}{L^2}\right)^2}\right].$$

(d) Knowing that

$$V_{\text{ef}}\,(r) = \frac{L^2}{2mr^2} - \frac{k}{r} - \frac{kJ_2 R^2}{2r^3},$$

it results that:

$$V_{\text{ef}}\,(r_0 + \Delta r) = \frac{L^2}{2m\,(r_0 + \Delta r)^2} - \frac{k}{r_0 + \Delta r} - \frac{kJ_2 R^2}{2\,(r_0 + \Delta r)^3};$$

$$V_{\text{ef}}\,(r_0 + \Delta r) = \frac{L^2}{2mr_0^2}\left(1 + \frac{\Delta r}{r_0}\right)^{-2} - \frac{k}{r_0}\left(1 + \frac{\Delta r}{r_0}\right)^{-1}$$

$$- \frac{kJ_2 R^2}{2r_0^3}\left(1 + \frac{\Delta r}{r_0}\right)^{-3};$$

$$\frac{\Delta r}{r_0} << 1;$$

$$(1 + \alpha)^n \cong 1 + n\alpha + \frac{1}{2}n(n-1)\alpha^2; \quad \alpha << 1;$$

$$V_{\text{ef}}(r_0 + \Delta r) \cong \frac{L^2}{2mr_0^2}\left(1 - 2\frac{\Delta r}{r_0} + 3\frac{(\Delta r)^2}{r_0^2}\right)$$

$$-\frac{k}{r_0}\left(1 - \frac{\Delta r}{r_0} + \frac{(\Delta r)^2}{r_0^2}\right) - \frac{kJ_2R^2}{2r_0^3}\left(1 - 3\frac{\Delta r}{r_0} + 6\frac{(\Delta r)^2}{r_0^2}\right);$$

$$V_{\text{ef}}(r_0 + \Delta r) \cong \left(\frac{L^2}{2mr_0^2} - \frac{k}{r_0} - \frac{kJ_2R^2}{2r_0^3}\right)$$

$$+\left(-\frac{L^2}{mr_0^3} + \frac{k}{r_0^2} + \frac{3kJ_2R^2}{2r_0^4}\right)(\Delta r) + \left(\frac{3L^2}{2mr_0^4} - \frac{k}{r_0^3} - \frac{3kJ_2R^2}{r_0^5}\right)(\Delta r)^2;$$

$$\left(\frac{L^2}{2mr_0^2} - \frac{k}{r_0} - \frac{kJ_2R^2}{2r_0^3}\right) = V_{\text{ef}}(r_0);$$

$$\left(-\frac{L^2}{mr_0^3} + \frac{k}{r_0^2} + \frac{3kJ_2R^2}{2r_0^4}\right) = \frac{1}{r_0^2}\frac{2kmr_0^2 - 2L^2r_0 + 3kmJ_2R^2}{2mr_0^2} = 0;$$

$$V_{\text{ef}}(r_0 + \Delta r) \cong V_{\text{ef}}(r_0) + \left(\frac{3L^2}{2mr_0^4} - \frac{k}{r_0^3} - \frac{3kJ_2R^2}{r_0^5}\right)(\Delta r)^2;$$

$$r_0 = \frac{L^2}{2km}\left[1 + \sqrt{1 - 6J_2R^2\left(\frac{km}{L^2}\right)^2}\right];$$

$$\sqrt{1 - 6J_2R^2\left(\frac{km}{L^2}\right)^2} = \left[1 - 6J_2R^2\left(\frac{km}{L^2}\right)\right]^{\frac{1}{2}}; \quad 6J_2R^2\left(\frac{km}{L^2}\right) << 1;$$

$$(1 + \alpha)^n \cong 1 + n\alpha + \frac{1}{2}n(n-1)\alpha^2; \quad \alpha << 1;$$

$$\sqrt{1 - 6J_2R^2\left(\frac{km}{L^2}\right)^2} = \left[1 - 6J_2R^2\left(\frac{km}{L^2}\right)\right]^{\frac{1}{2}}$$

$$\cong 1 - 3J_2R^2\left(\frac{km}{L^2}\right)^2 - \frac{9}{2}J_2^2R^4\left(\frac{km}{L^2}\right)^4$$

$$\cong 1 - 3J_2R^2\left(\frac{km}{L^2}\right)^2;$$

$$r_0 \cong \frac{L^2}{2km}\left[2 - 3J_2R^2\left(\frac{km}{L^2}\right)^2\right];$$

$$r_0 \cong \frac{L^2}{km}\left[1 - \frac{3}{2}J_2R^2\left(\frac{km}{L^2}\right)^2\right];$$

$$\left(\frac{3L^2}{2mr_0^4} - \frac{k}{r_0^3} - \frac{3kJ_2R^2}{r_0^5}\right) = \frac{3L^2k^4m^4}{2mL^8}\left[1 - \frac{3}{2}J_2R^2\left(\frac{km}{L^2}\right)^2\right]^{-4}$$

$$- \frac{kk^3m^3}{L^6}\left[1 - \frac{3}{2}J_2R^2\left(\frac{km}{L^2}\right)^2\right]^{-3}$$

$$- \frac{3kJ_2R^2k^5m^5}{L^{10}}\left[1 - \frac{3}{2}J_2R^2\left(\frac{km}{L^2}\right)^2\right]^{-5};$$

$$\left(\frac{3L^2}{2mr_0^4} - \frac{k}{r_0^3} - \frac{3kJ_2R^2}{r_0^5}\right) \cong \frac{3k^4m^3}{2L^6}\left[1 + 4\frac{3}{2}J_2R^2\left(\frac{km}{L^2}\right)^2\right]$$

$$- \frac{k^4m^3}{L^6}\left[1 + 3\frac{3}{2}J_2R^2\left(\frac{km}{L^2}\right)^2\right]$$

$$- \frac{3k^6m^5J_2R^2}{L^{10}}\left[1 + 5\frac{3}{2}J_2R^2\left(\frac{km}{L^2}\right)^2\right];$$

$$\left(\frac{3L^2}{2mr_0^4} - \frac{k}{r_0^3} - \frac{3kJ_2R^2}{r_0^5} \right)$$

$$= \frac{k^4m^3}{L^6} \left[\frac{3}{2} + 6\frac{3}{2}J_2R^2 \left(\frac{km}{L^2} \right)^2 - 1 - \frac{9}{2}J_2R^2 \left(\frac{km}{L^2} \right)^2 \right]$$

$$- \frac{3k^6m^5J_2R^2}{L^{10}} \left[1 + \frac{15}{2}J_2R^2 \left(\frac{km}{L^2} \right)^2 \right];$$

$$\left(\frac{3L^2}{2mr_0^4} - \frac{k}{r_0^3} - \frac{3kJ_2R^2}{r_0^5} \right) \cong \frac{k^4m^3}{L^6} \left[\frac{1}{2} + \frac{9}{2}J_2R^2 \left(\frac{km}{L^2} \right)^2 \right]$$

$$- \frac{3k^6m^5J_2R^2}{L^{10}};$$

$$\left(\frac{3L^2}{2mr_0^4} - \frac{k}{r_0^3} - \frac{3kJ_2R^2}{r_0^5} \right)$$

$$= \frac{k^4m^3}{L^6} \left[\frac{1}{2} + \frac{9}{2}J_2R^2 \left(\frac{km}{L^2} \right)^2 - 3J_2R^2 \left(\frac{km}{L^2} \right)^2 \right];$$

$$\left(\frac{3L^2}{2mr_0^4} - \frac{k}{r_0^3} - \frac{3kJ_2R^2}{r_0^5} \right) = \frac{k^4m^3}{2L^6} \left[1 + 3J_2R^2 \left(\frac{km}{L^2} \right)^2 \right];$$

$$V_{\text{ef}}(r_0 + \Delta r) \cong V_{\text{ef}}(r_0) + \left(\frac{3L^2}{2mr_0^4} - \frac{k}{r_0^3} - \frac{3kJ_2R^2}{r_0^5} \right) (\Delta r)^2;$$

$$V_{\text{ef}}(r_0 + \Delta r) = V_{\text{ef}}(r_0) + \frac{k^4m^3}{L^6} \left[1 + 3J_2R^2 \left(\frac{km}{L^2} \right)^2 \right] \frac{(\Delta r)^2}{2}.$$

Under these conditions, the total energy of the Sun–planet system for $r = r_0 + \Delta r$, is

$$E_{r_0 + \Delta r} = \frac{p_{\text{rad}}^2}{2m} + V_{\text{ef}}(r_0 + \Delta r);$$

$$E_{r_0 + \Delta r} = \frac{p_{\text{rad}}^2}{2m} + V_{\text{ef}}(r_0) + \frac{k^4 m^3}{L^6}\left[1 + 3J_2 R^2 \left(\frac{km}{L^2}\right)^2\right]\frac{(\Delta r)^2}{2}.$$

For a harmonic oscillator, the total energy and period of its oscillations are given by expressions

$$E = \frac{p^2}{2m} + k_0 \frac{x^2}{2};$$

$$T = 2\pi \sqrt{\frac{m}{k_0}}.$$

Comparing the two expressions of the total energies, it results that:

$$k_0 = \frac{k^4 m^3}{L^6}\left[1 + 3J_2 R^2 \left(\frac{km}{L^2}\right)^2\right];$$

$$T_{\text{osc}} = 2\pi \frac{L^3}{k^2 m}\sqrt{\frac{1}{1 + 3J_2 R^2 \left(\frac{km}{L^2}\right)^2}} = 2\pi \frac{L^3}{k^2 m}\left[1 + 3J_2 R^2 \left(\frac{km}{L^2}\right)^2\right]^{-\frac{1}{2}};$$

$$3J_2 R^2 \left(\frac{km}{L^2}\right)^2 << 1;$$

$$(1 + \alpha)^n \cong 1 + n\alpha + \frac{1}{2}n(n-1)\alpha^2; \quad \alpha << 1;$$

$$T_{\text{osc}} \cong 2\pi \frac{L^3}{k^2 m}\left(1 - \frac{3}{2}\frac{k^2 m^2 R^2}{L^4}J_2\right).$$

For a planet that evolves around the Sun, in an elliptical orbit with semi-axes a and b, respectively, it is demonstrated that the kinetic moment and the period of revolution are given by the expressions

$$L = mb\sqrt{\frac{KM_S}{a}}; \quad T_{\text{rev}} = 2\pi\sqrt{\frac{a^3}{KM_S}}.$$

In particular, if the orbit is approximately circular $(a \approx b)$, it follows that:

$$2\pi \frac{L^3}{k^2 m} = 2\pi \frac{L^3}{(KmM_S)^2 m} = T_{\text{rev}};$$

$$T_{\text{osc}} = T_{\text{rev}}\left(1 - \frac{3}{2}\frac{k^2 m^2 R^2}{L^4}J_2\right); \quad T_{\text{osc}} < T_{\text{rev}}.$$

Particular case: $J_2 = 0 \rightarrow T_{\text{osc}} = T_{\text{rev}}$.

Problem 2. Intercontinental Ballistic Missiles

Suppose an intercontinental ballistic missile must be launched at the miminum speed from the geographical North Pole of the Earth to reach a point on the Earth's Equator. In that case, its trajectory must be a sector of the ellipse represented in Figure 2.1, with the center of the Earth (O) in one of its foci (F_2), and the other focus (F_1) being midway between the launch point and the landing point of the rocket.

(a) *Determine* the elements of the vector representing the initial speed required for this launch. The following are known: the radius of the Earth, R, and the gravitational acceleration on the ground, g_0.

We know that: **(1)** the sum of the distances from every point of the ellipse to the two foci is constant, $2a$, representing the length of the axis of the ellipse; **(2)** the tangent between a point on an ellipse is perpendicular to the bisector of the angle formed by the directions that pass through that point and through the foci of the ellipse (the optical property of the ellipse); **(3)** the total energy of the rocket–Earth system, when the rocket evolves on an ellipse with semi-major axis a, having the Earth in one of its foci, is $E = -K\frac{mM}{2a}$, where K is the constant of universal attraction, m is the mass of the rocket, and M is the mass of the Earth.

(b) *Determine* the speed of the rocket at point A, representing the peak of the elliptical orbit and its altitude at that moment.

(c) *Study*, under the same conditions, the scenario represented in Figure 2.2, where the intercontinental ballistic missile must reach the geographic South Pole of the Earth, evolving on a sector of another ellipse, having the center of the Earth (O) in one of its foci (F_2).

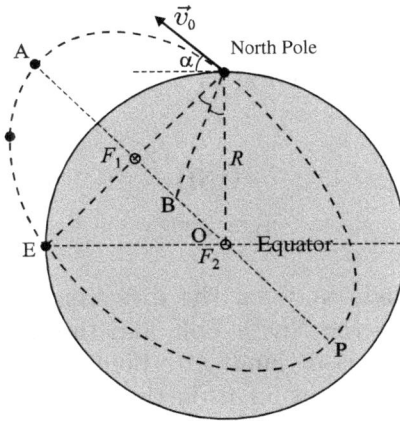

Fig. 2.1

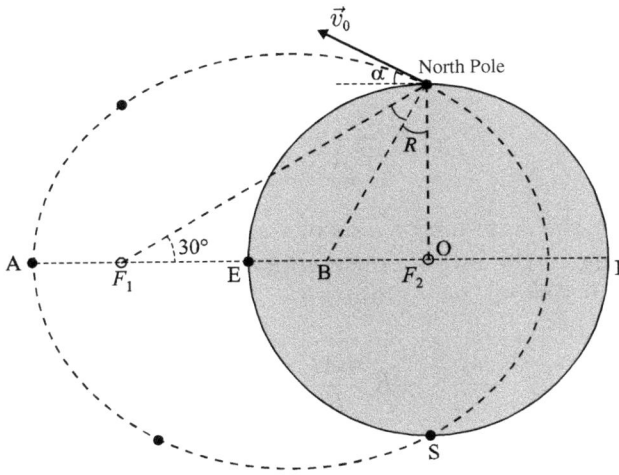

Fig. 2.2

(d) From the two poles of the Earth, two ballistic missiles are launched simultaneously, in the same plane, with the same initial speeds, oriented horizontally. After time t, the rockets reach their maximum distance from each other. Determine this distance. *We know*: the acceleration of free fall at the surface of the Earth, g_0, and the radius of the Earth, R.

Solution

(a) From the geometry of Figure 2.1, since ONF_1 is an isosceles triangle, it follows that:

$$\angle ONF_1 = 45°; \quad \angle ONB = \angle BNF_1 = 22.5°;$$

$$\alpha = 22.5°,$$

representing the angle between the direction of the launch of the ballistic missile from the North Pole and the direction of the local horizon, so that the missile lands at a point on the equator.

The North Pole being a point on the ellipse, in accordance with the definition of the ellipse, it results that:

$$NF_1 + NF_2 = 2a;$$

$$R\frac{\sqrt{2}}{2} + R = 2a;$$

$$a = \frac{R}{2}\left(1 + \frac{\sqrt{2}}{2}\right),$$

representing the semi-major axis of the ellipse.

According to the law of conservation of mechanical energy of the rocket–Earth system, we obtain:

$$\frac{mv_0^2}{2} - K\frac{mM}{R} = -K\frac{mM}{2a};$$

$$v_0 = \sqrt{\frac{2KM}{R(\sqrt{2}+1)}};$$

$$g_0 = K\frac{M}{R^2}; \quad v_0 = \sqrt{\frac{2g_0 R}{\sqrt{2}+1}},$$

representing the rocket's speed at the moment of launch from the North Pole in order for it to reach, under the specified conditions, a point on the equator.

(b) Since the apogee A is a point on the ellipse, it results that:

$$AF_1 + AF_2 = 2a; \quad AF_2 = R + h; \quad AF_1 = AF_2 - R\frac{\sqrt{2}}{2};$$

$$h = \frac{R}{2}(\sqrt{2} - 1),$$

representing the maximum altitude of the rocket in its ballistic flight from the North Pole to the equator.

According to the law of conservation of mechanical energy, it results that:

$$\frac{mv_{min}^2}{2} - K\frac{mM}{R+h} = -K\frac{mM}{2a};$$

$$v_{min} = 2\sqrt{\frac{g_0 R}{(\sqrt{2}+1)(\sqrt{2}+2)}}.$$

(c) From the geometry of Figure 2.2, it follows that:

$$\angle ONF_1 = 60°; \quad \angle ONB = \angle BNF_1 = 30°;$$

$$\alpha = 30°; \quad NF_1 + NF_2 = 2a;$$

$$a = \frac{3}{2}R.$$

From the law of conservation of mechanical energy, it follows that:

$$v_0 = 2\sqrt{\frac{g_0 R}{3}};$$

$$h = \frac{R}{2}(\sqrt{3}+1);$$

$$v_{min} = \sqrt{\frac{2g_0 R(3 - \sqrt{3})}{3(3 + \sqrt{3})}}.$$

(d) The two rockets will move on identical elliptical orbits, each orbit being tangent to the surface of the Earth at one of the geographic poles, and the center of the Earth being in one of the foci of the two ellipses, as shown in Figure 2.3. Each rocket is launched from the perigee of the ellipse. The maximum distance between the two rockets is achieved when each rocket reaches the apogee of its orbit, which happens after a time $t = T/2$.

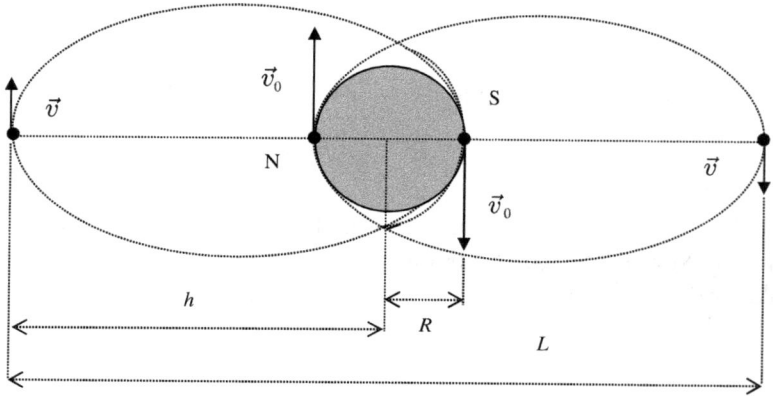

Fig. 2.3

For the movement of each rocket on its elliptical trajectory, according to Kepler's third law, we can write that

$$T^2 = k \left(\frac{h + R}{2} \right)^3.$$

This relationship proves that the period of movement is the same for any other rocket that moves on any other ellipse with the same major semi-axis, regardless of its minor semi-axis. Thus, for the movement of a rocket on a circular trajectory whose radius is

$$r = \frac{h + R}{2},$$

we can have:

$$T = \frac{2\pi r}{v} = \frac{2\pi r}{\sqrt{\frac{KM}{r}}} = \frac{2\pi r}{R} \sqrt{\frac{r}{g_0}};$$

$$r = \sqrt[3]{\frac{R^2 g_0 T^2}{4\pi^2}} = \frac{h + R}{2}; \quad h = \sqrt[3]{\frac{2R^2 g_0 T^2}{\pi^2}} - R,$$

so that the maximum distance between the rockets is

$$L = 2h = 2 \left(\sqrt[3]{\frac{4R^2 g_0 t}{\pi^2}} - R \right).$$

Problem 3. An Electromagnetic Pendulum

In the images in Figure 3.1, a multiplier coil (2) with mass m is represented, having identical circular turns, each with radius r, suspended at the level of the diameter or horizontally with the help of two vertical conducting rods, each having length L. The north pole area of a U-shaped permanent magnet (1) is located in the center of the coil. The width of each pole of the magnet (the width of the U-shaped magnetic strip) is 1, and the induction of the magnetic field of the magnet, considered uniform throughout the space between the poles, is represented by the vertical vector $\vec{B}$. The ends of the coil conductor are connected to the lower ends of the rods, and their upper ends are connected to the two terminals of the device shown. A direct voltage generator is connected to the terminals (S) of the device. After closing the switch (4), when the device's mobile frame is in the equilibrium position, the intensity of the current through the turns of the coil is I.

(a) *Demonstrate* that the small oscillations of the mobile frame in relation to the equilibrium position are harmonic.

(b) *Determine* the period of the small oscillations of the mobile frame in relation to the equilibrium position. The gravitational acceleration, g, is known.

(c) *Particular case*: $B = 0$.

In the space between the magnet poles, the sectors of the coil turns can be considered linear and horizontal. During the oscillations, the coil does not move out from between the magnet's poles. Neglect the electromagnetic induction, the non-uniformity of the magnetic field between the magnet's poles, the magnetic field around the magnet's poles, the masses of the coil suspension rods, and the forces of friction.

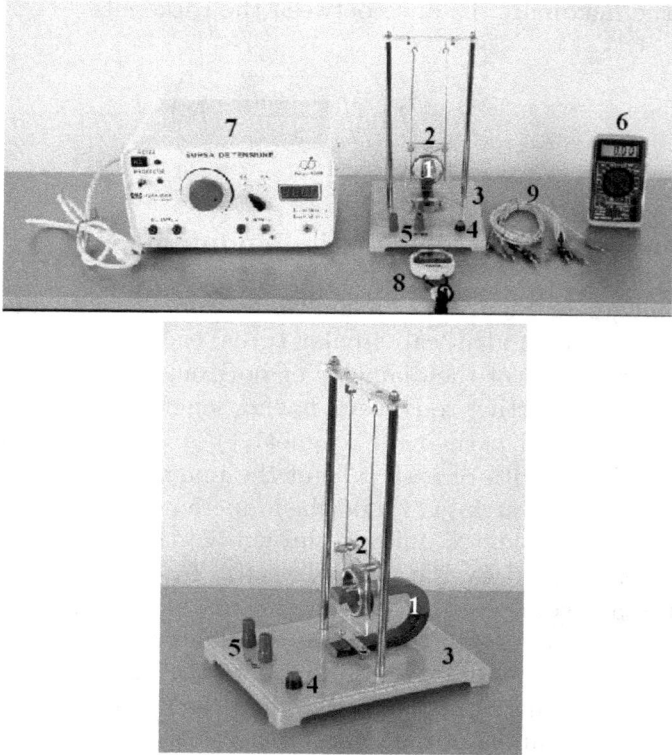

Fig. 3.1

Solution

(a) When the electric current passes through the turns of the multiplier coils, the mobile frame is moved from the vertical position due to the inertia of the electromagnetic forces. It will settle in an equilibrium position, as shown in Figure 3.2, when the resultant moment of the forces acting on the mobile frame is null.

For this equilibrium position, in relation to point 0, it follows that:

$$F_{\text{em}} \left(L + r \right) \cos \alpha_0 = GL \sin \alpha_0;$$

$$\tan \alpha_0 = \frac{F_{\text{em}} \left(L + r \right)}{mgL}.$$

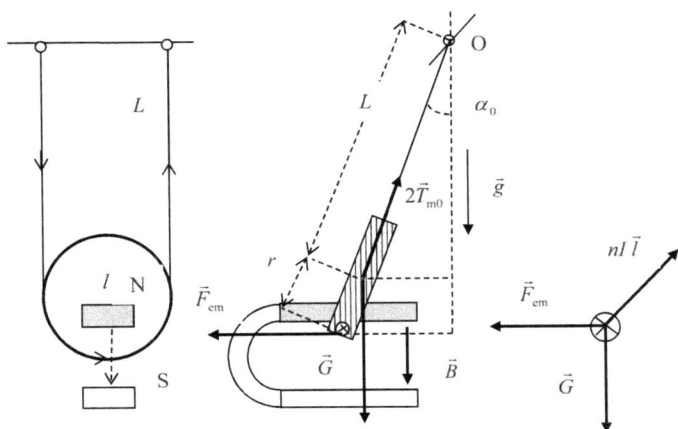

Fig. 3.2

If the moving coil is displaced from the equilibrium position by a small angle a, as indicated in Figure 3.3, and then released, it will not remain there but will return to the equilibrium position, because the torques of the two forces, $\vec{F}_{em}$ and $\vec{G}$, in relation to the suspension point O are no longer equal in modulus and are in the opposite direction. It is easy to verify that, in this position, $M_{\vec{G}} > M_{\vec{F}_{em}}$.

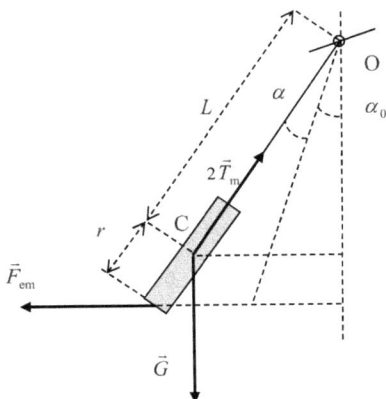

Fig. 3.3

As a result, the resultant moment of the forces acting on the mobile frame is no longer zero. Consequently, the mobile frame will return to the equilibrium position, exceed it due to inertia, and then oscillate relative to the initial equilibrium position.

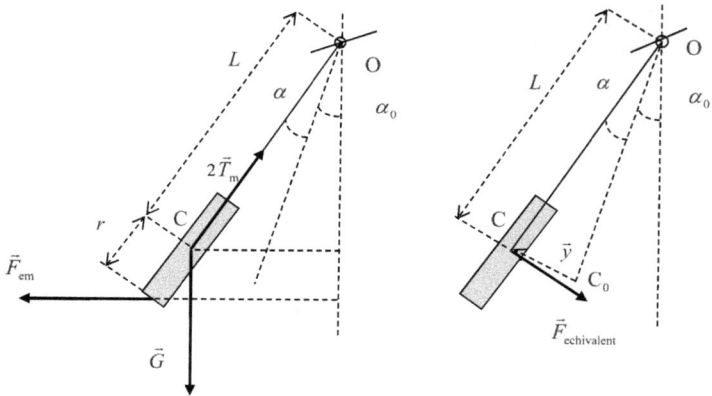

Fig. 3.4

In order to establish the type of oscillatory movement, we will replace the forces acting on the mobile system ($\vec{F}_{em}$, $\vec{G}$, $2\vec{T}_m$), with a single force, called the equivalent force, $\vec{F}_{equivalent}$, whose effect should be the same as that of the forces acting on the mobile frame. We will consider that the equivalent force acts at the center of mass of the mobile system (approximately the center of the multiplier) and is perpendicular to the arm OC, as indicated in Figure 3.4.

So, we will look for the modulus of the force $\vec{F}$, in such a way that we have:

$$FL + F_{em}\,(L + r)\cos\,(\alpha + \alpha_0) = GL\sin\,(\alpha + \alpha_0);$$

$$FL + F_{em}\,(L + r)(\cos\alpha\cos\alpha_0 - \sin\alpha\sin\alpha_0)$$
$$= GL\,(\sin\alpha\cos\alpha_0 + \cos\alpha\sin\alpha_0);$$
$$FL + [F_{em}\,(L + r)\cos\alpha_0 - GL\sin\alpha_0]\cos\alpha$$

$$= [F_{em}\,(L + r)\sin\alpha_0 + GL\cos\alpha_0]\sin\alpha;$$

$$F_{em}\,(L + r)\cos\alpha_0 - GL\sin\alpha_0 = 0;$$
$$FL = [F_{em}\,(L + r)\sin\alpha_0 + GL\cos\alpha_0]\sin\alpha; \quad \sin\alpha \approx \alpha;$$

$$FL \approx [F_{\text{em}}\,(L+r)\sin\alpha_0 + GL\cos\alpha_0]\,\alpha; \quad y \approx L\alpha;$$

$$F = \frac{F_{\text{em}}\,(L+r)\sin\alpha_0 + GL\cos\alpha_0}{L^2}y;$$

$$k = \frac{F_{\text{em}}\,(L+r)\sin\alpha_0 + GL\cos\alpha_0}{L^2};$$

$$F = ky = F_{\text{equivalent}}; \quad \vec{F} = k\vec{y} = -\vec{F}_{\text{equivalent}}; \quad \vec{F}_{\text{equivalent}} = -k\vec{y}.$$

This proves that the mobile frame's motion is a harmonic oscillatory motion.

(b)

$$k = m\omega^2 = m\frac{4\pi^2}{T^2};$$

$$\frac{F_{\text{em}}\,(L+r)\sin\alpha_0 + GL\cos\alpha_0}{L^2} = m\frac{4\pi^2}{T^2};$$

$$\tan\alpha_0 = \frac{F_{\text{em}}\,(L+r)}{mgL};$$

$$\sin\alpha_0 = \frac{\tan\alpha_0}{\sqrt{1+\tan^2\alpha_0}}; \quad \cos\alpha_0 = \frac{1}{\sqrt{1+\tan^2\alpha_0}};$$

$$\frac{F_{\text{em}}\,(L+r)}{L^2}\frac{\tan\alpha_0}{\sqrt{1+\tan^2\alpha_0}} + \frac{mg}{L}\frac{1}{\sqrt{1+\tan^2\alpha_0}} = m\frac{4\pi^2}{T^2};$$

$$\frac{F_{\text{em}}\,(L+r)}{L^2}\tan\alpha_0 + \frac{mg}{L} = m\frac{4\pi^2}{T^2}\sqrt{1+\tan^2\alpha_0};$$

$$\frac{F_{\text{em}}\,(L+r)}{L^2}\frac{F_{\text{em}}\,(L+r)}{mgL} + \frac{mg}{L} = m\frac{4\pi^2}{T^2}\sqrt{1+\frac{F_{\text{em}}^2\,(L+r)^2}{m^2g^2L^2}};$$

$$\frac{mg}{L}\left[1+\frac{F_{\text{em}}^2\,(L+r)^2}{m^2g^2L^2}\right] = m\frac{4\pi^2}{T^2}\sqrt{1+\frac{F_{\text{em}}^2\,(L+r)^2}{m^2g^2L^2}};$$

$$\sqrt{1+\frac{F_{\text{em}}^2\,(L+r)^2}{m^2g^2L^2}} = \frac{4\pi^2 L}{T^2 g}; \quad \frac{F_{\text{em}}^2\,(L+r)^2}{m^2g^2L^2} = \frac{16\pi^4 L^2}{T^4 g^2} - 1;$$

$$F_{em} = nIlB;$$

$$T = 2\pi L \left(\sqrt[4]{\frac{m^2}{n^2 I^2 l^2 (L+r)^2 B^2 + m^2 g^2 L^2}} \right).$$

(c)

$$B = 0; \quad T = 2\pi \sqrt{\frac{L}{g}}.$$

Problem 4. A Linear Accelerator

In a linear accelerator for electrified particles, its n coaxial conductive cylinders are connected to the poles of a high-frequency electric generator (v = constant), as shown in Figure 4.1.

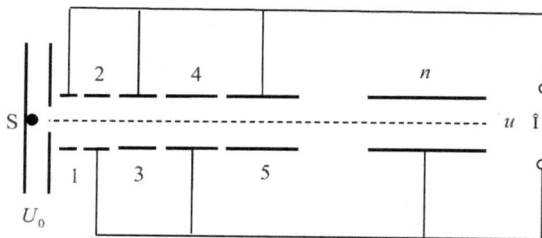

Fig. 4.1

The ions resulting from an electric discharge, S, are considered at rest, with mass m_0 and electric charge q. They are then accelerated in an electric field under constant voltage U_0 and injected into the first cylinder along its axis.

The lengths of the accelerator cylinders are calculated in such a way that, reaching the space between any two neighboring cylinders, the electrified particle will find there identical acceleration conditions, under the alternating electric voltage, whose polarity must change when the electrified particle crosses the distance between any two neighboring acceleration intervals.

The length of any acceleration interval is negligible in relation to the lengths of the cylinders, and the duration of the particle's passage through any acceleration interval is infinitely small. Under these conditions, the passage of the particle through the acceleration interval is equivalent to crossing a region where there is a constant potential difference, U_0.

(a) *Determine* the relationship between the period of electric voltage maintained by the generator and the duration of crossing each cylinder, as well as the length of each cylinder, if the speeds acquired by the accelerated electrified particles are relativistic.

(b) *Calculate* the limit speed of the accelerated electrified particles and the limit length of the last cylinder if $n \to \infty$.

(c) *Establish* the expression of the energy acquired by the electrified particle on each acceleration interval if its length is D and the phases of the acceleration voltage when the particle is at the entrance to the acceleration interval and in the middle of the acceleration interval are zero and φ_0, respectively. *Particular cases*: $D \to 0$, $\varphi_0 \neq 0$; $D \to 0$, $\varphi_0 = 0$. Neglect the weight of the particle and the deviations from the rectilinear trajectory.

(d) *Find* the length of the accelerator in the non-relativistic scenario for $D \to 0$.

Solution

(a) The relationship between the period of electrical oscillations of the generator and the duration of the passing of each cylinder is

$$\frac{T}{2} = t = \frac{l_1}{v_1} = \frac{l_2}{v_2} = \cdots = \frac{l_n}{v_n},$$

where $(l_1, l_2, l_3 \ldots \ldots l_n)$ are the lengths of the cylinders and $(v_1, v_2, v_3 \ldots \ldots \ldots v_n)$ are the speeds of the particles along each cylinder.

(b) The speed of the electrified particle, v_1, upon entering the first cylinder, accelerated under the constant voltage U_0 immediately after its excitement in the electric discharge S, is calculated in accordance with the law of energy conservation, as follows:

$$m_1 c^2 - m_0 c^2 = qU_0; \quad m_0 c^2 \left(\frac{1}{\sqrt{1 - \beta_1^2}} - 1 \right) = qU_0; \quad \beta_1 = \frac{v_1}{c};$$

$$v_1 = c \frac{\sqrt{q^2 U_0^2 + 2qU_0 m_0 c^2}}{qU_0 + m_0 c^2}.$$

Within each cylinder, the electric field intensity is zero; therefore, the motion of the particle inside each cylinder is uniform (i.e., with constant velocity). Specifically, for the motion of the charged particle inside the first cylinder, we can conclude:

$$l_1 = v_1 \frac{T}{2} = \frac{v_1}{2v},$$

where v is the frequency of the alternating voltage maintained by the generator;

$$l_1 = \frac{c}{2v} \frac{\sqrt{q^2 U_0^2 + 2qU_0 m_0 c^2}}{qU_0 + m_0 c^2}.$$

The speed of the electrified particle, v_2, upon entering cylinder 2, accelerated under voltage U_0 immediately after exiting cylinder 1, is calculated in accordance with the law of conservation of energy as follows:

$$m_2 c^2 - m_1 c^2 = qU_0; \quad m_2 c^2 - m_0 c^2 = 2qU_0;$$

$$m_0 c^2 \left(\frac{1}{\sqrt{1 - \beta_2^2}} - 1 \right) = 2qU_0; \quad \beta_2 = \frac{v_2}{c};$$

$$v_2 = c \frac{\sqrt{2^2 q^2 U_0^2 + 2 \cdot 2qU_0 m_0 c^2}}{2qU_0 + m_0 c^2}.$$

For the length of the cylinder 2, it results that:

$$l_2 = v_2 \frac{T}{2} = \frac{v_2}{2v};$$

$$l_2 = \frac{c}{2v} \frac{\sqrt{2^2 q^2 U_0^2 + 2 \cdot 2qU_0 m_0 c^2}}{2qU_0 + m_0 c^2}.$$

In a similar way, v_3 and l_3 are calculated, as well as, by generalization, v_n and l_n, respectively. Then, for $n \to \infty$, we calculate the required limit values. It results that:

$$v_3 = c \frac{\sqrt{3^2 q^2 U_0^2 + 2 \cdot 3qU_0 m_0 c^2}}{3qU_0 + m_0 c^2};$$

$$l_3 = \frac{c}{2v} \frac{\sqrt{3^2 q^2 U_0^2 + 2 \cdot 3qU_0 m_0 c^2}}{3qU_0 + m_0 c^2};$$

....

$$v_n = c \frac{\sqrt{n^2 q^2 U_0^2 + 2 \cdot nqU_0 m_0 c^2}}{nqU_0 + m_0 c^2};$$

$$l_n = \frac{c}{2v} \frac{\sqrt{n^2 q^2 U_0^2 + 2 \cdot nqU_0 m_0 c^2}}{nqU_0 + m_0 c^2};$$

$$v_{max} = \lim_{n \to \infty} v_n = c; \quad l_{max} = \lim_{n \to \infty} l_n = \frac{c}{2v} = \frac{\lambda}{2}.$$

(c) Let dW be the variation of the kinetic energy of the electrified particle achieved over the distance dx in the acceleration interval between two neighboring cylinders, as shown in Figure 4.2:

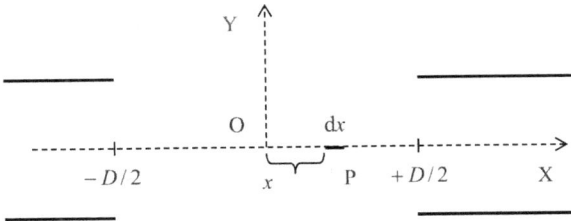

Fig. 4.2

$$dW = qE_x dx,$$

where E_x is the electric field intensity of the electromagnetic wave propagating along the acceleration interval.

If at time $t = 0$ the wavefront of the electromagnetic wave is at the beginning of the acceleration interval and at time $t = t_0$ the wavefront is at half the distance between the cylinders, then the wavefront will be at the point $P(x)$ at time t, such that:

$$x = c(t - t_0); \quad t = \frac{x}{c} + t_0.$$

Since the tension between the cylinders varies over time according to the law $u = U_0 \cos \omega t$, then corresponding to the moment t_0, we have:

$$u_0 = U_0 \cos \omega t_0 = U_0 \cos \varphi_0; \quad \phi_0 = \omega t0; \quad t_0 \frac{\varphi_0}{\omega};$$

$$t = \frac{x}{c} + \frac{\varphi_0}{\omega}.$$

The intensity of the electric field at the point $P(x)$ varies over time according to the following law:

$$E_x = \frac{u}{D} = \frac{U_0}{D} \cos \omega t; \quad E_x = \frac{U_0}{D} \cos \omega \left(\frac{x}{c} + \frac{\varphi_0}{\omega} \right);$$

$$E_x = \frac{U_0}{D} \cos \omega \left(\frac{2\pi x}{\lambda} + \frac{\varphi_0}{\omega} \right); \quad E_x = \frac{U_0}{D} \cos \omega \left(\frac{\pi x}{l_n} + \frac{\varphi_0}{\omega} \right).$$

It results that:

$$dW = \frac{qU_0}{D} \cos \omega \left(\frac{\pi x}{l_n} + \frac{\varphi_0}{\omega} \right) dx;$$

$$\Delta W = \frac{qU_0}{D} \int_{-D/2}^{+D/2} \cos \left(\frac{\pi x}{l_n} + \varphi_0 \right) dx;$$

$$\Delta W = \frac{qU_0}{D} \frac{l_n}{\pi} \left[\cos \varphi_0 \sin \frac{\pi x}{l_n} \Big|_{-D/2}^{+D/2} + \sin \varphi_0 \cos \frac{\pi x}{l_n} \Big|_{-D/2}^{+D/2} \right];$$

$$\Delta W = qU_0 \frac{\sin \frac{\pi D}{2l_n}}{\frac{\pi D}{2l_n}} \cos \varphi_0.$$

Particular cases:

(1) $D \to 0$, $\varphi_0 \neq 0$; $\Delta W = qU_0 \cos \varphi_0$;
(2) $D \to 0$, $\varphi_0 = 0$; $\Delta W = qU_0$.

(d) If we admit that $c \to \infty$, the non-relativistic variant from the previous results gives us:

$$v_1 = \sqrt{\frac{2qU_0}{m_0}}; \quad l_1 = \frac{1}{v}\sqrt{\frac{qU_0}{2m_0}};$$

$$v_2 = \sqrt{\frac{2 \cdot 2qU_0}{m_0}}; \quad l_2 = \frac{1}{v}\sqrt{\frac{2qU_0}{2m_0}};$$

$$v_3 = \sqrt{\frac{3 \cdot 2qU_0}{m_0}}; \quad l_3 = \frac{1}{v}\sqrt{\frac{3qU_0}{2m_0}};$$

$$\cdots\cdots\cdots\cdots\cdots\cdots\cdots\cdots\cdots$$

$$v_n = \sqrt{\frac{n \cdot 2qU_0}{m_0}} = \sqrt{n}v_1; \quad l_n = \frac{1}{v}\sqrt{\frac{nqU_0}{2m_0}} = \sqrt{n}l_1;$$

$$L = l_1 + l_2 + \cdots + l_n;$$

$$L = \frac{1}{v}\sqrt{\frac{qU_0}{2m_0}}\left(1 + \sqrt{2} + \sqrt{3} + \cdots + \sqrt{n}\right);$$

$$L = \frac{1}{v}\sqrt{\frac{qU_0}{2m_0}}\sum_{n=1}^{N}\sqrt{n}.$$

Problem 5. Spherical Space Probe

A spherical space probe, with one hemisphere perfectly reflecting and the other hemisphere perfectly absorbing, is in a vacuum, very far from the Sun and any other planet. Positions 1, 2, and 3 in Figure 5.1 represent different orientations of the space probe with respect to the direction of the incident sunlight (the shaded hemisphere is the absorbing one).

(a) *Determine* the momentum and kinetic moment transmitted to the sphere by a photon, with frequency ν, incident at an angle θ. The probe is oriented as shown in drawings 1 and 2 of Figure 5.1. *We know*: the radius of the sphere, R; Planck's constant, h; and the speed of light in vacuum, c.

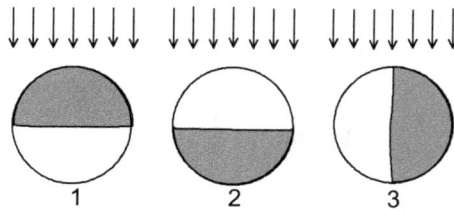

Fig. 5.1

(b) *Justify* the behavior of the space probe, oriented as in drawings 1 and 2 of Figure 5.1, if it is rotated with a small angle around a fixed axis relative to the Sun, perpendicular to the direction of incident light, passing through the center of the sphere and then released from rest. The center of mass of the probe is at the center of the sphere.

(c) *Determine* the force acting on the sphere in drawings 1 and 2. *We know*: the average power of solar radiation per unit area of a surface at normal incidence, E_0.

(d) *Qualitatively describe* the non-relativistic movement of the space probe under the action of solar light, if it is free and initially at rest with respect to the Sun and has the orientation presented in drawing 3. The gravitational actions of the probe, the photoelectric effect, and the emission of electromagnetic radiation are neglected.

Solution

(a) The variation in the impulse of the incident photon on the absorbing face of the sphere is:

$$\Delta \vec{p} = 0 - \vec{p}_i = -\vec{p}_i; \quad \Delta p = p_i = h/\lambda = h v/c.$$

The impulse transmitted to the sphere by a photon, incident at a point on the absorbing face of the probe (Figure 5.2), is equal in magnitude and identical in orientation to the impulse of the incident photon:

$$\vec{p} = -\Delta \vec{p} = \vec{p}_i; \quad p = p_i = h/\lambda = h v/c.$$

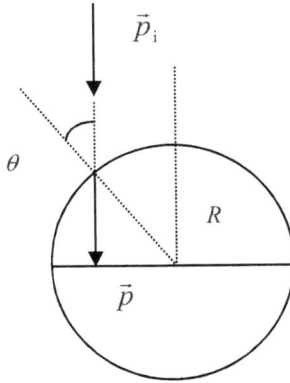

Fig. 5.2

As a result, the kinetic moment transmitted to the probe by the incident photon is:

$$\vec{L} = \vec{R} \times \vec{p};$$

$$L = Rp \sin\left(180^0 - \theta\right) = Rp \sin\theta = \frac{Rh}{\lambda}\sin\theta = \frac{Rh\nu}{c}\sin\theta.$$

The variation in the momentum of a photon, incident on the reflecting face of the probe (Figure 5.3), is:

$$\Delta\vec{p} = \vec{p}_r - \vec{p}_i;$$

$$\Delta p = 2p_i \cos\theta = 2\frac{h}{\lambda}\cos\theta = 2\frac{h\nu}{c}\cos\theta.$$

In this case, the impulse transmitted to the sphere is:

$$\vec{p} = -\Delta\vec{p}; \quad p = 2\frac{h}{\lambda}\cos\theta = 2\frac{h\nu}{c}\cos\theta.$$

The kinetic momentum transmitted to the sphere by the incident photon on the reflecting surface of the space probe is

$$\vec{L} = \vec{R} \times \vec{p}.$$

(b) If the sphere is rotated, compared to the initial position (1), with a small angle $\Delta\alpha$, as shown in Figure 5.4, where sectors I (reflective

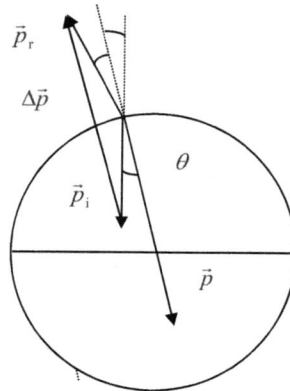

Fig. 5.3

sector) and III (absorbing sector) are symmetrical with respect to
the diameter of the sphere parallel to the solar rays, and sector II
(absorbing sector) is symmetrical with respect to the diameter of
the sphere parallel to the solar rays, then the resultant forces acting
on the three sectors have the orientations represented in the figure,
so that only the component $\vec{F}_3$ (light pressure force acting on the
absorbing sector III) has a rotation effect.

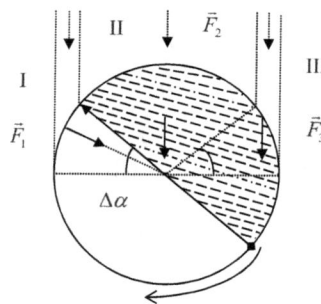

Fig. 5.4

As a result, the probe will continue its accelerated rotation in the
direction imposed by the initial rotation, evolving as indicated by
the sequences in Figure 5.5 until it reaches position 2.

Due to inertia, the sphere will continue to rotate in the same
direction before slowing to a stop. The final position of the probe

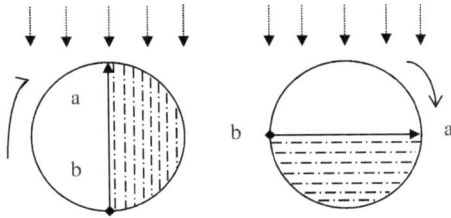

Fig. 5.5

will represent a rotation of 180° after releasing the probe from the initial position. Figure 5.6 presents the justification for the slowing motion.

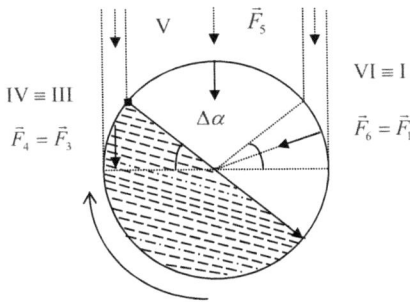

Fig. 5.6

After stopping, the probe will rotate in the opposite direction by 180°, returning to the initial position, at rest. If the sphere is rotated relative to the initial position (2) by a small angle $\Delta\alpha$, as shown in Figure 5.7, where sectors I (absorbing sector) and III (reflecting sector) are symmetrical with respect to the diameter of the sphere parallel to the sun's rays, and sector II (reflective sector) is symmetrical with respect to the diameter of the sphere parallel to the sun's rays, then the resultant forces acting on the three sectors have the orientations represented in the drawing, so that only the component $\vec{F}_1$ (light pressure force acting on the absorbent sector I) has a rotation effect.

After release, the probe will begin to rotate rapidly in the opposite direction to the direction imposed by the initial rotation, evolving towards position 2, which it will overcome due to gravity. After this, the sphere will continue to rotate in the same direction but will slow

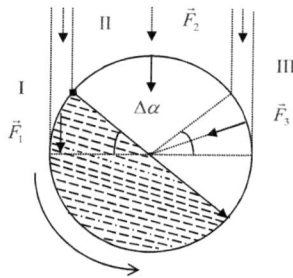

Fig. 5.7

to a stop. The final position of the probe will represent a rotation with an angle of $2\Delta\alpha$ compared to the initial position. The justification for the slowing movement is presented in Figure 5.8.

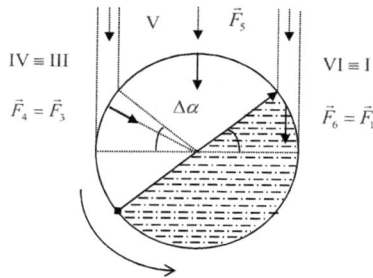

Fig. 5.8

After stopping, the probe will rotate in the opposite direction with an angle of $2\Delta a$, returning to the initial position, at rest.

(c) N_1 photons with frequency ν_1, N_2 photons with frequency ν_2, etc., arrive on the elementary spherical sector from the reflecting surface of the probe (Figure 5.9) corresponding to position (2), under the same angle of incidence, θ. The total impulse transmitted to the sphere from the direction of light propagation is:

$$p_{//} = N_1 2\frac{h\nu_1}{c}\cos^2\theta + N_2 2\frac{h\nu_2}{c}\cos^2\theta + \cdots$$

$$= \frac{2}{c}(N_1 h\nu_1 + N_2 h\nu_2 + \cdots)\cos^2\theta$$

$$= \frac{2}{c}E_{\theta;\theta+d\theta}\cos^2\theta,$$

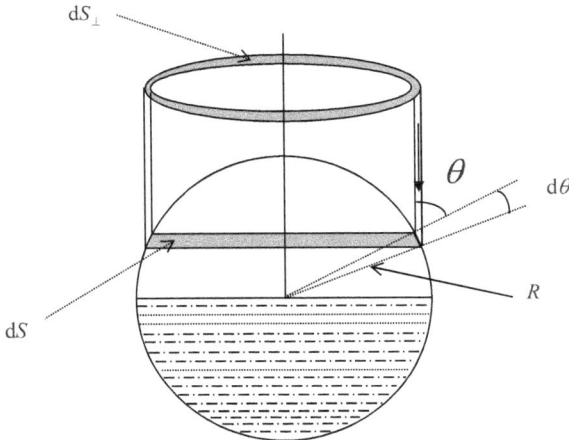

Fig. 5.9

where $E_{\theta;\theta+d\theta}$ is the energy of all photons incident on the elementary spherical sector considered, whose area is

$$dS = 2\pi R^2 \sin\theta d\theta.$$

The cross-section of the light beam incident on the elementary spherical sector considered has the surface area

$$dS_\perp = dS \cos\theta.$$

Considering the definition of the average power of solar radiation per unit area of a surface at normal incidence, it follows that:

$$E_0 = \frac{E_{\theta;\theta+d\theta}}{dS_\perp dt};$$

$$E_{\theta;\theta+d\theta} = E_0 dS_\perp dt;$$

$$p_{//} = \frac{2}{c} E_0 dS_\perp dt \cos^2\theta;$$

$$p_{//} = \frac{4\pi R^2 E_0}{c} \cos^3\theta \sin\theta d\theta dt.$$

In these conditions, the elementary force that acts on the sphere as a result of the photons incident on the elementary sector considered is:

$$dF = \frac{p_{//}}{dt} = \frac{4\pi R^2 E_0}{c} \cos^3 \theta \sin \theta d\theta;$$

$$dF = -\frac{4\pi R^2 E_0}{c} \cos^3 \theta d(\cos \theta).$$

The force acting on the sphere as a result of the photons incident on the reflecting hemisphere of the space probe will be:

$$F = -\frac{4\pi R^2}{c} E_0 \int_0^{\pi/2} \cos^3 \theta d(\cos \theta);$$

$$F = -\frac{\pi R^2}{c} E_0 \cos^4 \theta \Big|_0^{\pi/2};$$

$$F = \frac{E_0}{c} \pi R^2.$$

Similarly, if N_1 photons with frequency ν_1, N_2 photons with frequency ν_2, etc., arrive on the elementary spherical sector from the absorbing surface of the probe (Figure 5.10), corresponding to position (1), under the same angle of incidence, θ, then the total impulse transmitted to the sphere along the direction of light propagation is:

$$p_{//} = N_1 \frac{h\nu_1}{c} + N_2 \frac{h\nu_2}{c} + \cdots$$

$$= \frac{1}{c} (N_1 h\nu_1 + N_2 h\nu_2 + \cdots)$$

$$= \frac{1}{c} E_{\theta;\theta+d\theta},$$

where $E_{\theta;\theta+d\theta}$ is the energy of all photons incident on the elementary spherical sector considered, whose area is

$$dS = 2\pi R^2 \sin \theta d\theta.$$

The cross-section of the light beam incident on the elementary spherical sector considered has the surface area

$$dS_\perp = dS \cos \theta.$$

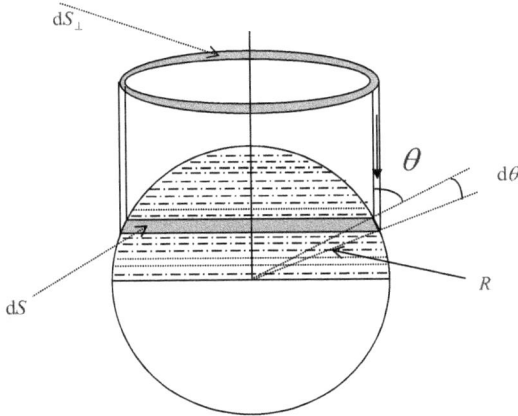

Fig. 5.10

Considering the definition of the average power of solar radiation per unit area of a surface at normal incidence, it results that:

$$E_0 = \frac{E_{\theta;\theta+\mathrm{d}\theta}}{\mathrm{d}S_\perp \mathrm{d}t};$$

$$E_{\theta;\theta+\mathrm{d}\theta} = E_0 \mathrm{d}S_\perp \mathrm{d}t;$$

$$p_{//} = \frac{1}{c} E_0 \mathrm{d}S_\perp \mathrm{d}t;$$

$$p_{//} = \frac{2\pi R^2 E_0}{c} \cos\theta \sin\theta \mathrm{d}\theta \mathrm{d}t.$$

Under these conditions, the elementary force acting on the sphere as a result of the photons incident on the considered elementary sector is:

$$\mathrm{d}F = \frac{p_{//}}{\mathrm{d}t} = \frac{2\pi R^2 E_0}{c} \cos\theta \sin\theta \mathrm{d}\theta;$$

$$\mathrm{d}F = -\frac{2\pi R^2 E_0}{c} \cos\theta \mathrm{d}(\cos\theta).$$

The force acting on the sphere as a result of the photons incident on the absorbing hemisphere of the space probe is:

$$F = -\frac{2\pi R^2}{c} E_0 \int_0^{\pi/2} \cos\theta \mathrm{d}(\cos\theta);$$

$$F = -\frac{\pi R^2}{c} E_0 \cos^2 \theta \Big|_0^{\pi/2} ;$$

$$F = \frac{E_0}{c} \pi R^2.$$

(d) The probe will rotate around the axis that passes through its center and is perpendicular to the direction of the sun's rays, periodically changing its direction of rotation and permanently moving away from the Sun, oscillating either side of the direction of its forward movement, in the order of the sequences shown in Figure 5.11. At the extreme positions on either side, the static angular velocity is

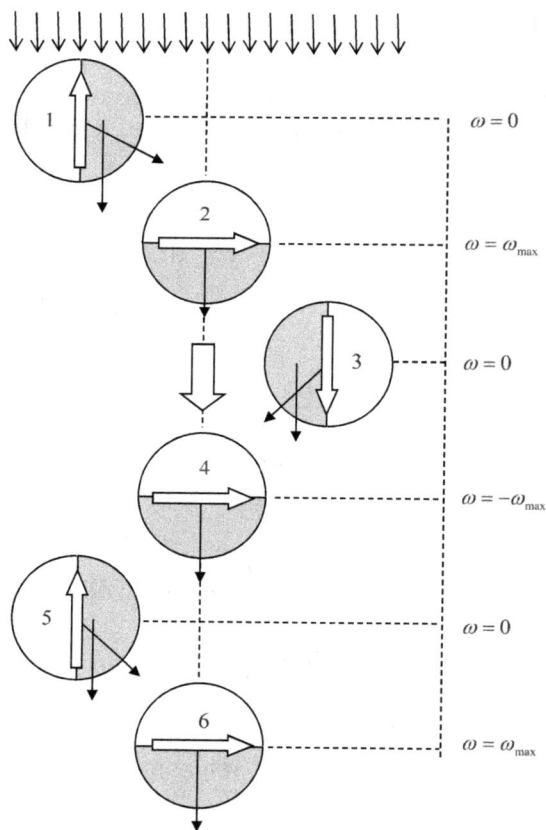

Fig. 5.11

zero, and the direction of rotation changes. In the middle positions, the angular speed of the probe is maximum. This behavior of the probe results from the orientation of the light pressure forces, which act either on the reflective sector or on the absorbing sector of the sphere, when exposed to the light of the Sun.

Problem 6. The Movement of Train Wheels

A train is moving at a velocity $\vec{v} =$ constant, while the wheels of its carriages roll without slipping on the railway tracks. The wheel rim has an outer radius R, and the flange on the rim (the protruding part of the rim, acting as a guide) has a height h (Figure 6.1).

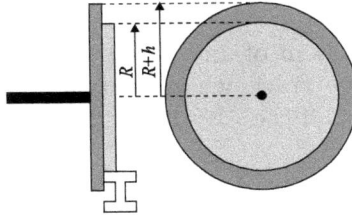

Fig. 6.1

(a) Using the velocity composition rule, *demonstrate* that, while rolling, each point of the wheel executes an instantaneous rotational movement in relation to the point on the rail with which the wheel makes instantaneous contact.

(b) *Identify*: **(1)** the points of the wheel whose instantaneous velocities in relation to the ground indicate instantaneous movements in the same direction as the train's movement, calculating this velocity for the point where its value is maximum; **(2)** the points of the wheel whose instantaneous velocities in relation to the ground indicate instantaneous movements in the opposite direction to the movement of the train, calculating this velocities for the point where its value is maximum; **(3)** the points of the wheel whose instantaneous velocities relative to the ground indicate instantaneous movements perpendicular to the direction of the train's moevement, calculating this velocity for the point where its value is maximum.

(c) *Compare* the instantaneous velocity relative to the points on the
ground with which instantaneous contact is made: (1) in front
of the vertical plane of the point of contact between the wheel
and the rail, above the horizontal plane of the point of contact
between the wheel and the rail, and at the points located in
front of the same vertical plane but below the specified horizontal plane; (2) behind the vertical plane of the point of contact
between the wheel and the rail, below the horizontal plane of the
point of contact between the wheel and the rail, and at the points
located behind the same vertical plane but above the specified
horizontal plane.

Solution

(a) Let XOY be a system of axes fixed to the railway track, where
O is the point on the track where instantaneous contact is made
with the rolling rim, and the axes OX and OY have the orientations
represented in Figure 6.2.

The system X'Y' in the center of the wheel is in translational
motion with a velocity relative to the fixed system OXY.

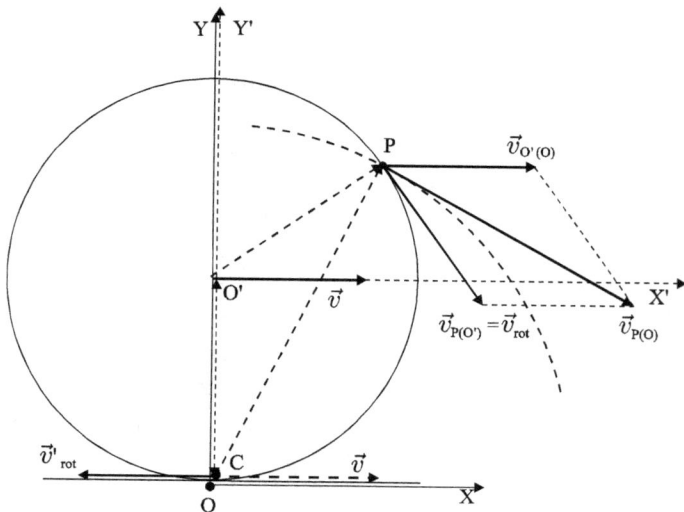

Fig. 6.2

A point P on the wheel is in uniform circular motion relative to the system O'X'Y (relative motion) with the velocity

$$\vec{v}_{\text{rel}} = \vec{v}_{P(O')} = \vec{v}_{\text{rot}}$$

and in absolute motion (transport motion) relative to the fixed system OXY with velocity

$$\vec{v}_{\text{abs}} = \vec{v}_{P(O)}.$$

Thus, the velocity of the mobile system's movement is

$$\vec{v}_{\text{tr}} = \vec{v}_{O'(O)} = \vec{v},$$

such that:

$$\vec{v}_{\text{abs}} = \vec{v}_{\text{rel}} + \vec{v}_{\text{tr}};$$

$$\vec{v}_{P(O)} = \vec{v}_{P(O')} + \vec{v}_{O'(O)};$$

$$\vec{v}_{P(O')} = \vec{v}_{\text{rot}} = \vec{\omega} \times \overrightarrow{O'P},$$

where $\vec{\omega}$ is the angular velocity of the wheel's rotational motion relative to the O'X'Y' system for all points of the wheel (rigid solid body).

In particular, for point C on the wheel, in contact with point O on the rail, we have:

$$\vec{v}_{C(O)} = \vec{v}_{C(O')} + \vec{v}_{O'(O)};$$

$$\vec{v}_{C(O')} = \vec{v}_{\text{rot}} = \vec{\omega} \times \overrightarrow{O'C};$$

$$\overrightarrow{OC'} = -\overrightarrow{CO'} = -\overrightarrow{OO'};$$

$$\vec{v}_{C(O')} = -\vec{\omega} \times \overrightarrow{OO'}.$$

Since the wheel rolls on the rail without slipping,

$$\vec{v}_{C(O)} = 0.$$

It results that:

$$0 = \vec{v}_{\text{rot}} + \vec{v};$$

$$\vec{v}_{\text{rot}}' = \vec{\omega} \times \overrightarrow{O'C} = -\vec{v};$$

$$\vec{v}_{\text{rot}} = \vec{\omega} \times \overrightarrow{O'P};$$

$$O'C = O'P;$$

$$v_{\text{rot}} = v_{\text{rot}}' = \omega R = v,$$

The module of the tangential (rotational) velocity of a point on the wheel rim is the same for all points (referring to the mobile system) and is equal to the modulus of the velocity of the center of the wheel in relation to the fixed system. Thus,

$$0 = -\vec{\omega} \times \overrightarrow{OO'} + \vec{v}_{O'(O)};$$

$$\vec{v}_{O'(O)} = \vec{\omega} \times \overrightarrow{OO'};$$

$$\vec{v}_{P(O)} = \vec{\omega} \times \overrightarrow{O'P} + \vec{\omega} \times \overrightarrow{OO'};$$

$$\vec{v}_{P(O)} = \vec{\omega} \times \left(\overrightarrow{OO'} + \overrightarrow{O'P} \right);$$

$$\vec{v}_{P(O)} = \vec{\omega} \times \overrightarrow{OP}.$$

This proves that the instantaneous movement of the point P, relative to O, is circular, corresponding to a circle with the instantaneous radius OP, so that $\vec{v}_{P(O)} \perp \overrightarrow{OP}$, and an angular velocity equal to the angular velocity of the wheel's rotation in relation to X'O'Y'.

(b) The wheel's uniform rolling motion without slipping is equivalent at any time to the composition of the translational motion of a "locked" wheel (in which the velocity of each point of the wheel is $\vec{v}$) with the uniform rotational motion of the wheel about its fixed axis (in which the angular velocity of each point is $\omega = \frac{v}{R}$), as indicated by Figure 6.3. Thus, the velocity of each point of the wheel during rolling is the result of the composition of the velocities of that point corresponding to the combined movements.

For example:

$$\vec{v}_A = \vec{v} + \vec{v}_{\text{rot, A}};$$

$$v_A = v + v_{\text{rot,A}} = v + \omega R_A = 2v;$$

$$\vec{v}_{\mathrm{B}} = \vec{v} + \vec{v}_{\mathrm{rot,B}}; \quad \vec{v}_{\mathrm{B}} \perp \overrightarrow{\mathrm{CB}};$$

$$v_{\mathrm{B}} = \sqrt{v^2 + v_{\mathrm{rot,B}}^2} = v\sqrt{2};$$

$$\vec{v}_{\mathrm{C}} = \vec{v} + \vec{v}_{\mathrm{rot,C}};$$

$$v_{\mathrm{C}} = v - v_{\mathrm{rot,C}} = 0.$$

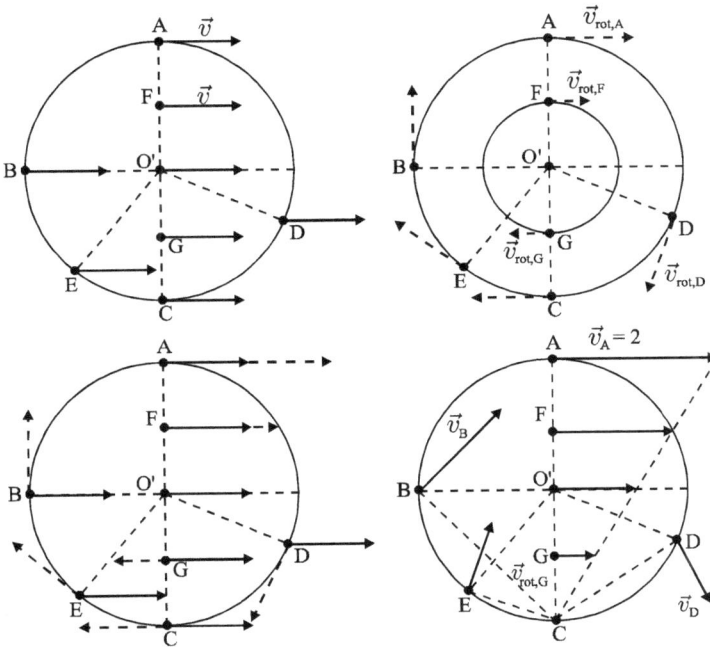

Fig. 6.3

For the points that belong to the rim of the wheel, as shown in Figure 6.4, the instantaneous velocities relative to the ground result from the same composition of the velocities of the two simultaneous movements.

For example:

$$\vec{v}_{\mathrm{H}} = \vec{v} + \vec{v}_{\mathrm{rot,H}};$$

$$v_{\mathrm{H}} = v + v_{\mathrm{rot,H}}; \quad v_{\mathrm{rot,H}} = \omega R_{\mathrm{H}};$$

$$R_H = R + h;$$

$$v_{\mathrm{H}} = 2v + v\frac{h}{R};$$

$$\vec{v}_{\mathrm{K}} = \vec{v} + \vec{v}_{\mathrm{rot,\,K}};$$

$$v_{\mathrm{K}} = v_{\mathrm{rot,K}} - v; \quad v_{\mathrm{rot,K}} = \omega R_{\mathrm{K}};$$

$$R_{\mathrm{K}} = R + h; \quad v_{\mathrm{K}} = v\frac{h}{R};$$

$$\vec{v}_{\mathrm{M}} = \vec{v} + \vec{v}_{\mathrm{rot,\,M}};$$

$$\vec{v}_{\mathrm{M}} = \vec{\omega} \times \overrightarrow{\mathrm{CM}}; \quad \vec{v}_{\mathrm{M}} \perp \overrightarrow{\mathrm{CM}};$$

$$v_{\mathrm{M}} = \omega \mathrm{CM}; \quad \omega = \frac{v}{R};$$

$$\mathrm{CM} = \sqrt{h\,(2R + h)}; \quad v_{\mathrm{M}} = \frac{v}{R}\sqrt{h\,(2R + h)};$$

$$\vec{v}_{\mathrm{N}} = \vec{v} + \vec{v}_{\mathrm{rot,N}};$$

$$\vec{v}_{\mathrm{N}} = \vec{\omega} \times \overrightarrow{\mathrm{CN}}; \quad v_{\mathrm{N}} = \omega \mathrm{CN};$$

$$\mathrm{CN} = \sqrt{h(2R + h)}; \quad v_{\mathrm{N}} = \frac{v}{R}\sqrt{h(2R + h)}.$$

(1) The velocities of all points located on the common instantaneous vertical plane of the wheel center (O′) and the point of contact with the rail (C) above point C are oriented in the same direction as vector $\vec{v}$, which means that the instantaneous movements of these points are in the direction of the train's movement. Among the velocities of these points, the highest is the velocity of point H ($v_{\mathrm{H}} = v_{\max}$).

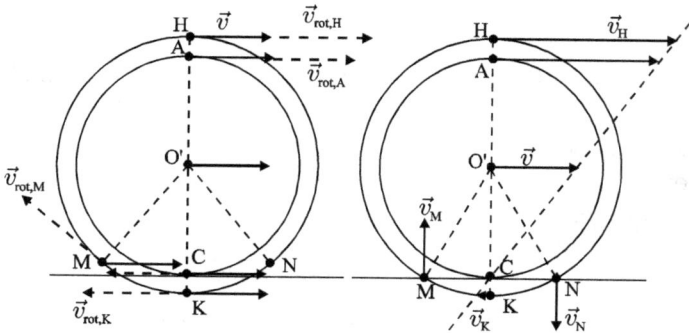

Fig. 6.4

(2) The velocities of all points of the wheel located in the common instantaneous vertical plane of the center of the wheel and the contact point with the rail, below point C, are oriented in opposite directions to vector $\vec{v}$, meaning that the instantaneous movements of these points in relation to the ground are in the opposite direction to the movement of the train. Among the velocities of these points, the highest is at point K $(v_K = v_{max})$.

(3) The velocities of all points of the wheel located in the instantaneous horizontal plane of the point of contact with the rail are vectors whose directions are vertical, which means that the instantaneous movements of these points in relation to the ground are perpendicular to the direction of the train's movement. Among the velocities of these points, the velocities of points M and N are highest $(v_M = v_N = v_{max})$, so that $\vec{v}_M = -\vec{v}_N$.

(c) If the wheel is divided into four sectors, as shown in Figure 6.5, then from the previous analyses, it results that: **(1)** the horizontal components of the velocities of the points in regions I and II in relation to the ground have opposite orientations (in region I to the right, in region II to the left), and the orientations of the vertical components are identical (vertically down); **(2)** the horizontal components of the velocities of the points in regions III and IV in relation to the ground have opposite orientations (in region III to the left, and in region IV to the right), and the vertical components are identical (vertically upwards).

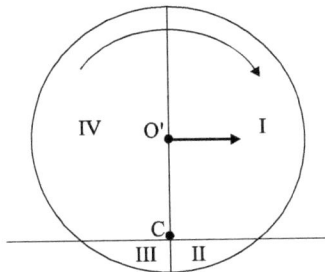

Fig. 6.5

Observations:

- The velocities of the points on the borders of sectors I and IV and sectors II and III, in relation to the ground, have horizontal directions.
- The velocities of the points on the borders of sectors I and II and sectors III and IV, in relation to the ground, have vertical directions.

Chapter 7

International Pre-Olympic Physics Contest 2011, Satu Mare, Romania

Problem 1. What Newton Didn't Know!

The revolution movement of the Moon around the Earth is accompanied by a rotation movement of the Moon around its own axis, the directions of the two movements being identical. The ellipse in Figure 1.1 represents the trajectory of the center of the Moon in its movement resulting from the gravitational interaction with the planet Earth. In the focus F_1 of this ellipse, we find the center of the Earth.

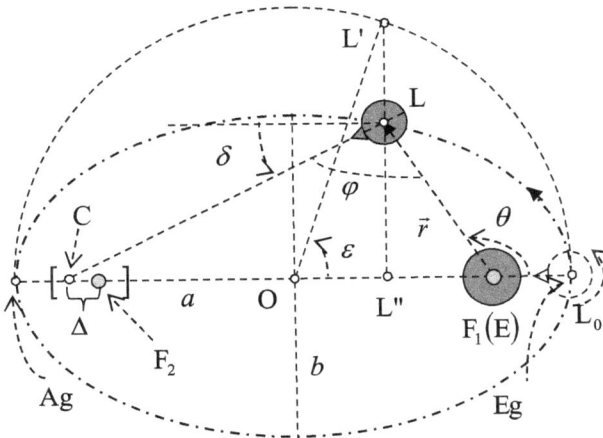

Fig. 1.1

In the initial position, L_0, the center of the Moon coincides with the perigee of the ellipse. After a certain time, the center of the Moon reaches the ellipse in position L, and the vector radius $\vec{r}$ of the center of the Moon rotates by an angle of θ. In the same time interval, the Moon rotates around its axis (perpendicular to the plane of the drawing) with an angle δ, highlighted in the drawing as the angle by which the reference axis from the Moon's orbital section has rotated, in the plane of the Moon's orbit. As a result, the direction of the reference axis in the plane of the orbital section of the Moon intersects the major axis of the ellipse at a point C near its focus F_2.

Conclusion: When the Moon moves around the Earth, the point C moves along the major axis of the ellipse, oscillating to either side of the focus F_2, thus highlighting the role of the second focus of the ellipse, F_2, a role that Newton did not know about!

(a) *Identify* and *explain* the optical phenomenon in which the Moon is involved, the observer being on Earth (in the focus F_1 of the ellipse), a phenomenon resulting from the simultaneous occurence of the two movements of the Moon, specified in the statement of the problem.
   Determine the intervals of the values of the distance Δ, to the left and right of the focus F_2, if Δ represents the distance from point C to the focus F_2, at a certain moment during the movement of the Moon's center from the perigee to the apogee. *Analyze* the symmetry/asymmetry of these intervals in relation to the focus F_2 and *interpret* the result.

(b) *Localize* on the ellipse the position L_{F_2} of the center of the Moon for which the direction of the reference axis from the plane of the Moon's orbital section intersects the major axis of the ellipse in its focus F_2.

Given that: (1) the relationship between the angles θ and ε, highlighted in Figure 1.1, is $\cos \theta = \frac{\cos \varepsilon - e}{1 - e \cos \varepsilon}$, where $e = \sqrt{1 - b^2/a^2}$ is the numerical eccentricity of the ellipse; (2) the area of the surface described by the vector radius of the center of the Moon until the center of the Moon has reached the position L is $S = \frac{ab}{2}(\varepsilon - e \sin \varepsilon)$, where a and b are the two semi-axes of the ellipse; (3) the major semi-axis of the ellipse is $a = 384.400$ km; (**4**) the average angular velocity during the movement of the Moon on the large ellipse around the Earth, ω, is equal to the angular velocity corresponding to the rotation of the Moon around its own axis, Ω; (**5**) because the numerical eccentricity of the ellipse is very small, $e \approx 0.0549$, we will accept the following approximation:

$$f(e) = \sin\varepsilon\sqrt{1 - e^2}\cot(\varepsilon - e\sin\varepsilon) - e - \cos\varepsilon$$

$$\approx \frac{\cos\varepsilon}{2}e^2 - \frac{1}{3}\left(\frac{1}{2} - 2\cos^2\varepsilon\right)e^3.$$

Solution

(**a**) We will first analyze the situation from the statement of the problem, when the Moon is on an elliptical orbit, at a moment when the angle δ is small and the point C is to the left of the focus F_2, as shown in Figure 1.2.

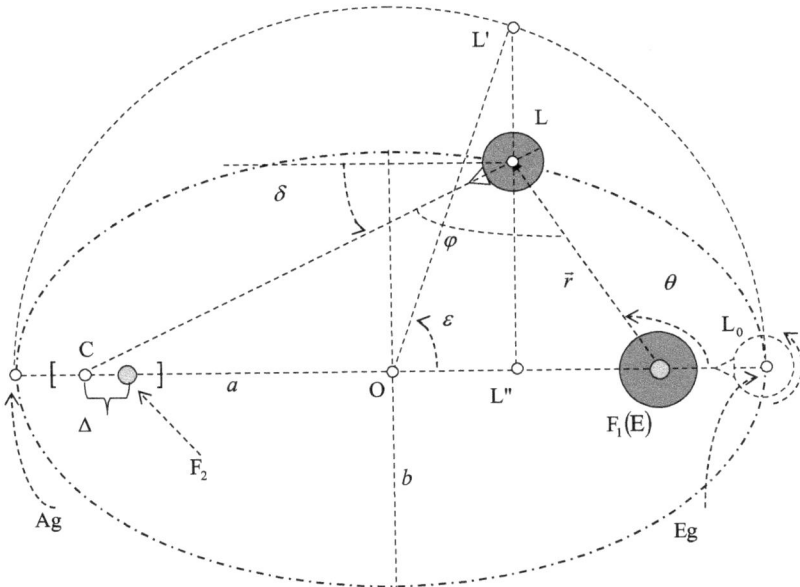

Fig. 1.2

If in the time interval t, during which the center of the Moon moves on the large ellipse from position L_0 to position L, the area of the surface described by the vector radius of the center of the Moon is

$$S = \frac{ab}{2}(\varepsilon - e\sin\varepsilon),$$

then, according to Kepler's second law, it results that:

$$\frac{S}{S_0} = \frac{t}{T},$$

where T is the period of the Moon's rotation around the Earth, and S_0 is the surface area of the ellipse;

$$S_0 = \pi ab;$$

$$\frac{\frac{ab}{2}(\varepsilon - e\sin\varepsilon)}{\pi ab} = \frac{t}{T};$$

$$t = \frac{T(\varepsilon - e\sin\varepsilon)}{2\pi},$$

representing the duration of the movement of the center of the Moon from position L_0 to position L.

If the average angular speed during the movement of the Moon on the large ellipse around the Earth is

$$\omega = \frac{2\pi}{T},$$

and Ω is the angular velocity corresponding to the rotation of the Moon around its axis, knowing that the two angular velocities are equal ($\omega = \Omega$), then the angle with which, in time t, the Moon uniformly rotates around its axis (equal to the rotational angle of the reference axis in the plane of the Moon's orbital section) is

$$\delta = \Omega t = \omega t = \frac{2\pi}{T}\frac{T(\varepsilon - e\sin\varepsilon)}{2\pi} = \varepsilon - e\sin\varepsilon.$$

From the triangle LCF_1, using the theorem of sines, it follows that:

$$\frac{d_{CF_1}}{\sin\varphi} = \frac{d_{LF_1}}{\sin(\theta - \varphi)};$$

$$\delta + \varphi = \theta; \quad d_{LF_1} = r;$$

$$d_{CF_1} = r\frac{\sin\varphi}{\sin(\theta - \varphi)} = r\frac{\sin\varphi}{\sin\delta},$$

such that, for the interval $\Delta = d_{\text{CF}_2}$, which interests us,

$$\Delta = d_{\text{CF}_2} = d_{\text{CF}_1} - d_{\text{F}_1\text{F}_2}.$$

To calculate the distance between the foci of the ellipse, $d_{\text{F}_1\text{F}_2}$, we use the equation of the ellipse in plane polar coordinates and the drawing in Figure 1.3, resulting in:

$$r = \frac{p}{1 + e \cos \theta}; \quad \theta = 0;$$

$$r = r_{\min} = \frac{p}{1 + e}; \quad \theta = \pi; \quad r = r_{\max} = \frac{p}{1 - e};$$

$$r_{\min} + r_{\max} = 2a = \frac{p}{1 + e} + \frac{p}{1 - e} = \frac{2p}{1 - e^2};$$

$$p = a(1 - e^2);$$

$$r_{\min} = \frac{p}{1 + e} = a(1 - e);$$

$$c = a - r_{\min} = ae;$$

$$d_{\text{F}_1\text{F}_2} = 2c = 2ae.$$

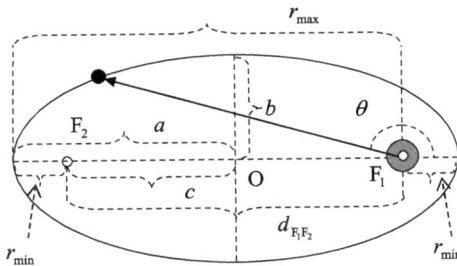

Fig. 1.3

In these conditions, for the interval $\Delta = d_{\text{CF}_2}$, it results that:

$$\Delta = d_{\text{CF}_1} - d_{\text{F}_1\text{F}_2} = r\frac{\sin \varphi}{\sin \delta} - 2ae;$$

$$\Delta = \frac{p}{1 + e \cos \theta} \frac{\sin \varphi}{\sin \delta} - 2ae = \frac{a(1 - e^2)}{1 + e \cos \theta} \frac{\sin \varphi}{\sin \delta} - 2ae;$$

$$\frac{\Delta}{a} = \frac{(1 - e^2)}{1 + e \cos \theta} \frac{\sin \varphi}{\sin \delta} - 2e; \quad \varphi = \theta - \delta;$$

$$\frac{\Delta}{a} = \frac{(1-e^2)}{1+e\cos\theta}\frac{\sin(\theta-\delta)}{\sin\delta} - 2e;$$

$$\frac{\Delta}{a} = \frac{(1-e^2)}{1+e\cos\theta}(\sin\theta\cot\delta - \cos\theta) - 2e;$$

$$\delta = \varepsilon - e\sin\varepsilon;$$

$$\cot\delta = \cot(\varepsilon - e\sin\varepsilon);$$

$$\cos\theta = \frac{\cos\varepsilon - e}{1 - e\cos\varepsilon}; \quad \sin\theta = \sqrt{1-\cos^2\theta} = \frac{\sin\varepsilon}{1 - e\cos\varepsilon}\sqrt{1-e^2};$$

$$1 + e\cos\theta = \frac{1-e^2}{1-e\cos\varepsilon};$$

$$\frac{\Delta}{a} = -e + \sin\varepsilon\sqrt{1-e^2}\cot\delta - \cos\varepsilon;$$

$$\frac{\Delta}{a} = \sin\varepsilon\sqrt{1-e^2}\cot(\varepsilon - e\sin\varepsilon) - e - \cos\varepsilon;$$

$$\frac{\Delta}{a} \approx \frac{\cos\varepsilon}{2}e^2 - \frac{1}{3}\left(\frac{1}{2} - 2\cos^2\varepsilon\right)e^3;$$

$$\varepsilon = 0; \quad e = 0.0549;$$

$$\frac{\Delta}{a} = \frac{1}{2}0.00301401 - \frac{1}{3}\left(\frac{1}{2} - 2\right)0.000165469;$$

$$\frac{\Delta}{a} = \frac{1}{2}0.00301401 + \frac{1}{2}0.000165469 = 0.00158974;$$

$$a = 384.400\,\text{km};$$

$$\Delta = 611.09\,\text{km},$$

representing the maximum distance between point C, to the left, and focus F, for $\varepsilon \to 0$;

$$\Delta_{\text{max,left}} = 611.09\,\text{km}.$$

Let us now analyze the situation represented in Figure 1.4, when the Moon is on an elliptical orbit, at a time when the angle δ is large, and point C is to the right of the focus F_2.

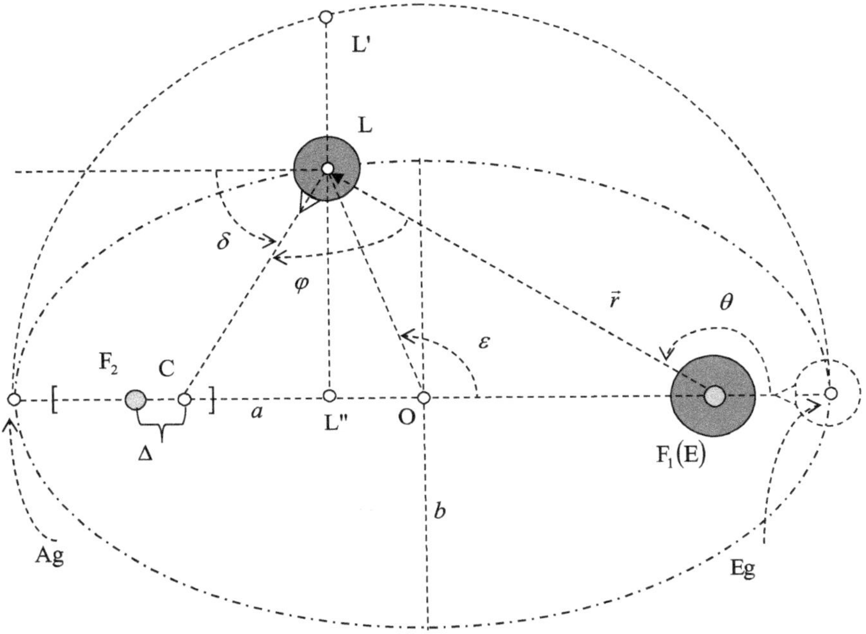

Fig. 1.4

In these conditions, for the interval $\Delta = d_{CF_2}$, it results that:

$$\Delta = d_{F_1F_2} - d_{CF_1} = 2ae - r\frac{\sin\varphi}{\sin\delta};$$

$$\Delta = 2ae - \frac{p}{1+e\cos\theta}\frac{\sin\varphi}{\sin\delta} = 2ae - \frac{a(1-e^2)}{1+e\cos\theta}\frac{\sin\varphi}{\sin\delta};$$

$$\frac{\Delta}{a} = 2e - \frac{(1-e^2)}{1+e\cos\theta}\frac{\sin\varphi}{\sin\delta}; \quad \varphi = \theta - \delta;$$

$$\frac{\Delta}{a} = 2e - \frac{(1-e^2)}{1+e\cos\theta}\frac{\sin(\theta-\delta)}{\sin\delta};$$

$$\frac{\Delta}{a} = 2e - \frac{(1-e^2)}{1+e\cos\theta}(\sin\theta\cot\delta - \cos\theta);$$

$$\delta = \varepsilon - e\sin\varepsilon;$$

$$\cot\delta = \cot(\varepsilon - e\sin\varepsilon);$$

$$\cos\theta = \frac{\cos\varepsilon - e}{1 - e\cos\varepsilon}; \quad \sin\theta = \sqrt{1 - \cos^2\theta} = \frac{\sin\varepsilon}{1 - e\cos\varepsilon}\sqrt{1 - e^2};$$

$$1 + e\cos\theta = \frac{1 - e^2}{1 - e\cos\varepsilon};$$

$$\frac{\Delta}{a} = e - \sin\varepsilon\sqrt{1 - e^2}\cot\delta + \cos\varepsilon;$$

$$\frac{\Delta}{a} = e - \sin\varepsilon\sqrt{1 - e^2}\cot(\varepsilon - e\sin\varepsilon) + \cos\varepsilon;$$

$$\frac{\Delta}{a} = -\left[-e + \sin\varepsilon\sqrt{1 - e^2}\cot(\varepsilon - e\sin\varepsilon) - \cos\varepsilon\right];$$

$$\frac{\Delta}{a} \approx -\frac{\cos\varepsilon}{2}e^2 + \frac{1}{3}\left(\frac{1}{2} - 2\cos^2\varepsilon\right)e^3;$$

$$\frac{\Delta}{a} \approx -\left[\frac{\cos\varepsilon}{2}e^2 - \frac{1}{3}\left(\frac{1}{2} - 2\cos^2\varepsilon\right)e^3\right];$$

$$\varepsilon = \pi; \quad e = 0.0549;$$

$$\frac{\Delta}{a} = \frac{1}{2}0.00301401 + \frac{1}{3}\left(\frac{1}{2} - 2\right)0.000165469;$$

$$\frac{\Delta}{a} = \frac{1}{2}0.00301401 - \frac{1}{2}0.000165469 = 0.001424271;$$

$$a = 384.400\,\text{km};$$

$$\Delta = 547.48\,\text{km},$$

representing the maximum distance between point C, to the right, and the focus F_2, for $\varepsilon \to \pi$;

$$\Delta_{\text{max,right}} = 547.48\,\text{km};$$

$$\Delta_{\text{max,right}} < \Delta_{\text{max,left}}.$$

This proves the asymmetry of the two intervals near the focus F_2, as illustrated in Figure 1.5.

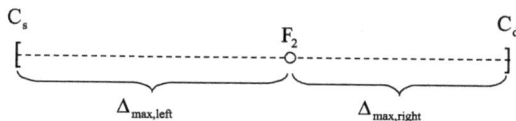

Fig. 1.5

(b) If the direction of the reference axis from the plane of the Moon's orbital section intersects the major axis of the ellipse in its focus F_2, it means that

$$\frac{\Delta}{a} \approx \frac{\cos \varepsilon}{2} e^2 - \frac{1}{3}\left(\frac{1}{2} - 2\cos^2 \varepsilon\right) e^3 = 0,$$

from which it results that:

$$4e \cos^2 \varepsilon + 3 \cos \varepsilon - e = 0;$$

$$\cos \varepsilon = \frac{-3 \pm \sqrt{9 + 16e^2}}{8e};$$

$$\cos \varepsilon = \frac{-3 \pm 3.008}{0.4392};$$

$$\cos \varepsilon = \frac{-3 + 3.008}{0.4392} = 0.018214;$$

$$\varepsilon \approx 89^0;$$

$$\cos \theta = \frac{\cos \varepsilon - e}{1 - e \cos \varepsilon};$$

$$\cos \theta \approx -0.036722721;$$

$$\theta \approx 92.15^0.$$

Analyze the phenomenon referred to in this problem and the scenarios wherein the observer is in the focus F_2 or the center O of the ellipse.

Problem 2. Geostationary Satellite

The circular orbit of a geostationary satellite lies in the plane of the equator. Due to a brief malfunction or shutdown of one of the satellite's thrusters, its orbital speed decreases by an amount Δv, which is small compared to the satellite's original geostationary velocity. As a result, the orientation of the velocity vector changes.
    Determine:

(a) the new rotation period of the satellite, T;
(b) the maximum possible value of Δv, so that the satellite does not fall to the surface of the Earth.

Given: T_0, the period of the geostationary satellite; g_0, the gravitational acceleration on the ground; and R, the radius of the Earth. *It is known that*: $(1 + x)^n \approx 1 + nx$, if $x \ll 1$.

Solution

(a) The geostationary evolution of the satellite on the orbit with radius r, highlighted in Figure 2.1, makes the angular velocity of the satellite equal to the angular velocity of the Earth's rotation (the external satellite's rotation period is equal to the Earth's rotation period). The force of gravitational attraction exerted by the Earth on the satellite is centripetal, so it results that:

$$K\frac{mM}{r^2} = \frac{mv^2}{r}; \quad K\frac{M}{r} = v^2;$$

$$g_0 = K\frac{M}{R^2}; \quad T_0 = \frac{2\pi r}{v}; \quad \frac{g_0 R^2}{r} = \frac{4\pi^2 r^2}{T_0^2};$$

$$r = \sqrt[3]{\frac{g_0 R^2 T_0^2}{4\pi^2}}; \quad v = \frac{2\pi}{T_0}r; \quad v = \sqrt[3]{\frac{2\pi g_0 R^2}{T_0}}.$$

After reducing the magnitude of the velocity vector by the amount $\Delta v \ll v$, the satellite will evolve on an elliptical orbit. The point where the speed correction was achieved represents the apogee of the new elliptical orbit, in whose distant focus is the center of the Earth.

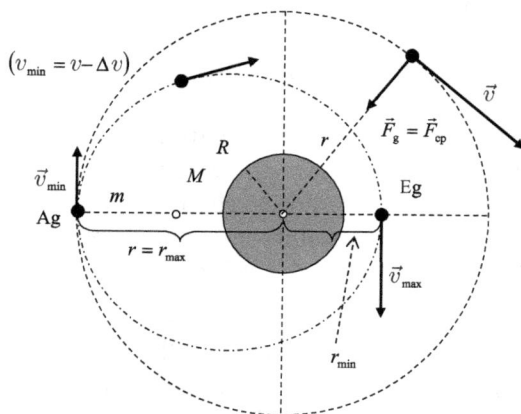

Fig. 2.1

In accordance with the laws of conservation of kinetic momentum and mechanical energy, it follows that:

$$v_{\min} r_{\max} = v_{\max} r_{\min};$$

$$(v - \Delta v) r = v_{\max} r_{\min};$$

$$-K\frac{mM}{r_{\max}} + \frac{mv_{\min}^2}{2} = -K\frac{mM}{r_{\min}} + \frac{mv_{\max}^2}{2};$$

$$-K\frac{M}{r} + \frac{(v - \Delta v)^2}{2} = -K\frac{M}{r_{\min}} + \frac{v_{\max}^2}{2};$$

$$r + r_{\min} = 2a;$$

$$a = \frac{r + r_{\min}}{2},$$

representing the major semi-axis of the ellipse.

In accordance with Kepler's third law, if T is the period of rotation of the satellite on the new orbits, it results that:

$$\frac{T^2}{T_0^2} = \frac{a^3}{r^3}; \quad \frac{T^2}{T_0^2} = \frac{(r + r_{\min})^3}{(2r)^3};$$

$$-K\frac{M}{r} + \frac{(v - \Delta v)^2}{2} = -K\frac{M}{r_{\min}} + \frac{v_{\max}^2}{2}; \quad K\frac{M}{r} = v^2;$$

$$-v^2 + \frac{(v - \Delta v)^2}{2} = -K\frac{M}{r}\frac{r}{r_{\min}} + \frac{v_{\max}^2}{2};$$

$$-v^2 + \frac{(v - \Delta v)^2}{2} = -v^2\frac{r}{r_{\min}} + \frac{v_{\max}^2}{2};$$

$$(v - \Delta v)\frac{r}{r_{\min}} = v_{\max};$$

$$-v^2 + \frac{(v - \Delta v)^2}{2} = -v^2\frac{r}{r_{\min}} + \frac{1}{2}(v - \Delta v)^2\frac{r^2}{r_{\min}^2};$$

$$v^2\left(\frac{r}{r_{\min}} - 1\right) = -\frac{(v - \Delta v)^2}{2} + \frac{1}{2}(v - \Delta v)^2\frac{r^2}{r_{\min}^2};$$

$$v^2\left(\frac{r}{r_{\min}} - 1\right) = \frac{1}{2}(v - \Delta v)^2\left(\frac{r^2}{r_{\min}^2} - 1\right);$$

$$v^2 = \frac{1}{2}(v - \Delta v)^2\left(\frac{r}{r_{\min}} + 1\right);$$

$$\frac{r + r_{\min}}{2r_{\min}} = \frac{v^2}{(v - \Delta v)^2}; \quad \frac{r + r_{\min}}{2r} = \frac{v^2}{(v - \Delta v)^2}\frac{r_{\min}}{r};$$

$$\frac{T^2}{T_0^2} = \left(\frac{r + r_{\min}}{2r}\right)^3;$$

$$-K\frac{M}{r} + \frac{(v - \Delta v)^2}{2} = -K\frac{M}{r_{\min}} + \frac{v_{\max}^2}{2};$$

$$-K\frac{M}{R^2}\frac{R^2}{r} + \frac{(v - \Delta v)^2}{2} = -K\frac{M}{R^2}\frac{R^2}{r_{\min}} + \frac{v_{\max}^2}{2};$$

$$-g_0\frac{R^2}{r} + \frac{(v - \Delta v)^2}{2} = -g_0\frac{R^2}{r_{\min}} + \frac{v_{\max}^2}{2};$$

$$(v - \Delta v)\frac{r}{r_{\min}} = v_{\max};$$

$$-g_0\frac{R^2}{r} + \frac{(v - \Delta v)^2}{2} = -g_0\frac{R^2}{r_{\min}} + \frac{1}{2}(v - \Delta v)^2\frac{r^2}{r_{\min}^2};$$

$$g_0 R^2\left(\frac{1}{r_{\min}} - \frac{1}{r}\right) = \frac{1}{2}(v - \Delta v)^2\left(\frac{r^2}{r_{\min}^2} - 1\right);$$

$$g_0 R^2\frac{r - r_{\min}}{rr_{\min}} = \frac{1}{2}(v - \Delta v)^2\frac{r^2 - r_{\min}^2}{r_{\min}^2};$$

$$g_0 R^2\frac{1}{r} = \frac{1}{2}(v - \Delta v)^2\frac{r + r_{\min}}{r_{\min}};$$

$$g_0 R^2\frac{1}{r} = \frac{1}{2}(v - \Delta v)^2\left(\frac{r}{r_{\min}} + 1\right);$$

$$\frac{r}{r_{\min}} + 1 = \frac{2g_0 R^2\frac{1}{r}}{(v - \Delta v)^2};$$

$$v^2 = K\frac{M}{r} = K\frac{M}{R^2}\frac{R^2}{r} = g_0\frac{R^2}{r};$$

$$\frac{r}{r_{\min}} + 1 = \frac{2v^2}{(v - \Delta v)^2};$$

$$\frac{r}{r_{\min}} = \frac{2v^2}{(v - \Delta v)^2} - 1;$$

$$\frac{r}{r_{\min}} = \frac{2v^2 - v^2 + 2v\Delta v - (\Delta v)^2}{(v - \Delta v)^2} = \frac{v^2 + 2v\Delta v - (\Delta v)^2}{(v - \Delta v)^2};$$

$$\frac{r}{r_{\min}} \approx \frac{v^2 + 2v\Delta v}{(v - \Delta v)^2} = \frac{v^2 \left(1 + 2\frac{\Delta v}{v}\right)}{(v - \Delta v)^2};$$

$$\frac{r_{\min}}{r} = \frac{(v - \Delta v)^2}{v^2 \left(1 + 2\frac{\Delta v}{v}\right)};$$

$$\frac{r + r_{\min}}{2r} = \frac{v^2}{(v - \Delta v)^2} \frac{r_{\min}}{r};$$

$$\frac{r + r_{\min}}{2r} = \frac{v^2}{(v - \Delta v)^2} \frac{(v - \Delta v)^2}{v^2 \left(1 + 2\frac{\Delta v}{v}\right)};$$

$$\frac{r + r_{\min}}{2r} = \frac{1}{1 + 2\frac{\Delta v}{v}} = \left(1 + 2\frac{\Delta v}{v}\right)^{-1};$$

$$\frac{r + r_{\min}}{2r} \approx 1 - 2\frac{\Delta v}{v};$$

$$\frac{T^2}{T_0^2} = \left(\frac{r + r_{\min}}{2r}\right)^3;$$

$$\frac{T^2}{T_0^2} = \left(1 - 2\frac{\Delta v}{v}\right)^3; \quad T^2 = T_0^2 \left(1 - 2\frac{\Delta v}{v}\right)^3;$$

$$T = T_0 \sqrt{\left(1 - 2\frac{\Delta v}{v}\right)^3} = T_0 \left(1 - 2\frac{\Delta v}{v}\right)^{3/2};$$

$$T \approx T_0 \left(1 - 3\frac{\Delta v}{v}\right);$$

$$v = \sqrt[3]{\frac{2\pi g_0 R^2}{T_0}};$$

$$T = T_0 \left(1 - \sqrt[3]{\frac{T_0}{2\pi g_0 R^2}} 3\Delta v\right).$$

(b) The admissible extreme situation is the one represented in Figure 2.2, when $r_{min} = R$. In accordance with the laws of conservation of kinetic momentum and mechanical energy, it results that:

$$(v - \Delta v)r = v_{max}R;$$

$$-K\frac{M}{r} + \frac{(v - \Delta v)^2}{2} = -K\frac{M}{R} + \frac{v_{max}^2}{2};$$

$$-K\frac{M}{R^2}\frac{R^2}{r} + \frac{(v - \Delta v)^2}{2} = -K\frac{M}{R^2}R + \frac{v_{max}^2}{2};$$

$$-g_0\frac{R^2}{r} + \frac{(v - \Delta v)^2}{2} = -g_0R + \frac{v_{max}^2}{2};$$

$$v_{max} = (v - \Delta v)\frac{r}{R};$$

$$g_0R\left(1 - \frac{R}{r}\right) = \frac{1}{2}(v - \Delta v)^2\left(\frac{r^2}{R^2} - 1\right);$$

$$2g_0R\left(\frac{r - R}{r}\right) = (v - \Delta v)^2\left(\frac{r^2 - R^2}{R^2}\right);$$

$$(v - \Delta v)^2 = \frac{2g_0R^3}{r(r + R)}; \quad \Delta v = v - R\sqrt{\frac{2g_0R}{r(r + R)}}.$$

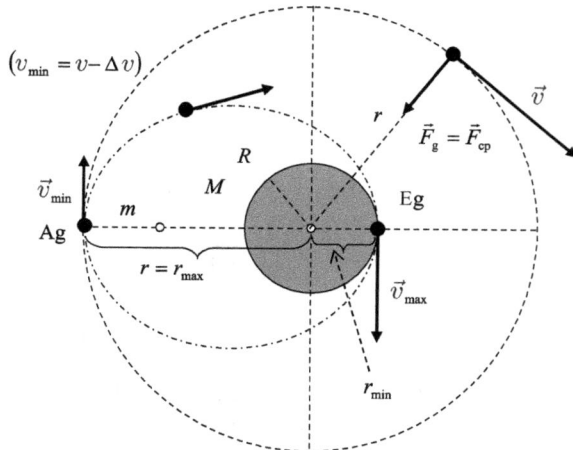

Fig. 2.2

Appendix: Geostationary Satellites

To achieve such an evolution, the satellite is first raised to the altitude of the injection point, where it is given the horizontal speed $\vec{v}$ necessary for maintaining the satellite on a circular orbit with the radius r around the Earth.

In these conditions, from Figure 2.3, it results that:

$$F = K\frac{mM}{r^2} = \frac{mv^2}{r}; \quad v = \sqrt{k\frac{M}{r}};$$

$$E = \frac{mv^2}{2} - K\frac{mM}{r} = -K\frac{mM}{2r} < 0;$$

$$g = K\frac{M}{r^2}; \quad v = \sqrt{gr}; \quad r = R + h;$$

$$g = K\frac{M}{(R+h)^2}\frac{R^2}{R^2}; \quad g_0 = K\frac{M}{R^2};$$

$$g = g_0\frac{R^2}{(R+h)^2}; \quad v = R\sqrt{\frac{g_0}{R+h}};$$

$$T = \frac{2\pi r}{v},$$

where T is the time required for one complete rotation on the circular orbit (the sidereal revolutionary period of the satellite);

$$T = \frac{2\pi(R+h)}{R}\sqrt{\frac{R+h}{g_0}},$$

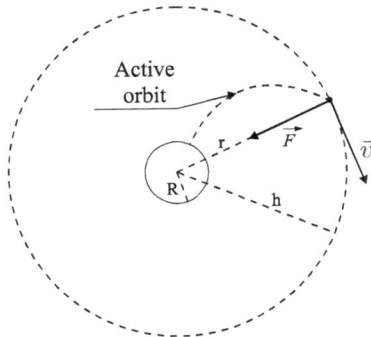

Fig. 2.3

where g_0 is the gravitational acceleration at the ground;

$$T = \frac{2\pi}{\omega}; \quad \omega = \frac{R}{R+h}\sqrt{\frac{g_0}{R+h}}.$$

We can define the *period of synodic revolution* of the satellite, T_s, as the time between two consecutive passes of the satellite across the zenith of an observer located on the equatorial line:

$$T_s = \frac{2\pi}{\omega \pm \omega_0},$$

where ω_0 is the angular speed of the Earth's rotation;

$$T_s = \frac{2\pi}{\frac{R}{R+h}\sqrt{\frac{g_0}{R+h}} \pm \omega_0},$$

where the sign "$+$" corresponds to the case when the direction of the satellite's rotation is opposite to the Earth's rotation, and the sign "$-$" corresponds to the case when the satellite's and Earth's directions of rotation are identical.

A particularly interesting case is when the satellite revolves in a circle around the Earth in the same direction as the rotation of the Earth, at a certain height, so that

$$\frac{R}{R+h}\sqrt{\frac{g_0}{R+h}} = \omega_0.$$

Therefore, the satellite's angular rotational speed should be equal to the Earth's angular rotational speed.

Such a satellite is called a *geostationary satellite*, and it appears to be fixed above a point on the surface of the Earth.

The period of synodic revolution of a geostationary satellite is $T_s \to \infty$.

To be geostationary, the satellite must evolve at the altitude

$$h = \sqrt[3]{\frac{g_0 R^2}{\omega_0^2}} - R \approx 35800\,\text{km},$$

so that its tangential speed must be

$$v = R\sqrt{\frac{g_0}{R+h}} \approx 3.08 \, \text{km/s}.$$

Problem 3. Cylindrical Planet

One of the planets of a nameless star is a very long, homogeneous cylinder. The average density of the planet and its radius are the same as those of the Earth, so its rotation period around its axis is the same as the Earth's period of rotation.

(a) *Determine* the value of the first cosmic velocity of the satellite that revolves around this planet, $v_{\text{I,planet}}$, given that the first cosmic velocity of a satellite that revolves in a circular orbit around the Earth, very close to its surface, is $v_{\text{I,Earth}} = 7.9 \, \text{km/s}$.

(b) *Determine* the value of the altitude at which a stationary satellite for telecommunications must revolve around this planet, knowing that the radius of the circular orbit of a geostationary satellite for telecommunications is $r_{\text{E}}^{*} = 42.170 \, \text{km}$, and the radius of the Earth is $R_{\text{E}} = 6.370 \, \text{km}$.

(c) It is shown that the speed required for a satellite to escape from the gravitational field of the cylinder-shaped planet (second cosmic speed) is

$$v_{\text{II,planet}} = \sqrt{2} \ln \frac{H}{R} v_{\text{I,planet}},$$

where H is the length of the planet and R is the radius of the planet.

The satellite is launched from a point on the planet's surface, located midway along one of its meridional generators, in the plane of the cross-section of a cylinder tangent to the planet.

  Calculate the length, H, of the planet, knowing that the gravitational potential energy of the satellite–cylindrical planet system is equal to the gravitational potential energy of the satellite–Earth system when the same satellite is launched, for the same purpose, from the Earth's surface. It is known that the gravitational acceleration at the level of the Earth's surface is $g_{0\text{E}} = 9.8 \, \text{ms}^{-2}$.

  Given: $\ln(2.28) \approx 0.82$, or $e^{0.82} = (2.71)^{0.82} = \sqrt[100]{e^{82}} = \sqrt[100]{(2.71)^{82}} \approx 2.28$.

Solution

(a) The gravitational field around a plate in the shape of a very long cylinder has cylindrical symmetry, so that the direction of the intensity of the gravitational field vector $\vec{\Gamma} = \vec{g}$, at those points located far from the ends of the cylinder, towards its interior, is perpendicular to the axis of the cylinder, as shown in Figure 3.1, and its modulus depends only on the distance to this axis.

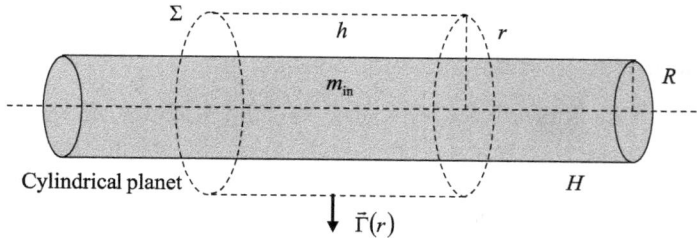

Fig. 3.1

Gauss's theorem for the electric field generated by an electrified body has the form

$$\Phi = \oint_{\Sigma} \vec{E} \cdot d\vec{S} = \frac{q_{\text{int}}}{\varepsilon_0},$$

so that the flow of the electric field through a closed surface surrounding an electrified body is equal to the ratio of the electric charge inside the surface and the dielectric permittivity of a vacuum.

Analogously, this theorem can also be written for the gravitational field of a cylinder-shaped planet with the radius R and the length H, so that the flow of the gravitational field through the closed surface Σ, represented in Figure 3.1, is

$$\Phi = \oint_{\Sigma} \vec{\Gamma} \cdot d\vec{S} = 4\pi K m_{\text{int}},$$

where $\vec{\Gamma}$ is the intensity of the gravitational field at the distance r from the central axis of the planet, K is the constant of universal attraction, and m_{int} is the part of the mass of the planet located inside the surface Σ.

Due to the symmetry, the intensity of the gravitational field is perpendicular to the cylinder Σ's generator. The flux of the gravitational field through the bases of the cylinder Σ is null.

In these conditions, it results that:

$$\Phi = 2\pi r h \Gamma = 4\pi K m_{\text{in}}; \quad m_{\text{in}} = \rho \pi R^2 h,$$

where ρ is the average density of the planet ($\rho = \rho_E$);

$$2\pi r h \Gamma = 4\pi K \rho \pi R^2 h;$$

$$\Gamma(r) = \frac{2\pi \rho K R^2}{r} = g(r).$$

As a result, the speed of a satellite evolving around this planet in a circular orbit with the radius r (the first cosmic speed) is obtained as follows:

$$mg = \frac{m v_{I,\text{planet}}^2}{r}; \quad v_{I,\text{planet}} = \sqrt{2\pi \rho K R^2},$$

an expression independent of the value of the radius of the orbit and therefore true also for the evolution of the satellite in a very low orbit, very close to the surface of the planet ($r = R$);

$$\rho = \rho_E = \frac{M_E}{V_E};$$

$$v_{I,\text{planet}} = \sqrt{2\pi \frac{M_E}{V_E} K R^2}; \quad V_E = \frac{4\pi R_E^3}{3},$$

where V_E is the volume of the Earth;

$$v_{I,\text{planet}} = \sqrt{\frac{3}{2} \frac{M_E}{R_E^3} K R^2}; \quad R = R_E; \quad v_{I,\text{planet}} = \sqrt{\frac{3}{2} K \frac{M_E}{R_E}}.$$

The first cosmic speed, in the case of the evolution of the satellite around the Earth, very close to its surface, is

$$v_{I,\text{Earth}} = \sqrt{K \frac{M_E}{R_E}}; \quad R = R_E.$$

It results that:

$$v_{I,\text{planet}} = \sqrt{\frac{3}{2} K \frac{M_E}{R_E}} = \sqrt{\frac{3}{2}} v_{I,\text{Earth}};$$

$$v_{I,\text{planet}} = 1.22 \cdot 7.9 \,\text{km/s} = 9.63 \,\text{km/s}.$$

(b) It is known that, to be geostationary, a satellite must evolve on a circular orbit whose radius is:

$$r_{\mathrm{E}}^* = \sqrt[3]{\frac{g_{0\mathrm{E}} R_{\mathrm{E}}^2}{\omega_0^2}},$$

where $g_{0\mathrm{E}}$ is the gravitational acceleration at the level of the Earth's surface and ω_0 is the angular velocity corresponding to the rotation of the Earth around its axis;

$$\omega_0 = \frac{2\pi}{T_0},$$

where T_0 is the period of the Earth's rotation around its axis $(T_0 = 24\,\mathrm{h})$;

$$g_{0\mathrm{E}} = K \frac{M_{\mathrm{E}}}{R_{\mathrm{E}}^2}; \quad r_{\mathrm{E}}^* = \sqrt[3]{\frac{g_{0\mathrm{E}} R_{\mathrm{E}}^2}{\omega_0^2}};$$

$$r_{\mathrm{E}}^* = \sqrt[3]{\frac{K M_{\mathrm{E}} T_0^2}{4\pi^2}},$$

for which

$$r_{\mathrm{E}}^{*3} = \frac{K M_{\mathrm{E}} T_0^2}{4\pi^2}.$$

On the other hand, for a satellite to evolve synchronously around the cylindrical planet, the radius of its orbit must be:

$$r_0 = \frac{T_0 v}{2\pi} = \frac{T_0}{2\pi} \sqrt{2\pi \rho K R^2} = R \sqrt{\frac{T_0^2 \rho K}{2\pi}};$$

$$\rho = \rho_{\mathrm{E}} = \frac{M_{\mathrm{E}}}{V_{\mathrm{E}}} = \frac{3 M_{\mathrm{E}}}{4\pi R_{\mathrm{E}}^3}; \quad R = R_{\mathrm{E}};$$

$$r_0 = \sqrt{\frac{3}{2} \frac{K M_{\mathrm{E}} T_0^2}{4\pi^2} \frac{1}{R_{\mathrm{E}}}};$$

$$r_0 = \sqrt{\frac{3}{2} \frac{r_{\mathrm{E}}^{*3}}{R_{\mathrm{E}}}};$$

$$r_0 = r_{\mathrm{E}}^* \sqrt{\frac{3}{2} \frac{r_{\mathrm{E}}^*}{R_{\mathrm{E}}}}; \quad r_0 = 1.33 \cdot 10^8 \,\mathrm{m};$$

$$h_0 = r_0 - R = r_0 - R_E;$$

$$h_0 = r_0 - R_E = r_E^* \sqrt{\frac{3}{2} \frac{r_E^*}{R_E}} - R_E = 1.27 \cdot 10^8 \text{ m}.$$

(c) To launch the satellite so that it escapes from the gravitational field of the planet, the satellite must evolve on a parabolic arc in whose focus is located the center of mass of the planet, so that the total energy of the system meets the following condition:

$$E_{\text{kin}} + E_{\text{pot}} = 0;$$

$$\frac{m v_{\text{II,planet}}^2}{2} + E_{\text{pot}} = 0;$$

$$v_{\text{II,planet}} = \sqrt{2} \ln \frac{H}{R} v_{\text{I,planet}}; \quad v_{\text{I,planet}} = 9.63 \text{ km/s}.$$

It results that:

$$E_{\text{pot}} = -m \left(\ln \frac{H}{R} \right)^2 v_{\text{I,planet}}^2;$$

$$R = R_E; \quad E_{\text{pot}} = -K \frac{m M_E}{R_E};$$

$$\left(\ln \frac{H}{R_E} \right)^2 v_{\text{I,planet}}^2 = K \frac{M_E}{R_E};$$

$$\ln \frac{H}{R_E} = \frac{\sqrt{K \frac{M_E}{R_E}}}{v_{\text{I,planet}}} = \frac{\sqrt{K \frac{M_E}{R_E^2} R_E}}{v_{\text{I,planet}}};$$

$$\ln \frac{H}{R_E} = \frac{\sqrt{g_{0E} R_E}}{v_{\text{I,planet}}}; \quad g_{0E} = 9.8 \text{ ms}^{-2}; \quad v_{\text{I,planet}} = 9.63 \text{ km/s};$$

$$\ln \frac{H}{R_E} = 0.82; \quad H = R_E e^{0.82}; \quad H = R_E \sqrt[100]{(2.71)^{82}};$$

$$e^{0.82} = \sqrt[100]{(2.71)^{82}} \approx 2.28; \quad \ln(2.28) \approx 0.82;$$

$$H = 2.28 \cdot 6370 \text{ km} = 14523.6 \text{ km}.$$

Problem 4. Automatic Space Station Hit by a Meteorite

A meteorite, which was approaching the Earth along a straight line passing through the center of the Earth, as shown in Figure 4.1, hits an automatic space station rotating around the Earth in a circular orbit with radius R. After the impact, the meteorite remains incorporated into the space station and is forced to evolve around the Earth in a new closed orbit, so that the minimum distance from the center of the Earth is $R/2$.

Fig. 4.1

(a) *Specify* the shape of the orbit of the space station after the impact with the meteorite and *determine*: **(1)** the speed of the meteorite before hitting the station; **(2)** the minimum and maximum speeds of the station on the new orbit after the impact with the meteorite; **(3)** the maximum distance of the station from the center of the Earth on the new orbit.

   Given: M, the mass of the Earth; K, the constant of universal attraction; m_1, the mass of the meteorite; and m_2, the mass of the space station.

   Establish the relationship between the two masses, m_1 and m_2, so that the proposed scenario is possible.

(b) *Determine* the minimum speed that the meteorite should have at the moment of impact with the station so that, after the impact, the space station evolves in an open orbit in relation to the Earth.

Specify the shape of the station's orbit after impact.

Determine the minimum distance between the station and the Earth after the impact, as well as the maximum speed of the station after the impact with the meteorite.

(c) *Determine:* **(1)** the angle at the center described by the position vector of the meteorite–station assembly, from the moment of impact until the moment when, evolving on the open orbit, the assembly passes at the minimum distance from the center of the Earth; **(2)** the duration of the evolution of the assembly on the specified sector of the open orbit.

Solution

(a) In accordance with the notation in Figure 4.2 ($\vec{v}_1$, velocity of the meteorite before the impact; $\vec{v}_2$, velocity of the station before

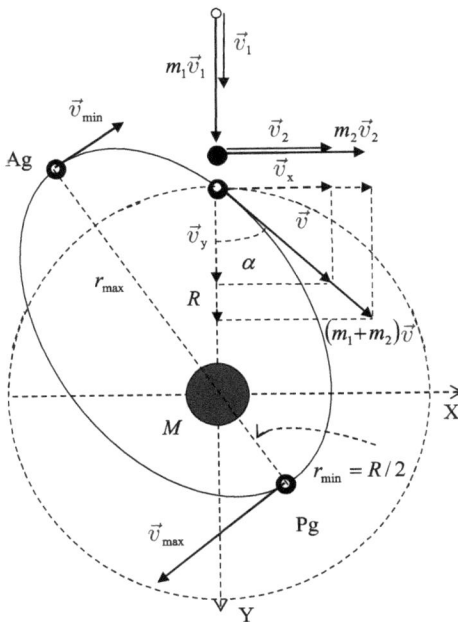

Fig. 4.2

collision; $v_2 = \sqrt{K\frac{M}{R}}$; $\vec{v}$, velocity of the assembly immediately after the collision), using the laws of conservation of momentum, energy, and kinetic momentum, it results that:

$$m_1\vec{v}_1 + m_2\vec{v}_2 = (m_1 + m_2)\vec{v};$$

$$m_1 v_1 = (m_1 + m_2)v_y = (m_1 + m_2)v\cos\alpha;$$

$$m_2 v_2 = (m_1 + m_2)v_x = (m_1 + m_2)v\sin\alpha;$$

$$\tan\alpha = \frac{m_2 v_2}{m_1 v_1};$$

$$v_x = \frac{m_2 v_2}{m_1 + m_2}; \quad v_2 = \sqrt{K\frac{M}{R}};$$

$$v_x = \frac{m_2}{m_1 + m_2}\sqrt{K\frac{M}{R}}; \quad v_y = \frac{m_1 v_1}{m_1 + m_2};$$

$$m_1^2 v_1^2 + m_2^2 v_2^2 = (m_1 + m_2)^2 v^2;$$

$$-K\frac{(m_1 + m_2)M}{R} + \frac{1}{2}(m_1 + m_2)v^2$$

$$= -K\frac{(m_1 + m_2)M}{\frac{R}{2}} + \frac{1}{2}(m_1 + m_2)v_{\max}^2 < 0;$$

$$-K\frac{M}{R} + \frac{1}{2}v^2 = -K\frac{M}{\frac{R}{2}} + \frac{1}{2}v_{\max}^2; \quad v^2 = v_{\max}^2 - 2K\frac{M}{R};$$

$$-K\frac{(m_1 + m_2)M}{R} + \frac{1}{2}(m_1 + m_2)(v_x^2 + v_y^2)$$

$$= -K\frac{(m_1 + m_2)M}{\frac{R}{2}} + \frac{1}{2}(m_1 + m_2)v_{\max}^2;$$

$$-K\frac{M}{R} + \frac{1}{2}(v_x^2 + v_y^2) = -K\frac{M}{\frac{R}{2}} + \frac{1}{2}v_{\max}^2;$$

$$(m_1 + m_2)vR\sin\alpha = (m_1 + m_2)v_{\max}r_{\min};$$

$$(m_1 + m_2)v_x R = (m_1 + m_2)v_{\max}\frac{R}{2}; \quad v_x = \frac{1}{2}v_{\max}; \quad v_{\max} = 2v_x;$$

$$-K\frac{M}{R} + \frac{1}{2}(v_x^2 + v_y^2) = -K\frac{2M}{R} + \frac{1}{2}v_{\max}^2;$$

$$-K\frac{M}{R} + \frac{1}{2}\left(v_x^2 + \frac{m_1^2 v_1^2}{(m_1 + m_2)^2}\right) = -K\frac{2M}{R} + \frac{1}{2}4v_x^2;$$

$$v_x = \frac{m_2}{m_1 + m_2}\sqrt{K\frac{M}{R}};$$

$$\frac{1}{2}\frac{m_1^2 v_1^2}{(m_1 + m_2)^2} = \frac{3}{2}\frac{m_2^2}{(m_1 + m_2)^2}K\frac{M}{R} - K\frac{M}{R};$$

$$\frac{1}{2}\frac{m_1^2 v_1^2}{(m_1 + m_2)^2} = K\frac{M}{R}\left(\frac{3m_2^2}{2(m_1 + m_2)^2} - 1\right);$$

$$v_1 = \frac{m_1 + m_2}{m_1}\sqrt{2K\frac{M}{R}\left(\frac{3m_2^2}{2(m_1 + m_2)^2} - 1\right)};$$

$$\frac{3m_2^2}{2(m_1 + m_2)^2} > 1; \quad m_1 < \left(\sqrt{\frac{3}{2}} - 1\right)m_2; \quad m_1 < 0.22m_2;$$

$$r_{\min}v_{\max} = r_{\max}v_{\min};$$

$$v_x = \frac{m_2}{m_1 + m_2}\sqrt{K\frac{M}{R}}; \quad v_{\max} = 2v_x;$$

$$v_{\max} = \frac{2m_2}{m_1 + m_2}\sqrt{K\frac{M}{R}}; \quad r_{\min} = \frac{R}{2};$$

$$-K\frac{(m_1 + m_2)M}{r_{\min}} + \frac{1}{2}(m_1 + m_2)v_{\max}^2$$
$$= -K\frac{(m_1 + m_2)M}{r_{\max}} + \frac{1}{2}(m_1 + m_2)v_{\min}^2;$$

$$-K\frac{M}{r_{\min}} + \frac{1}{2}v_{\max}^2 = -K\frac{M}{r_{\max}} + \frac{1}{2}v_{\min}^2; \quad r_{\max} = \frac{r_{\min}v_{\max}}{v_{\min}};$$

$$-K\frac{M}{r_{\min}} + \frac{1}{2}v_{\max}^2 = -K\frac{Mv_{\min}}{r_{\min}v_{\max}} + \frac{1}{2}v_{\min}^2;$$

$$v_{\min}^2 - 2\frac{KM}{r_{\min}v_{\max}}v_{\min} + 2K\frac{M}{r_{\min}} - v_{\max}^2 = 0; \quad v_{\min} = \frac{r_{\min}v_{\max}}{r_{\max}};$$

$$\left(v_{\max}^2 - 2K\frac{M}{r_{\min}}\right)r_{\max}^2 + 2KMr_{\max} - r_{\min}^2 v_{\max}^2 = 0;$$

$$v_{\max} = \frac{2m_2}{m_1 + m_2}\sqrt{K\frac{M}{R}}; \quad r_{\min} = \frac{R}{2};$$

$$r_{\max} = R\frac{\sqrt{(m_1 + m_2)^4 + 4m_2^2(m_2^2 - (m_1 + m_2)^2)} - (m_1 + m_2)^2}{4(m_2^2 - (m_1 + m_2)^2)}.$$

(b) In accordance with the notation in Figure 4.3 ($\vec{v}_{1,\min}$, velocity of the meteorite before the impact; $\vec{v}_2$, station's velocity before the collision; $v_2 = \sqrt{K\frac{M}{R}}$; $\vec{v}$, velocity of the assembly immediately after the collision), using the laws of conservation of momentum, energy, and kinetic momentum, it results that:

$$m_1\vec{v}_{1,\min} + m_2\vec{v}_2 = (m_1 + m_2)\vec{v};$$

$$m_1 v_{1,\min} = (m_1 + m_2)v_y = (m_1 + m_2)v\cos\alpha;$$

$$m_2 v_2 = (m_1 + m_2)v_x = (m_1 + m_2)v\sin\alpha;$$

$$\tan\alpha = \frac{m_2 v_2}{m_1 v_{1,\min}}; \quad v_x = \frac{m_2 v_2}{m_1 + m_2}; \quad v_2 = \sqrt{K\frac{M}{R}};$$

$$v_x = \frac{m_2}{m_1 + m_2}\sqrt{K\frac{M}{R}}; \quad v_y = \frac{m_1 v_{1,\min}}{m_1 + m_2};$$

$$m_1^2 v_{1,\min}^2 + m_2^2 v_2^2 = (m_1 + m_2)^2 v^2;$$

$$-K\frac{(m_1 + m_2)M}{R} + \frac{1}{2}(m_1 + m_2)v^2$$

$$= -K\frac{(m_1 + m_2)M}{r_{\min}} + \frac{1}{2}(m_1 + m_2)v_{\max}^2 = 0;$$

$$-K\frac{M}{R} + \frac{1}{2}v^2 = 0; \quad -K\frac{M}{r_{\min}} + \frac{1}{2}v_{\max}^2 = 0;$$

$$v = \sqrt{2K\frac{M}{R}}; \quad v_{\max} = \sqrt{2K\frac{M}{r_{\min}}};$$

$$m_1^2 v_{1,\min}^2 + m_2^2 \frac{KM}{R} = (m_1 + m_2)^2 \frac{2KM}{R};$$

$$v_{1,\min} = \frac{1}{m_1} \sqrt{\frac{KM}{R}(2(m_1 + m_2)^2 - m_2^2)};$$

$$(m_1 + m_2)vR\sin\alpha = (m_1 + m_2)v_{\max}r_{\min};$$

$$(m_1 + m_2)v_x R = (m_1 + m_2)v_{\max}r_{\min};$$

$$vR\sin\alpha = v_{\max}r_{\min};$$

$$v_x R = v_{\max}r_{\min}; \quad v_x = \frac{m_2}{m_1 + m_2}\sqrt{K\frac{M}{R}};$$

$$-K\frac{M}{r_{\min}} + \frac{1}{2}v_{\max}^2 = 0; \quad r_{\min} = \frac{m_2^2 R}{2(m_1 + m_2)^2};$$

$$v_{\max} = \frac{2(m_1 + m_2)}{m_2}\sqrt{\frac{KM}{R}}.$$

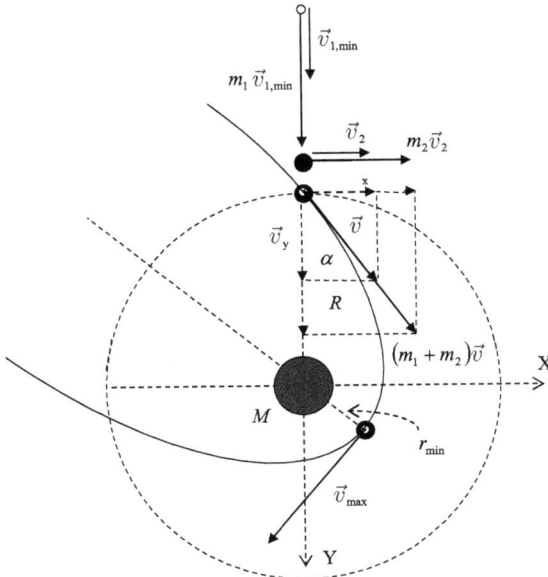

Fig. 4.3

(c)
(1) The satellite revolves on a parabolic trajectory, with the Earth in its focus, when it has to escape from the Earth's gravitational field.

Then, after reaching a very distant point, the speed of the satellite in relation to the Earth should be null.

The *parabola* describes the geometric locations of the points Q in a plane located at an equal distance from one fixed point P, called the focus, and a fixed straight line Δ, called the guiding line, as shown in Figure 4.4.

After the impact with the meteorite, occuring at point Q, as shown in Figure 4.4, the station–meteorite assembly with the speed $\vec{v}$, in order to escape from the terrestrial gravitational field, will begin its evolution on the parabola in whose focus is the Earth and whose equation in cartesian coordinates $(x; y)$ is $y^2 = 2px$, for which the parameter of the parabola is known, $p = 2r_{\min}$.

The *optical properties of the parabola* are as follows: all light rays emitted from the focus of a concave parabolic mirror, after reflection, will become parallel to the main optical axis, and reciprocally, the incident rays parallel to the main optical axis are reflected through the focus. As a result, the tangent to the parabola at the point Q is the bisector of the angle PQN.

Considering the definition of the parabola, it follows that:

$$QP = QN;$$

$$R = PB + QP\cos(\pi - \theta);$$

$$R = 2r_{\min} + R(-\cos\theta);$$

$$2r_{\min} = R(1 + \cos\theta); \quad r_{\min} = R\cos^2\frac{\theta}{2};$$

$$\cos\frac{\theta}{2} = \frac{m_2}{\sqrt{2(m_1 + m_2)}}.$$

When the station–meteorite assembly reaches point A, representing the vertex of the parabola, where $r = r_{\min}$, its speed will be $v = v_{\max}$, so that we have:

$$E = \frac{mv_{\max}^2}{2} - K\frac{mM}{r_{\min}} = 0; \quad v_{\max} = \sqrt{\frac{2KM}{r_{\min}}};$$

$$r_{\min}v_{\max} = Rv\sin\left(\frac{\pi}{2} - \frac{\theta}{2}\right) = Rv\cos\frac{\theta}{2};$$

$$R\cos^2\frac{\theta}{2}v_{\max} = Rv\cos\frac{\theta}{2}; \quad v_{\max} = \frac{v}{\cos\frac{\theta}{2}};$$

$$\cos\frac{\theta}{2} = \frac{v}{v_{\max}} = \frac{m_2}{\sqrt{2}(m_1 + m_2)}.$$

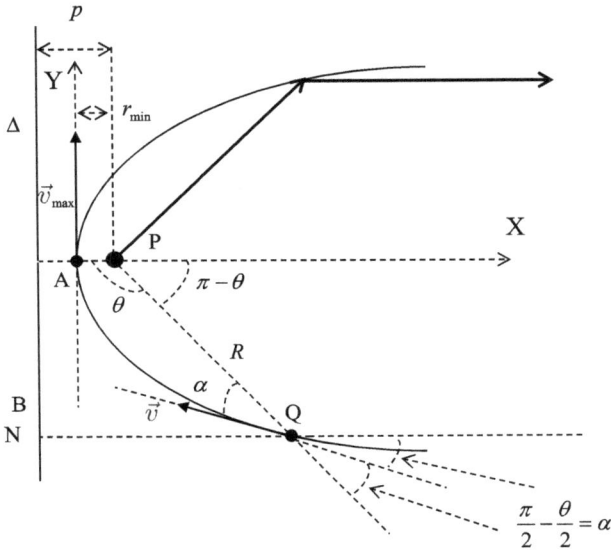

Fig. 4.4

With the same result, we can observe from Figure 4.5 that:

$$\alpha = \frac{\pi}{2} - \frac{\theta}{2}; \quad \frac{\theta}{2} = \frac{\pi}{2} - \alpha;$$

$$\cos\frac{\theta}{2} = \cos\left(\frac{\pi}{2} - \alpha\right) = \sin\alpha;$$

$$vR\sin\alpha = v_{\max}r_{\min}; \quad \sin\alpha = \frac{v_{\max}r_{\min}}{vR};$$

$$v_{\max} = \frac{2(m_1 + m_2)}{m_2}\sqrt{\frac{KM}{R}};$$

$$r_{\min} = \frac{m_2^2 R}{2(m_1 + m_2)^2}; \quad v = \sqrt{2K\frac{M}{R}};$$

$$\cos\frac{\theta}{2} = \frac{m_2}{\sqrt{2}(m_1 + m_2)}.$$

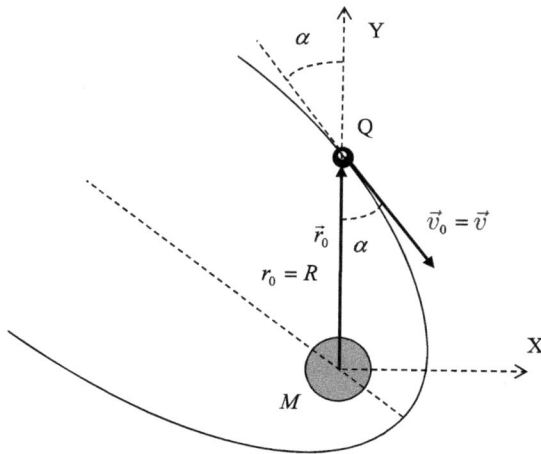

Fig. 4.5

(2) From the equation of the trajectory (of the conic) written in polar coordinates, for $e = 1$ (parabola), it results that:

$$r = \frac{p}{1 + e\cos\theta}; \quad r = \frac{p}{1 + \cos\theta};$$

$$r = \frac{p}{2\cos^2\frac{\theta}{2}} = \frac{p}{2}\left(1 + \tan^2\frac{\theta}{2}\right).$$

For the evolution on the parabola, taking place under the action of the central force of gravitational attraction, we have:

$$r^2\dot{\theta} = C = \sqrt{pKM}; \quad r^2\mathrm{d}\theta = C\mathrm{d}t;$$

$$\frac{p^2}{4}\left(1 + \tan^2\frac{\theta}{2}\right)^2\mathrm{d}\theta = C\mathrm{d}t;$$

$$\tan\frac{\theta}{2} = u; \quad \frac{1}{2}\frac{\mathrm{d}\theta}{\cos^2\frac{\theta}{2}} = \mathrm{d}u;$$

$$\mathrm{d}\theta = 2\cos^2\frac{\theta}{2}\mathrm{d}u = \frac{2}{1 + \tan^2\frac{\theta}{2}}\mathrm{d}u; \quad \mathrm{d}\theta = \frac{2}{1 + u^2}\mathrm{d}u;$$

$$\frac{p^2}{2C}(1 + u^2)\mathrm{d}u = \mathrm{d}t;$$

$$t - t_0 = \frac{p^2}{2c} \int_0^\pi (1 + u^2) du,$$

where t_0 is the time when the satellite passes through the point corresponding to r_{min}, for which $\theta = 0$ and $u = 0$, and t is the moment when the satellite's polar coordinates are r and θ;

$$t - t_0 = \frac{p^2}{2C} \left(u + \frac{u^3}{3} \right);$$

$$t_0 = 0; \quad t = \frac{p^2}{2C} \left(u + \frac{u^3}{3} \right),$$

where the parameter of the conic (parabola) is

$$p = \frac{r_0^2 v_0^2 \sin^2 \alpha}{KM},$$

with the values of the terms r_0, v_0 and α, represented in Figure 4.4, being associated with the conditions under which the injection (entry) onto the parabolic trajectory was achieved (the initiation of the parabolic movement);

$$p = \frac{R^2 v^2 \sin^2 \alpha}{KM} = \frac{R m_2^2}{(m_1 + m_2)^2},$$

an identical relationship to that obtained directly from the definition of the conic (parabola) parameter;

$$p = 2r_{min} = 2 \frac{m_2^2 R}{2(m_1 + m_2)^2} = \frac{m_2^2 R}{(m_1 + m_2)^2}.$$

Problem 5. Superficial Phenomena

A. Pulsating Liquid Film Bubble

At one end of a short, thin cylindrical tube, open at both ends, a spherical bubble is formed from a solution of soapy water, as shown in Figure 5.1, having mass m. The surface tension coefficient of the solution is σ. The bubble is uniformly electrified with electric charge q.

Given: the absolute electrical permittivity of air, ε_0, and the density of the soapy water solution, ρ.

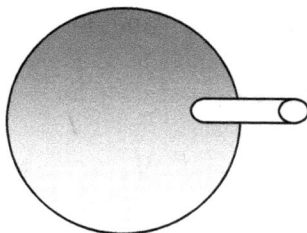

Fig. 5.1

Determine:

(a) the equilibrium radius of the bubble, R_0, if the tube remains open, and the period of the bubble's small pulsating oscillations, if during the pulsating oscillations, its spherical shape is preserved;

(b) the velocities of the n identical drops resulting from the "explosion" of the bubble, produced when it suddenly acquires the total load $Q = nq$.

B. The Narrow Space Between Coaxial Cylindrical Glasses

Into a cylindrical glass with the inner diameter D_i, containing water, another cylindrical glass with the outer diameter D_e, mass m, and walls of negligible thickness is introduced, mouth down, so that the two cylinders are perfectly coaxial.

(c) *Determine* the difference in water level in the system after releasing the inner glass if the space between the glasses is extremely narrow. The water wets the walls of the glasses perfectly.

Given: σ, ρ and g.

Particular case: $\sigma = 0$.

Solution

(a) First, we will calculate the pressure exerted by the electrostatic forces on the bubble. On the surface of the bubble, as shown in

Figure 5.2, we separate an elementary sector with the surface area ΔS, carrying the uniformly distributed electric charge Δq, and with the same density as the rest of the bubble's surface.

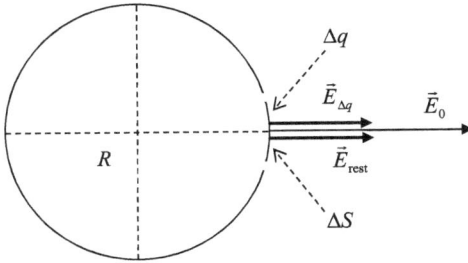

Fig. 5.2

The electrostatic force exerted by the electric charges over the rest of the bubble's surface on the charges of the elementary sector, represented in the drawing, whose orientation is justified by the symmetrical distribution of the electric charges, is:

$$F_e = E_{rest} \Delta q,$$

where E_{rest} is the intensity of the electric field generated by the rest of the electric charges on the surface of the balloon at the point where the elementary sector is located;

$$\Delta q = \frac{q \Delta S}{4 \pi R^2};$$

$$F_e = E_{rest} \frac{q \Delta S}{4 \pi R^2}.$$

If the elementary sector can be considered a plane sector with the surface area ΔS, carrying the electrical charge Δq, then the intensity of the electric field generated by this distribution is

$$E_{\Delta q} = \frac{1}{2} \frac{1}{\varepsilon_0} \frac{\Delta q}{\Delta S},$$

so that we have:

$$E_0 = E_{rest} + E_{\Delta q},$$

where E_0 is the intensity of the electric field generated by the electric charge distributed over the entire bubble;

$$E_{\text{rest}} = E_0 - E_{\Delta q} = \frac{q}{4\pi\varepsilon_0 R^2} - \frac{1}{2}\frac{1}{\varepsilon_0}\frac{\Delta q}{\Delta S} = \frac{1}{\varepsilon_0}\frac{q}{4\pi R^2} - \frac{1}{2}\frac{1}{\varepsilon_0}\frac{\Delta q}{\Delta S};$$

$$\frac{\Delta q}{\Delta S} = \frac{q}{4\pi R^2},$$

representing the density of the electric charge;

$$E_{\text{rest}} = \frac{1}{\varepsilon_0}\frac{q}{4\pi R^2} - \frac{1}{2}\frac{1}{\varepsilon_0}\frac{q}{4\pi R^2} = \frac{1}{2}\frac{q}{4\pi\varepsilon_0 R^2} = \frac{1}{2}E_0;$$

$$F_{\text{electrostatic}} = \frac{1}{2}\frac{q}{4\pi\varepsilon_0 R^2}\frac{q\Delta S}{4\pi R^2} = \frac{1}{32}\frac{q^2}{\varepsilon_0 R^4}\Delta S.$$

Thus, the electrostatic pressure exerted on the bubble, towards its exterior is

$$p_{\text{electrostatic}} = \frac{F_e}{\Delta S} = \frac{q^2}{32\pi^2\varepsilon_0 R^4}.$$

At the same time, the pressure from the superficial tensional forces towards the interior of the bubble is

$$p_{\text{superficial}} = \frac{4\sigma}{R}.$$

The two pressures will be balanced for a certain radius of the bubble, so that we obtain:

$$p_{\text{electrostatic,equilibrium}} = p_{\text{superficial,equilibrium}};$$

$$\frac{q^2}{32\pi^2\varepsilon_0 R_0^4} = \frac{4\sigma}{R_0};$$

$$R_0 = \sqrt[3]{\frac{q^2}{128\pi^2\varepsilon_0\sigma}}.$$

Figure 5.3 shows the bubble in its initial state, when the equilibrium radius between the pressure of the electrostatic forces and the pressure of the superficial tensional forces is R_0, as well as an elementary sector with a surface area ΔS, at first in the equilibrium position, when the resultant of the forces that act on it is

$$\vec{F}_{0,\text{superficial}} + \vec{F}_{0,\text{electrostatic}} = 0.$$

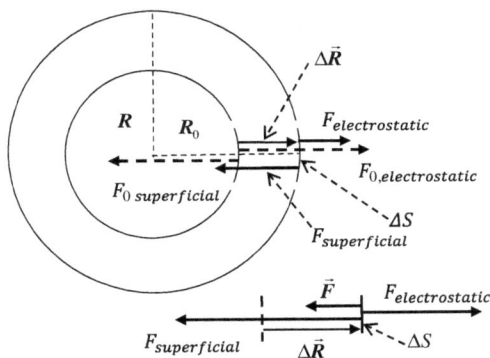

Fig. 5.3

In the same figure, the bubble is then represented at some moment during the oscillations when its radius is $R = R_0 + \Delta R$, and the forces acting on the same elementary sector, whose area has not changed significantly, are:

$$F_{\text{superficial}} = \frac{4\sigma}{R}\Delta S = \frac{4\sigma}{R_0 + \Delta R}\Delta S < F_{0,\text{superficial}};$$

$$F_{\text{electrostatic}} = \frac{q^2}{32\pi^2\varepsilon_0 R^4}\Delta S$$

$$= \frac{q^2}{32\pi^2\varepsilon_0(R_0 + \Delta R)^4}\Delta S < F_{0,\text{electrostatic}};$$

$$F_{\text{superficial}} > F_{\text{electrostatic}}.$$

Thus, if $\Delta R \ll R_0$, it results that:

$$F_{\text{superficial}} = \frac{4\sigma}{R_0 + \Delta R}\Delta S = \frac{4\sigma}{R_0}\left(1 + \frac{\Delta R}{R_0}\right)^{-1}\Delta S$$

$$\approx \frac{4\sigma}{R_0}\left(1 - \frac{\Delta R}{R_0}\right)\Delta S;$$

$$F_{\text{electrostatic}} = \frac{q^2}{32\pi^2\varepsilon_0(R_0 + \Delta R)^4}\Delta S = \frac{q^2}{32\pi^2\varepsilon_0 R_0^4}\left(1 + \frac{\Delta R}{R_0}\right)^{-4}\Delta S$$

$$\approx \frac{q^2}{32\pi^2\varepsilon_0 R_0^4}\left(1 - 4\frac{\Delta R}{R_0}\right)\Delta S;$$

$$\frac{q^2}{32\pi^2 \varepsilon_0 R_0^4} = \frac{4\sigma}{R_0};$$

$$F_{\text{electrostatic}} = \frac{4\sigma}{R_0}\left(1 - 4\frac{\Delta R}{R_0}\right)\Delta S;$$

$$F = F_{\text{superficial}} - F_{\text{electrostatic}}$$

$$= \frac{4\sigma}{R_0}\left(1 - \frac{\Delta R}{R_0}\right)\Delta S - \frac{4\sigma}{R_0}\left(1 - 4\frac{\Delta R}{R_0}\right)\Delta S;$$

$$F = \frac{12\sigma\Delta S}{R_0^2}\Delta R;$$

$$k = \frac{12\sigma\Delta S}{R_0^2};$$

$$F = k\Delta R;$$

$$\vec{F} = -k\Delta\vec{R}.$$

This proves that the oscillations of the elementary sector and, therefore, the oscillations of the entire bubble are harmonic oscillations.

If Δm is the mass of the considered elementary sector, it follows that:

$$k = \omega^2 \Delta m = \frac{4\pi^2}{T^2}\Delta m = \frac{12\sigma}{R_0^2}\Delta S;$$

$$\Delta m = \frac{m\Delta S}{4\pi R_0^2};$$

$$T = \sqrt{\frac{\pi m}{12\sigma}}.$$

(b) From the law of conservation of energy, it results that:

$$\frac{1}{2}\frac{n^2 q^2}{4\pi\varepsilon_0 R_0} + 2\sigma 4\pi R_0^2 = \frac{mv^2}{2} + n\frac{1}{2}\frac{\frac{q^2}{n^2}}{4\pi\varepsilon_0 r} + n\sigma 4\pi r^2;$$

$$\frac{1}{2}\frac{n^2 q^2}{4\pi\varepsilon_0 R_0} + 2\sigma 4\pi R_0^2 = \frac{mv^2}{2} + n\frac{1}{2}\frac{\frac{q^2}{n^2}}{4\pi\varepsilon_0 r} + n\sigma 4\pi r^2;$$

$$\frac{mv^2}{2} = \frac{1}{2}\frac{n^2q^2}{4\pi\varepsilon_0}\left(\frac{1}{R_0} - \frac{1}{n^3r}\right) - 4\pi\sigma(nr^2 - 2R_0^2);$$

$$m = nm_1 = n\rho\frac{4\pi r^3}{3};$$

$$r^3 = \frac{3m}{4\pi\rho n};$$

$$r = \sqrt[3]{\frac{3m}{4\pi\rho n}};$$

$$mv^2 = \frac{n^2q^2}{4\pi\varepsilon_0}\left(\frac{1}{R_0} - \frac{1}{n^3r}\right) - 8\pi\sigma(nr^2 - 2R_0^2);$$

$$v = \sqrt{\frac{n^2q^2}{4\pi\varepsilon_0 m}\left(\frac{1}{R_0} - \frac{1}{n^3r}\right) - \frac{8\pi\sigma}{m}(nr^2 - 2R_0^2)}.$$

(c) The space between the two glasses, which has capillary dimensions, makes the system uneven, as shown in Figure 5.4.

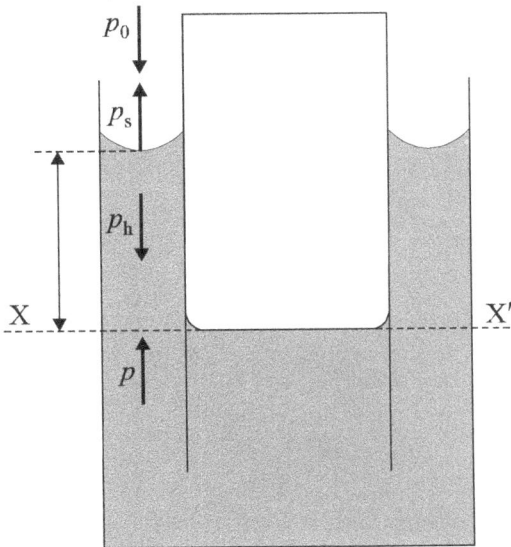

Fig. 5.4

At the level XX', from the condition of equilibrium of the pressures transmitted in the liquid, it results that

$$p_0 + p_h = p_s + p,$$

where p_0 is the atmospheric pressure, p_h is the hydrostatic pressure of the coloumn of liquid, p_s is the additional pressure of the curved superficial layer, and p is the pressure of the air inside the glass.

The free surface of the liquid between the two glasses is a concave surface identical to the inner surface of a cylindrical sector. An elementary sector of this surface can be considered as a cylindrical sector whose radius we calculate using the drawing in Figure 5.5:

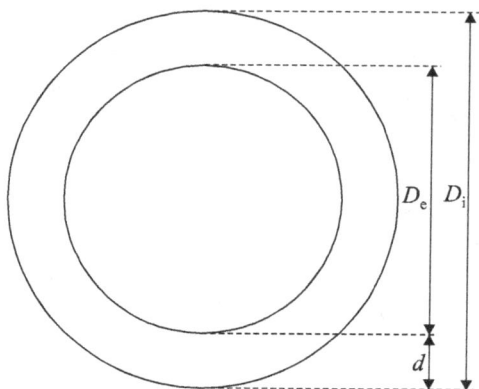

Fig. 5.5

$$2d + D_e = D_i; \quad d = 2r; \quad r = \frac{D_i - D_e}{4}.$$

As a result:

$$p_s = \frac{\sigma}{r} = \frac{4\sigma}{D_i - D_e};$$

$$p_0 + \rho g h = \frac{4\sigma}{D_i - D_e} + p.$$

Figure 5.6 represents the forces that act on the liquid, perpendicular to the contour of its free surface, as a result of the interaction with the walls of the glasses. Since the liquid wets the walls of the glass perfectly, these forces (the superficial tensional forces) are parallel to the walls, oriented vertically upwards. In accordance with the principle of reciprocal actions, a strong and opposite force will act from the liquid on the wall of the glass.

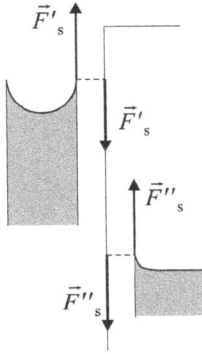

Fig. 5.6

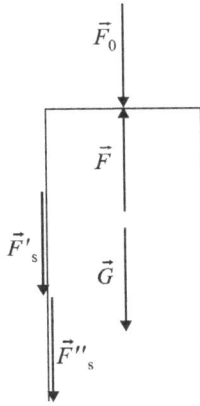

Fig. 5.7

The forces that act on the inner glass, ensuring its balance, are those represented in the drawing in Figure 5.7. It results that:

$$\vec{F}_0 + \vec{F} + \vec{G} + \vec{F}'_s + \vec{F}''_s = 0,$$

where $\vec{F}_0$ is the force determined by atmospheric pressure, $\vec{F}$ is the force determined by the air pressure inside the glass, and F'_s and $\vec{F}''_s$

are the reactions of superficial tensional forces,

$$F_0 + G + F'_s + F''_s = F;$$

$$p_0 S + mg + \sigma \pi D_e + \sigma \pi D_e = pS;$$

$$S = \frac{\pi D_e^2}{4};$$

$$p = p_0 + \frac{mg + 2\pi\sigma D_e}{\pi D_e^2/4}.$$

$$p = p_0 + \frac{4(mg + 2\pi\sigma D_e)}{\pi D_e^2}.$$

It results that:

$$h = \frac{4}{\rho g}\left(\frac{\sigma}{D_i - D_e} + \frac{mg + 2\pi\sigma D_e}{\pi D_e^2}\right);$$

$$\sigma = 0; \quad h = \frac{4m}{\pi\rho D_e^2}.$$

Problem 6. Magnetic Induction

A. Induction in a Ring

A conducting ring, with electrical resistance R, is placed in a magnetic field whose flux is variable over time according to the law $\Phi = \Phi(t) = \Phi_0 \cos \omega t$, where Φ_0 and ω are known quantities. The variable magnetic field is concentrated in a narrow area, as shown in Figure 6.1. The points M, N and P are divided into three identical sectors.

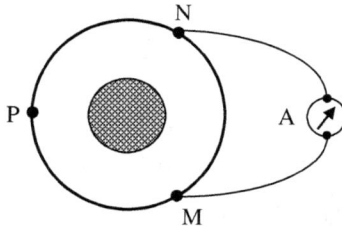

Fig. 6.1

(a) *Determine* the indication of the alternating current ammeter connected at points M and N, if the electrical resistance is r. The electrical resistance of the connecting conductors is negligible.

B. Induction in Two Rings

Two conducting rings, placed on a horizontal plane support, as shown in Figure 6.2, with mass m and radius r, are located inside a uniform magnetic field, whose magnetic induction vector, $\vec{B}_0$, is perpendicular to the horizontal plane of the rings. The points A and C at which the rings are touching each other, such that $\angle AOC = \alpha = \frac{\pi}{3}$, are points of very good electric contact.

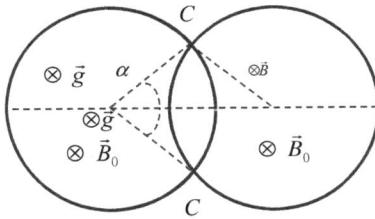

Fig. 6.2

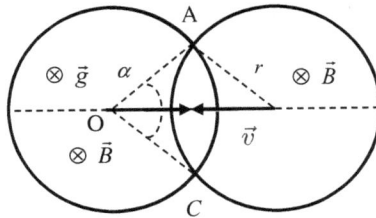

Fig. 6.3

(b) *Determine* the speed acquired by the ring after the rapid disconnection of the magnetic field. The electrical resistance of the piece of conductor from which each ring is made is R. Neglect the displacement of the rings during the disconnection of the magnetic field, the inductances of the rings, and any friction.

(c) Two rings are attached, placed on the same horizontal support. They move through translation towards each other over equal distances, in the direction of the line of their centers, as indicated by Figure 6.3, while the magnetic induction vector, $\vec{B}$, perpendicular to the plane of the rings, remains constant.

   Determine the orientations and values of the electromagnetic forces acting on each ring when their velocities are equal in modulus, (v), and of opposite directions, and $\alpha = \pi/3$. The inductances of the rings are neglected.

Solution

(a) The total electromotive voltage induced in the closed circuit MNPM of the conductive ring is

$$e = -\frac{d\Phi}{dt} = \Phi_0 \omega \sin \omega t.$$

Its distribution on the three identical sectors of the ring (MN, NP, PM) is, respectively, $e_{MN} = \frac{e}{3}$, $e_{NP} = \frac{e}{3}$, and $e_{PM} = \frac{e}{3}$, whose polarities are those represented in Figure 6.4.

   The total electromotive voltage induced in the closed circuit MNAM, which includes the MN sector of the ammeter and the two connecting conductors of the ammeter, must be zero because this contour is far from the region of the variable magnetic field. As the electromotive voltages induced in the identical sectors of the two connecting conductors are identical, $\frac{e_0}{2}$, with the polarities indicated in Figure 6.4, so that $e_{MNAM} = 0$, then $e_0 = \frac{e}{3}$.

   Accepting the instantaneous distribution of the currents through the sides of the network as represented in Figure 6.4, in accordance with Kirchhoff's laws, it results that:

$$3\frac{e}{3} = e = i_1 \frac{R}{3} + i_2 \left(\frac{R}{3} + \frac{R}{3} \right);$$

$$e_0 - \frac{e}{3} = \frac{e}{3} - \frac{e}{3} = 0 = i_3 r - i_1 \frac{R}{3};$$

$$i_2 = i_1 + i_3;$$

$$i_3 = \frac{3e}{9r + 2R} = \frac{3\Phi_0 \omega}{9r + 2R} \sin \omega t,$$

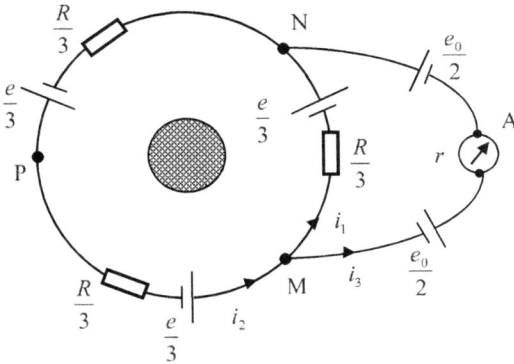

Fig. 6.4

representing the expression of the current intensity's instantaneous value through the side of the ammeter.

The indication of the ammeter, being the effective value of the intensity of the alternating current, results in:

$$i_3 = I_{3,\max} \sin \omega t; \quad I_{3,\max} = \frac{3\Phi_0 \omega}{9r + 2R};$$

$$I_3 = \frac{I_{3,\max}}{\sqrt{2}} = \frac{3\Phi_0 \omega}{\sqrt{2}(9r + 2R)}.$$

(b) When the external magnetic field is disconnected, the absolute value of the magnetic inductance varies over time from the initial value to the zero value. The variable magnetic field causes the appearance of an eddy electric field (with closed field lines), setting in orderly motion electric charge carriers in a free state, thus determining the appearance of inductive electric currents in the two conductive rings.

Let's analyze the contour AbCdA, which overlaps exactly the outline of the ring on the left, as shown in Figure 6.5. The electromotive induction voltage on this outline of the ring is

$$e = - \left(\frac{\Delta \Phi}{\Delta t} \right)_{\text{AbCdA}} = -S_{\text{AbCdA}} \frac{\Delta B}{\Delta t} = -\pi r^2 \frac{\Delta B}{\Delta t}.$$

According to Lentz's rule, the directions of the induction currents on the outline of this ring are aligned with the direction of rotation of the needles of a watch.

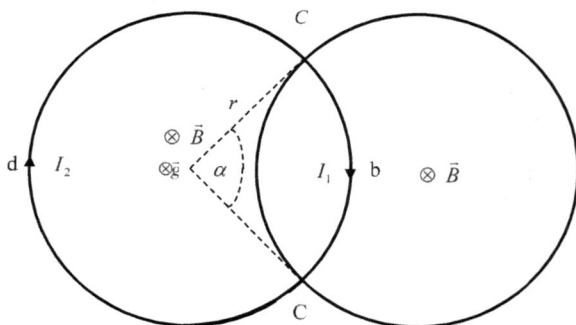

Fig. 6.5

For a certain moment $t \in \Delta t$, where Δt is the duration of the cancellation of the magnetic field, let $I_1(t)$ be the intensity of the current induced on the sector AbC of the specified outline and $I_2(t)$ the current intensity induced on the sector CdA of the same outline. In accordance with Kirchhoff's second theorem, for the outline AbCdA:

$$e = \frac{l_{\text{AbC}}}{2\pi r} RI_1(t) + \frac{l_{\text{CdA}}}{2\pi r} RI_2(t);$$

$$l_{\text{AbC}} = \alpha r = \frac{\pi}{3} r; \quad l_{\text{CdA}} = \frac{5\pi}{3} r;$$

$$I_1(t) + 5I_2(t) = -\frac{6\pi r^2}{R} \frac{\Delta B}{\Delta t}.$$

Similarly, for the outline AbCfA, represented in Figures 6.6 and 6.7, it results that:

$$e' = -\left(\frac{\Delta \Phi}{\Delta t}\right)_{\text{AbCfA}} = -S_{\text{AbCfA}} \frac{\Delta B}{\Delta t};$$

$$S_{\text{AbCfA}} = 2S_{\text{AbCA}} = 2(S_{\text{OAbCO}} - S_{\text{OACO}})$$

$$= 2\left(\frac{\alpha}{2} r^2 - r \sin \frac{\alpha}{2} r \cos \frac{\alpha}{2}\right);$$

$$S_{\text{AbCfA}} = 2r^2 \left(\frac{\alpha}{2} - \sin \frac{\alpha}{2} \cos \frac{\alpha}{2}\right);$$

$$e' = \frac{l_{\text{AbC}}}{2\pi r} RI_1(t) + \frac{l_{\text{CfA}}}{2\pi r} RI_1(t) = 2\frac{l_{\text{AbC}}}{2\pi r} RI_1(t) = \frac{l_{\text{AbC}}}{\pi r} RI_1(t)$$

$$= \frac{\alpha}{\pi} RI_1(t);$$

$$-2r^2 \left(\frac{\alpha}{2} - \sin \frac{\alpha}{2} \cos \frac{\alpha}{2} \right) \frac{\Delta B}{\Delta t} = \frac{\alpha}{\pi} R I_1(t); \quad \frac{\Delta B}{\Delta t} < 0;$$

$$I_1(t) = -\frac{6r^2}{R} \left(\frac{\alpha}{2} - \sin \frac{\alpha}{2} \cos \frac{\alpha}{2} \right) \frac{\Delta B}{\Delta t} > 0;$$

$$I_2(t) = -\frac{6r^2}{5R} \left(\pi - \frac{\alpha}{2} + \sin \frac{\alpha}{2} \cos \frac{\alpha}{2} \right) \frac{\Delta B}{\Delta t} > 0;$$

$$I_2(t) > I_1(t).$$

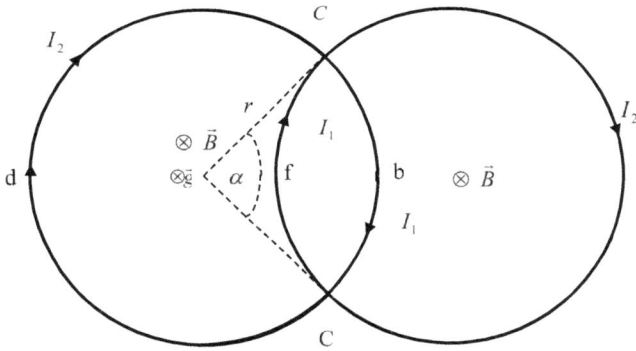

Fig. 6.6

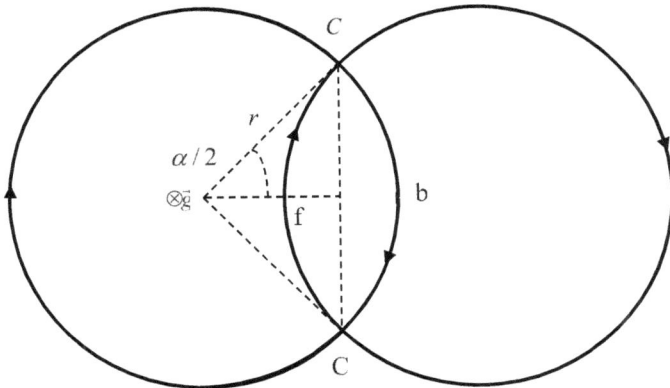

Fig. 6.7

For the very small duration (Δt) of the cancellation of the magnetic field, in accordance with Ampere's law, at every moment (t) on any elementary sector with a very small thickness (Δl) of each ring traversed by an electric current with the intensity $I(t)$, an electromagnetic force

$$\vec{F} = I(t)\Delta \vec{l} x \vec{B}(t)$$

will act, oriented in the direction of the radius of the ring to the exterior of the ring, whose value can be considered constant:

$$F = I(t)\Delta l B(t).$$

Considering the symmetry with respect to the center of the ring (I), any elemental sector on the arc EA of the ring (I), as shown in Figure 6.8, corresponds to an identical elementary sector on the arc CD of the same ring, through which the intensity of the current is the same, $I_2(t)$, so that the forces acting on the two elementary sectors in the same magnetic field are equal in magnitude and of opposite directions, their resultant being null. The resultant of the forces acting on the current from arcs EA and CD is null.

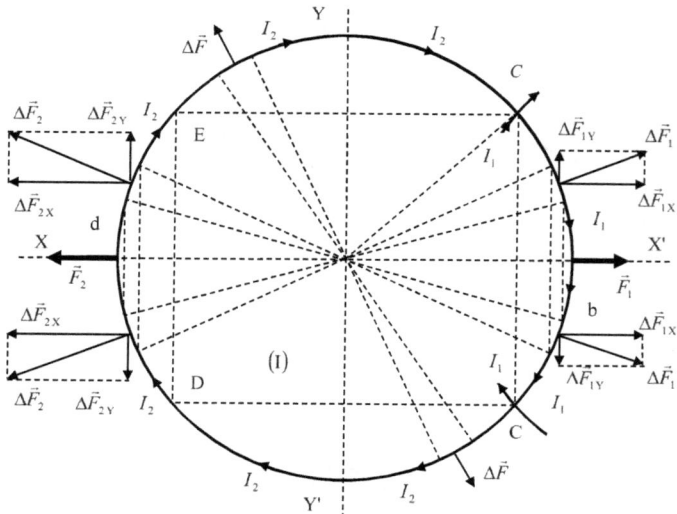

Fig. 6.8

Let us now analyze the forces that act upon the currents in the identical and symmetrical arcs AC and DE, which are traversed by currents of different intensities $(I_1(t) \neq I_2(t))$.

For reasons of symmetry, each of the arcs AC and DE are placed relative to the XX$'$ axis. Additionally, considering the symmetrical distribution of the intensity values $I_1(t)$ and $I_2(t)$ through each of the two arcs, relative to the XX$'$ axis, the resultant of the forces acting on each of the arcs AC and DE in the direction of the YY$'$ is null. Indeed, for two identical elementary sectors on the AC arc, located symmetrically with respect to the XX$'$ axis and traversed by a current with the intensity $I_1(t)$, the forces acting on them are identical, $\Delta \vec{F}_1$, and their components, $\Delta \vec{F}_{1Y}$, parallel to the YY$'$ axis, are equal in modulus but of opposite directions.

Similarly, for two identical elementary sectors on the DE arc, located symmetrically with respect to the XX$'$ axis and traversed by a current with the intensity $I_2(t)$, the forces acting on them are identical, $\Delta \vec{F}_2$, and their components, $\Delta \vec{F}_{2Y}$, parallel to the axis YY$'$, are equal in modulus but of opposite directions.

As a result, considering the forces $\vec{F}_1$ and $\vec{F}_2$, which act on the arcs AC and DE, respectively, traversed by electric currents with intensities $I_1(t)$ and $I_2(t)$, only the components $\Delta \vec{F}_{1X}$ and $\Delta \vec{F}_{2X}$ will be found:

$$\vec{F}_1 = \Sigma \Delta \vec{F}_{1X,k};$$

$$F_1 = \Sigma \Delta F_{1X,k} = \Sigma \Delta F_1 \cos \beta_k = \Sigma I_1(t) \Delta l_k B(t) \cos \beta_k$$

$$= I_1(t) B(t) \Sigma \Delta l_k \cos \beta_k;$$

$$F_1 = I_1(t) B(t) \Sigma \Delta y_k = I_1(t) B(t) (AC) = I_1(t) B(t) 2r \sin \frac{\alpha}{2};$$

$$\vec{F}_2 = \Sigma \Delta \vec{F}_{2X,k};$$

$$F_2 = \Sigma \Delta F_{2X,k} = \Sigma \Delta F_2 \cos \gamma_k = \Sigma I_2(t) \Delta l_k B(t) \cos \gamma_k$$

$$= I_2(t) B(t) \Sigma \Delta l_k \cos \gamma_k;$$

$$F_2 = I_2(t) B(t) \Sigma \Delta y_k = I_2(t) B(t) (DE) = I_2(t) B(t) 2r \sin \frac{\alpha}{2};$$

$$I_2(t) > I_1(t);$$

$$F_2 > F_1.$$

Thus, the resultant of the forces acting on ring (I) is:

$$F = F_2 - F_1 = 2rB(t)(I_2(t) - I_1(t));$$

$$F = -\frac{12r^3}{R}\sin\frac{\alpha}{2}\left[\frac{1}{5}\left(\pi - \frac{\alpha}{2} + \sin\frac{\alpha}{2}\cos\frac{\alpha}{2}\right) - \left(\frac{\alpha}{2} - \sin\frac{\alpha}{2}\cos\frac{\alpha}{2}\right)\right]$$

$$\times B(t)\frac{\Delta B(t)}{\Delta t};$$

$$F = -\frac{12r^3}{5R}\sin\frac{\alpha}{2}\left(\pi - 3\alpha + 6\sin\frac{\alpha}{2}\cos\frac{\alpha}{2}\right)B(t)\frac{\Delta B(t)}{\Delta t};$$

$$\frac{\Delta(B^2(t))}{\Delta t} = 2B(t)\frac{\Delta B(t)}{\Delta t};$$

$$F = -\frac{6r^3}{5R}\left(\pi - 3\alpha + 6\sin\frac{\alpha}{2}\cos\frac{\alpha}{2}\right)\sin\frac{\alpha}{2}\frac{\Delta(B^2(t))}{\Delta t}.$$

This force, acting on the ring for a very short time, causes a variation in the ring's momentum, so that:

$$m\Delta v = F\Delta t = -\frac{6r^3}{5R}\left(\pi - 3\alpha + 6\sin\frac{\alpha}{2}\cos\frac{\alpha}{2}\right)\sin\frac{\alpha}{2}\Delta(B^2(t));$$

$$\Delta v = v - 0 = v; \quad \Delta(B^2(t)) = 0 - B_0^2;$$

$$v = \frac{6r^3 B_0^2}{5mR}\left(\pi - 3\alpha + 6\sin\frac{\alpha}{2}\cos\frac{\alpha}{2}\right)\sin\frac{\alpha}{2};$$

$$\alpha = \frac{\pi}{3}; \quad v = \frac{9\sqrt{3}r^3 B_0^2}{10mR}.$$

(c) When a conductive ring moves through plane transitions in a uniform magnetic field, with magnetic induction $\vec{B}$, a Lorentz force acts on each free electron in its structure, $\vec{f}_L = q\vec{v}\times\vec{B}$, whose orientation is shown in Figure 6.9.

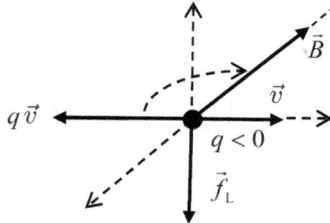

Fig. 6.9

Due to these forces, each elementary sector on the ring, having length Δl, can be considered a linear conductor, in which an electro-motive induction voltage appears.

Figure 6.10 shows two such elementary sectors located on the arc CmM of ring (I), symmetrical to the YY' axis, and two elementary sectors located on the arc NnA of the same ring, symmetrical to the same YY' axis. The polarities of the electromotive voltages induced in these elementary sectors, shown in the figure and justified by the tangential components of the Lorentz forces, prove that the total electromotive voltages induced in the arcs CmM and NnA are null.

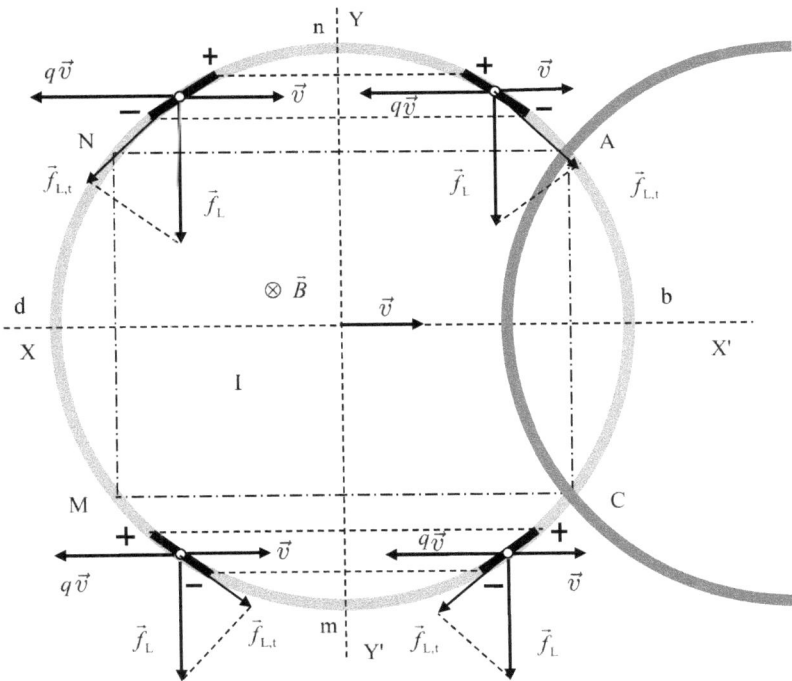

Fig. 6.10

Figure 6.11 shows two elementary sectors on the AbC arc and two elementary sectors on the MdN arc, symmetrical to the XX' axis. The polarities of the electromotive voltages induced in these elementary sectors, shown in the figure and justified by the tangential components of the Lorentz forces, prove that the total electromotive voltages induced in the arcs AbC and MdN are no longer zero; they add up arithmetically. The arcs AbC and MdN are symmetrical to the YY' axis, and each of them is symmetrical to the XX' axis, identical, and moves identically. As a result, the electromotive voltages induced in these arcs of ring (I) are identical.

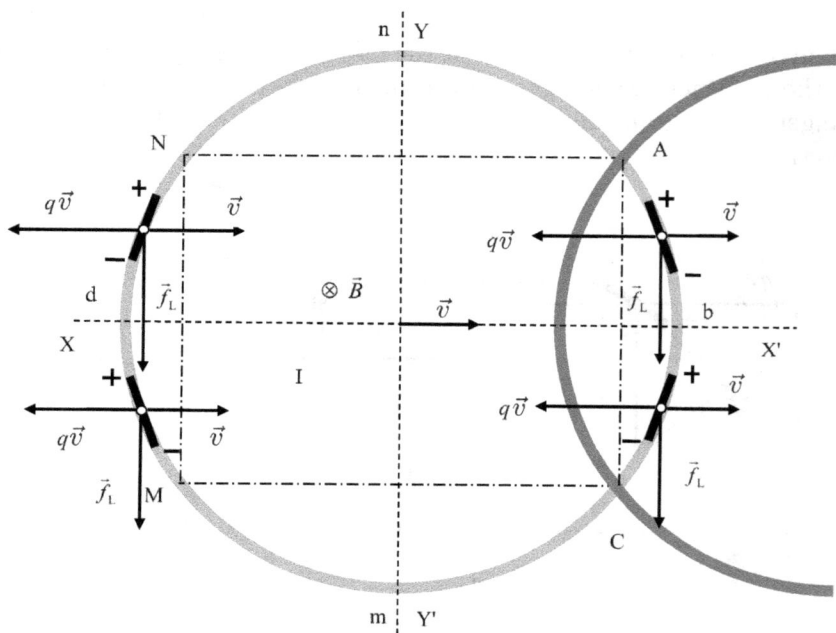

Fig. 6.11

Under these conditions, the distribution of the two identical total electromotive voltages of induction, $E_{AbC} = E_{CdA} = E$, on sectors AbC and CmMdNnA of ring (I), sectors whose electrical resistances are R_1 and R_2, respectively, is that represented in Figure 6.12, where the magnitudes and intensities of the induction currents are also noted.

It is easily demonstrated that:

$$R_1 = \frac{\alpha}{2\pi} R; \quad R_2 = \frac{2\pi - \alpha}{2\pi} R.$$

Knowing that the electromotive voltage of inductance in a coiled conductor with length l that is released with velocity $\vec{v}$ in a magnetic

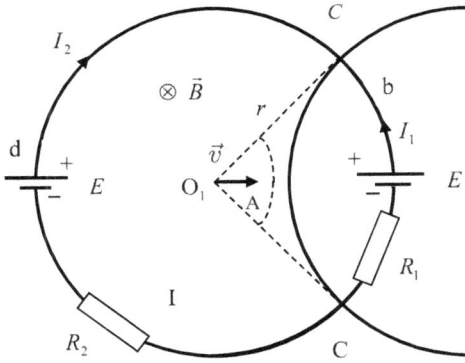

Fig. 6.12

field with induction $\vec{B} \perp \vec{v}$ is

$$e = -\frac{\Delta\Phi}{\Delta t} = -\frac{\Delta(\vec{B} \cdot \vec{S})}{\Delta t} = B\frac{\Delta S}{\Delta t},$$

based on Figure 6.13, it results that:

$$\Delta S = (l\cos\beta + v\Delta t)l\sin\beta - l^2\sin\beta\cos\beta = vl\Delta t\sin\beta;$$

$$e = Blv\sin\beta.$$

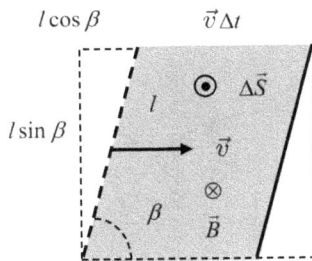

Fig. 6.13

Under these conditions, using Figure 6.14, it results that:

$$e_k = Bv\Delta l_k \sin\beta_k;$$

$$E = \Sigma e_k = Bv\Sigma\Delta l_k \sin\beta_k = Bv(\mathrm{AC});$$

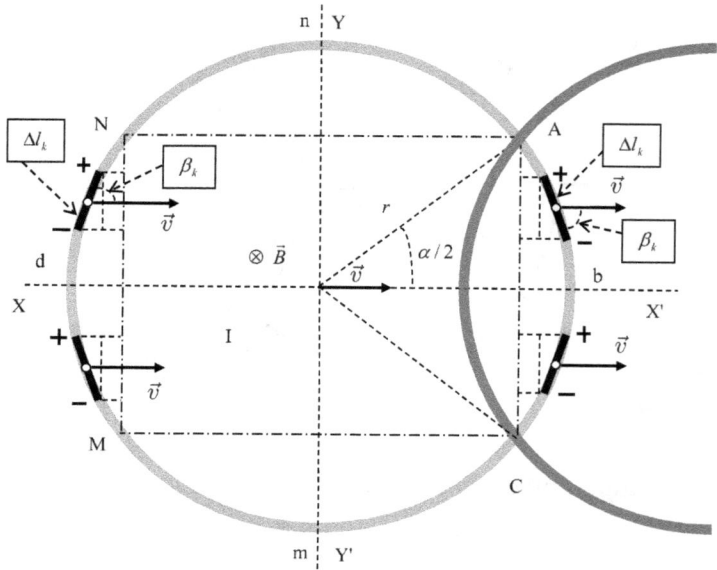

Fig. 6.14

$$\text{AC} = 2r \sin \frac{\alpha}{2};$$

$$E = 2Bvr \sin \frac{\alpha}{2}.$$

Following the same statements for ring (II) and noting the results in Figure 6.15, using Kirchhoff's theorems, we obtain:

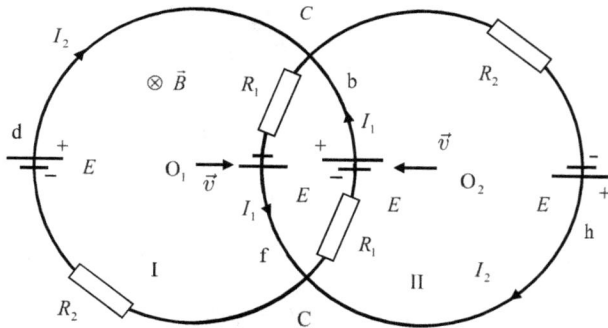

Fig. 6.15

$$2E = 2I_1 R_1; \quad I_1 = \frac{E}{R_1};$$

$$I_1 = \frac{4\pi B v r}{\alpha R} \sin \frac{\alpha}{2};$$

$$2E = I_1 R_1 + I_2 R_2;$$

$$I_2 = \frac{2E}{R_2} - I_1 \frac{R_1}{R_2};$$

$$I_2 = \frac{4\pi B v r}{(2\pi - \alpha)R} \sin \frac{\alpha}{2}.$$

With the electromagnetic forces that act on each of the four sectors of ring (I) being those shown in Figure 6.16, it is very easily demonstrated that, for reasons of symmetry,

$$\vec{F}_m = -\vec{F}_n,$$

so that their resultant is null.

In these conditions, the resultant of the forces acting on ring (I) is:

$$\vec{F}_I = \vec{F}_b + \vec{F}_d;$$

$$F_I = F_b + F_d;$$

$$F_I = (I_1 + I_2)Bl_{AB} = (I_1 + I_2)B2r \sin \frac{\alpha}{2};$$

$$F_I = \frac{16\pi^2 B^2 v r^2}{\alpha(2\pi - \alpha)R} \sin^2 \frac{\alpha}{2};$$

$$\alpha = \frac{\pi}{3}; \quad F_I = \frac{36}{5} \frac{B^2 v r^2}{R}.$$

Similarly, using Figure 6.17, it is proved that the force acting on ring (II) is $F_{II} = F_f + F_h = F_I$.

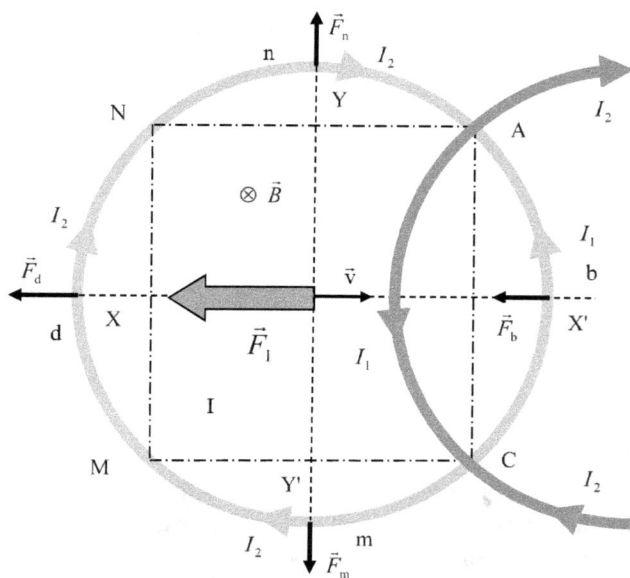

Fig. 6.16

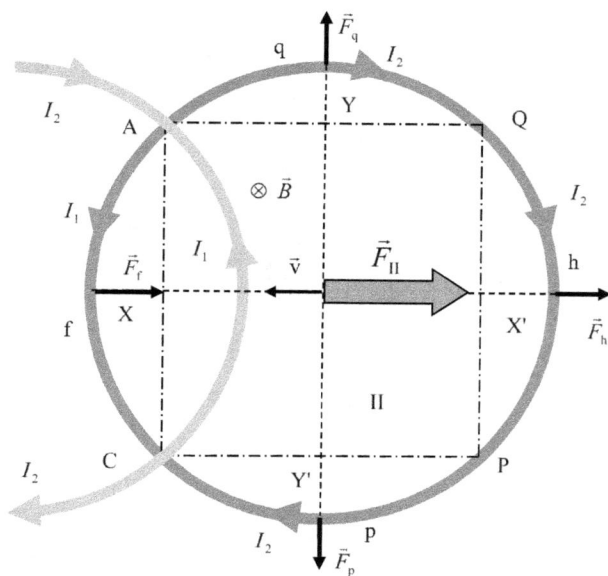

Fig. 6.17

Problem 7. Lens Between a Light Source and a Photodiode

A photodiode with the flat receiving surface $S_{pd} = 0.5\,\text{mm}^2$ is located in the focus of a convergent lens, arranged perpendicular to the main optical axis of the lens. On the other side of the lens, on the same main axis, there is a point source of light at a distance d from the lens.

Determine the ratio of the intensities of the currents through the photodiode if the light source located at a distance $d_1 = 1\,\text{m}$ from the lens moves, approaching the lens until it reaches the distance $d_2 = 0.3\,\text{m}$.

Given: the diameter of the lens, $D = 1\,\text{cm}$; the focal length of the lens, $f = 5\,\text{cm}$.

It is known that the intensity of the current through the photodiode is directly proportional to the incident radiation flow.

Solution

In the absence of the photodiode, the image of the source would be formed in S, as Figure 7.1 indicates, such that:

$$\frac{1}{d} + \frac{1}{p} = \frac{1}{f}; \quad p = \frac{fd}{d - f}.$$

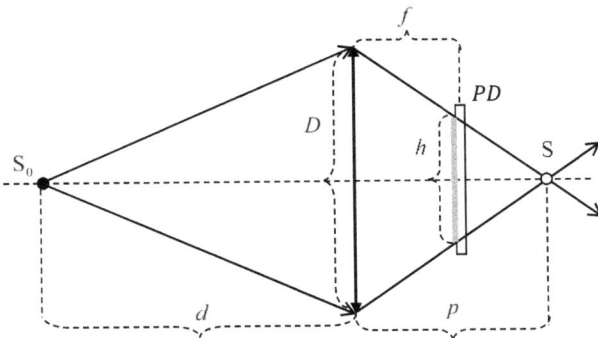

Fig. 7.1

Under these conditions, for the diameter of the light spot on the face of the photodiode and its surface area, from simple geometric considerations, we obtain:

$$h = \frac{fD}{d}; \quad S_{\text{spot}} = \frac{\pi h^2}{4} = \frac{\pi f^2 D^2}{4d^2}.$$

Thus, for the two distances of the source to the lens, we have:

$$f = 5 \text{ cm}; \quad D = 1 \text{ cm}; \quad d_1 = 1 \text{ m};$$

$$r_1 = \frac{h_1}{2} = \frac{fD}{2d_1} = 0.25 \text{ mm}; \quad S_{\text{spot},1} = \pi r_1^2 = \frac{\pi f^2 D^2}{4d_1^2} \approx 0.2 \text{ mm}^2;$$

$$S_{\text{pd}} = 0.5 \text{ mm}^2; \quad r_{\text{pd}} \approx 0.4 \text{ mm};$$

$$r_1 < r_{\text{pd}}; \quad S_{\text{spot},1} < S_{\text{pd}}.$$

This proves that, in this case, the entire flux of light falling on the lens will reach the surface of the photodiode, so that the intensity of the current through the photodiode, in this case, will be:

$$I_1 = k\Phi_1 = K S_{\text{spot},1};$$

$$f = 5 \text{ cm}; \quad D = 1 \text{ cm}; \quad d_2 = 0.3 \text{ m};$$

$$r_2 = \frac{h_2}{2} = \frac{fD}{2d_2} = 1.66 \text{ mm}; \quad S_{\text{spot},2} = \pi r_2^2 = \frac{\pi f^2 D^2}{4d_2^2} \approx 2.18 \text{ mm}^2;$$

$$r_2 > r_{\text{pd}}; \quad S_{spot,2} > S_{\text{pd}}.$$

This proves that, from the flow of light that falls on the lens, only a part, $\Phi_2 < \Phi_0$, corresponding to the surface area of the photodiode, S_{pd}, will reach the photodiode, so that the intensity of the current through the photodiode, in this case, will be

$$I_2 = k\Phi_2 = K S_{\text{pd}}.$$

In these conditions:

$$\frac{S_{\text{pd}}}{S_{\text{spot},2}} = \frac{0.5 \text{ mm}^2}{2.18 \text{ mm}^2} = \frac{1}{4.36};$$

$$\frac{I_2}{I_1} = \frac{k\Phi_2}{k\Phi_1} = \frac{K S_{\text{pd}}}{K S_{\text{spot},1}} = \frac{S_{\text{pd}}}{S_{\text{spot},1}} = \frac{S_{\text{pd}}}{S_{\text{spot},2}} \frac{S_{\text{spot},2}}{S_{\text{spot},1}};$$

$$\frac{I_2}{I_1} = \frac{1}{4.36} \frac{d_1^2}{d_2^2} \approx 2.54.$$

Problem 8. Two Transparent Plates

From the light intensity I_0, normally incident on a glass plate with very thin, semi-transparent parallel plane faces, the slide allows only I_0/k to pass, the rest being reflected. In the path of a parallel beam of monochromatic light of intensity I_0, two such semi-transparent plates are placed, their planes being parallel and close, as shown in the drawing in Figure 8.1.

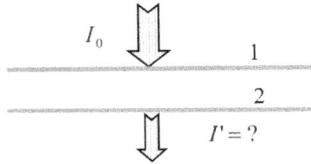

Fig. 8.1

Determine the intensity I' transmitted beyond the second plate, and *locate* and *calculate* the difference $I = I_0 - I'$. *Compare* intensities I' and I. Then, *verify* that $I + I' = I_0$. Neglect the absorption of light. It is known that $1 + q + q^2 + q^3 + \cdots + q^n = \frac{1}{1-q}$.

Solution

Using the representation in Figure 8.2, it results that:

$$i_0 = \frac{I_0}{k}; \quad I_1 = I_0 - i_0 = I_0 - \frac{I_0}{k} = I_0\left(1 - \frac{1}{k}\right);$$

$$I_1 = I_0\left(1 - \frac{1}{k}\right); \quad I'_1 = \frac{i_0}{k} = \frac{I_0}{k^2};$$

$$i_1 = i_0 - I'_1 = \frac{I_0}{k} - \frac{I_0}{k^2}; \quad i_1 = \frac{I_0}{k}\left(1 - \frac{1}{k}\right);$$

$$I_2 = \frac{i_1}{k} = \frac{I_0}{k^2}\left(1 - \frac{1}{k}\right); \quad i'_1 = i_1 - I_2;$$

$$i'_1 = \frac{I_0}{k}\left(1 - \frac{1}{k}\right)^2; \quad I'_2 = \frac{i'_1}{k}; \quad I'_2 = \frac{I_0}{k^2}\left(1 - \frac{1}{k}\right)^2;$$

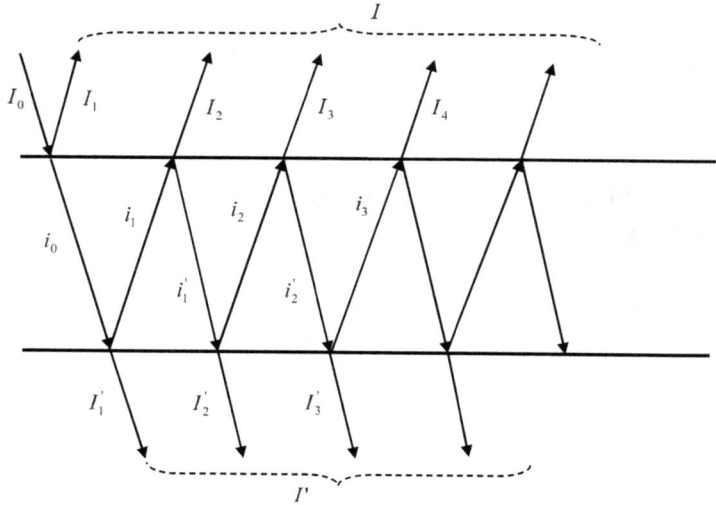

Fig. 8.2

$$i_2 = i_1' - I_2'; \quad i_2 = \frac{I_0}{k}\left(1 - \frac{1}{k}\right)^3;$$

$$I_3 = \frac{i_2}{k} = \frac{I_0}{k^2}\left(1 - \frac{1}{k}\right)^3; \quad i_2' = i_2 - I_3;$$

$$i_2' = \frac{I_0}{k}\left(1 - \frac{1}{k}\right)^4; \quad I_3' = \frac{i_2'}{k}; \quad I_3' = \frac{I_0}{k^2}\left(1 - \frac{1}{k}\right)^4;$$

$$i_3 = i_2' - I_3'; \quad i_3 = \frac{I_0}{k}\left(1 - \frac{1}{k}\right)^5;$$

$$I_4 = \frac{i_3}{k} = \frac{I_0}{k^2}\left(1 - \frac{1}{k}\right)^5;$$

$$\cdots\cdots\cdots\cdots\cdots\cdots\cdots\cdots\cdots\cdots\cdots\cdots$$

$$I' = I_1' + I_2' + I_3' + \cdots\cdots\cdots;$$

$$I' = \frac{I_0}{k^2} + \frac{I_0}{k^2}\left(1 - \frac{1}{k}\right)^2 + \frac{I_0}{k^2}\left(1 - \frac{1}{k}\right)^4 + \cdots\cdots\cdots;$$

$$I' = \frac{I_0}{k^2}\left[1 + \left(1 - \frac{1}{k}\right)^2 + \left(1 - \frac{1}{k}\right)^4 + \left(1 - \frac{1}{k}\right)^6 + \cdots\cdots\right];$$

$$q = \left(1 - \frac{1}{k}\right)^2;$$

$$1 + \left(1 - \frac{1}{k}\right)^2 + \left(1 - \frac{1}{k}\right)^4 + \left(1 - \frac{1}{k}\right)^6 + \cdots\cdots + \left(1 - \frac{1}{k}\right)^{2n}$$

$$= 1 + q + q^2 + q^3 + \cdots + q^n = \frac{1}{1 - q};$$

$$1 + \left(1 - \frac{1}{k}\right)^2 + \left(1 - \frac{1}{k}\right)^4 + \left(1 - \frac{1}{k}\right)^6 + \cdots\cdots$$

$$= \frac{1}{1 - \left(1 - \frac{1}{k}\right)^2} = \frac{k^2}{2k - 1};$$

$$I' = \frac{I_0}{2k - 1};$$

$$I = I_1 + I_2 + I_3 + I_4 + \cdots\cdots\cdots;$$

$$I = I_0\left(1 - \frac{1}{k}\right) + \frac{I_0}{k^2}\left(1 - \frac{1}{k}\right) + \frac{I_0}{k^2}\left(1 - \frac{1}{k}\right)^3$$

$$+ \frac{I_0}{k^2}\left(1 - \frac{1}{k}\right)^5 + \cdots\cdots\cdots;$$

$$I = I_0\left(1 - \frac{1}{k}\right) + \frac{I_0}{k^2}\left(1 - \frac{1}{k}\right)\left[1 + \left(1 - \frac{1}{k}\right)^2 + \left(1 - \frac{1}{k}\right)^4\right.$$

$$\left. + \left(1 - \frac{1}{k}\right)^6 + \cdots\cdots\right];$$

$$I = \frac{2I_0(k - 1)}{2k - 1}; \quad I' < I; \quad I + I' = I_0.$$

This is shown in Figure 8.3.

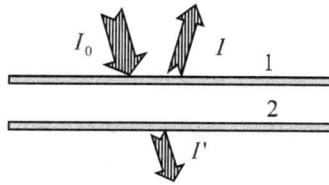

Fig. 8.3

Problem 9. The Flux of Light Beyond a Transparent Slide

The flux of radiation coming from a point-shaped light source is measured with the help of a photosensitive detector located at a distance l_0 from the source. Between the source and the photodetector, there is a transparent glass slide with flat and parallel faces, placed so that the planes of its faces are perpendicular to the right of the centers of the source and the detector. The refractive index of glass is n.

   Determine the thickness d of the transparent slide if the photodetector indication remains the same as in the absence of the slide. The reflective coefficient of light, under normal incidence, at a glass–air or air–glass boundary is $k = \frac{(n-1)^2}{(n+1)^2}$. It is known that $1 + q^2 + q^4 + \cdots + q^{2n} = \frac{1}{1-q^2}$.

Solution

After crossing the plate, the light flux reaching the photodetector is lower than the incident light flux on the slide due to light reflection at the air–glass and glass–air boundaries. However, this decrease also has a compensatory effect. Due to light refraction, the light source appears closer to the photodetector than in the absence of the plate.

   To begin with, using Figure 9.1, we will analyze the effect of light reflection on the two faces (a and b) of the slide.

   For the first reflection on the face of the slide at the air–glass boundary, according to the definition of the coefficient of reflection, we have:

$$k = \frac{\Phi_{r1,a}}{\Phi_{in1,a}}; \quad \Phi_{r1,a} = k\Phi_{in1,a},$$

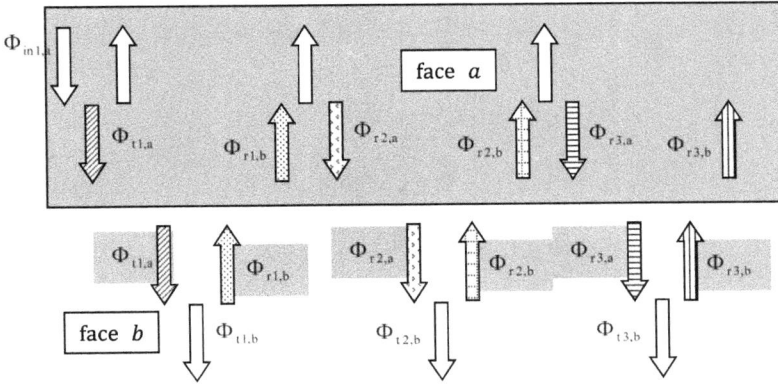

Fig. 9.1

where $\Phi_{in1,a}$ is the light flux recorded by the photodetector in the absence of a transparent slide, and $\Phi_{r1,a}$ is the light flux after the first reflection on face a;

$$\Phi_{t1,a} = \Phi_{in1,a} - \Phi_{r1,a} = \Phi_{in1,a}(1-k),$$

where $\Phi_{t1,a}$ is the flux transmitted during the first passage of light through face a.

Similar, corresponding to the first reflection on face b at the glass–air boundary, we have:

$$k = \frac{\Phi_{r1,b}}{\Phi_{in1,b}} = \frac{\Phi_{r1,b}}{\Phi_{t1,a}}; \quad \Phi_{r1,b} = k\Phi_{t1,a};$$

$$\Phi_{r1,b} = \Phi_{in1,a}k(1-k);$$

$$\Phi_{t1,b} = \Phi_{in1,b} - \Phi_{r1,b} = \Phi_{t1,a} - \Phi_{r1,b} = \Phi_{t1,a}(1-k);$$

$$\Phi_{t1,b} = \Phi_{in1,a}(1-k)^2.$$

Continuing this reasoning, for the second reflection produced on face a at the glass–air boundary, as well as for the following reflections on faces a and b, we have:

$$k = \frac{\Phi_{r2,a}}{\Phi_{in2,a}} = \frac{\Phi_{r2,a}}{\Phi_{r1,b}}; \quad \Phi_{r2,a} = k\Phi_{r1,b};$$

$$\Phi_{r2,a} = \Phi_{in1,a}k^2(1-k)$$

$$k = \frac{\Phi_{r2,b}}{\Phi_{in2,b}} = \frac{\Phi_{r2,b}}{\Phi_{r2,a}}; \quad \Phi_{r2,b} = k\Phi_{r2,a};$$

$$\Phi_{r2,b} = \Phi_{in1,a}k^3(1-k);$$

$$\Phi_{t2,b} = \Phi_{in2,b} - \Phi_{r2,b} = \Phi_{r2,a} - \Phi_{r2,b}$$

$$= \Phi_{in1,a}k^2(1-k) - \Phi_{in1,a}k^3(1-k);$$

$$\Phi_{t2,b} = \Phi_{in1,a}k^2(1-k)^2$$

$$k = \frac{\Phi_{r3,a}}{\Phi_{in3,a}} = \frac{\Phi_{r3,a}}{\Phi_{r2,b}}; \quad \Phi_{r3,a} = k\Phi_{r2,b};$$

$$\Phi_{r3,a} = \Phi_{in1,a}k^4(1-k);$$

$$k = \frac{\Phi_{r3,b}}{\Phi_{in3,b}} = \frac{\Phi_{r3,b}}{\Phi_{r3,a}}; \quad \Phi_{r3,b} = k\Phi_{r3,a};$$

$$\Phi_{r3,b} = \Phi_{in1,a}k^5(1-k);$$

$$\Phi_{t3,b} = \Phi_{in3,b} - \Phi_{t3,b} = \Phi_{r3,a} - \Phi_{r3,b}$$

$$= \Phi_{in1,a}k^4(1-k) - \Phi_{in1,a}k^5(1-k);$$

$$\Phi_{t3,b} = \Phi_{in1,a}k^4(1-k)^2;$$

$$\cdots \cdots \cdots \cdots \cdots \cdots \cdots \cdots \cdots \cdots \cdots \cdots$$

$$\Phi_t = \Phi_{t1,b} + \Phi_{t2,b} + \Phi_{t3,b} + \cdots \cdots \cdots;$$

$$\Phi_t = \Phi_{in1,a}(1-k)^2(1 + k^2 + k^4 + \cdots); \quad k < 1;$$

$$\Phi_t = \Phi_{in1,a}(1-k)^2\frac{1}{1-k^2} = \Phi_{in1,a}\frac{1-k}{1+k}.$$

This means that the light flux at the detector in the presence of the slide, due to the multiple reflections, is lower than the light flux that would reach the detector in the absence of the slide. We will now analyze the effect of the proximity of the source to the detector, an effect that is due to the refraction of light through the two faces

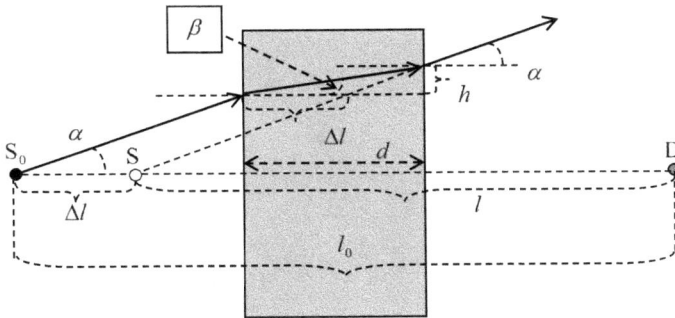

Fig. 9.2

(a and b) of the slide, using the drawing in Figure 9.2. For light rays departing at very small angles α from the source located in position S_0, using the law of refraction, as well as geometrical considerations, if d is the thickness of the transparent blade, it results that:

$$\sin \alpha = n \sin \beta; \quad \alpha \approx n\beta; \quad \beta = \frac{\alpha}{n};$$

$$\tan \beta = \frac{h}{d} \approx \beta; \quad h = \beta d = \frac{\alpha}{n} d;$$

$$\tan \alpha = \frac{h}{d - \Delta l} \approx \alpha; \quad h = \alpha(d - \Delta l);$$

$$\frac{\alpha}{n} d\alpha(d - \Delta l);$$

$$\Delta l = \left(1 - \frac{1}{n}\right) d.$$

Thus, in the presence of the slide, the source is in position S, closer than in the absence of the slide, at the distance $l = l_0 - \Delta l$ from the detector, which implies an increase in the light flux at the detector.

The energy that a point source of light emits per unit time, per unit solid angle, represents the *energy intensity* of the light source $\left(I_e = \frac{W}{\Delta t \cdot \Delta\Omega}\right)$.

The energy that arrives per unit time on a unit area of a body's surface under normal incidence represents the *illumination energy* that the light source produces on a certain surface $\left(E_e = \frac{W}{\Delta t \cdot \Delta A}\right)$. If $\frac{W}{\Delta t} = \Phi$ represents the *energy flux* of the light emitted by the source (representing the energy of the radiation that crosses under normal

incidence a given surface area, or the energy of the radiation that
reaches under normal incidence a given surface area), then:

$$I_e = \frac{\Phi}{\Delta\Omega};$$

$$E_e = \frac{\Phi}{\Delta A} = \frac{I_e \Delta\Omega}{\Delta A} = I_e \frac{\Delta\Omega}{\Delta A};$$

$$\Delta\Omega = \frac{\Delta A}{r^2}; \quad E_e = \frac{I_e}{r^2},$$

where r is the distance from the point source to the photodetector.

In the absence of the transparent slide, the illumination energy
produced by the light source, located at point S_0, on a given surface
area of the photodetector is:

$$E_{e0} = \frac{I_{e0}}{l_0^2},$$

where I_{e0} is the energy intensity of the light source located at point
S_0;

$$I_{e0} = \frac{\Phi_{in1,a}}{\Delta\Omega}; \quad E_{e0} = \frac{\Phi_{in1,a}}{l_0^2 \Delta\Omega}.$$

In the presence of the glass slide, the illumination energy that the
light source located at point S produces on a given surface area of
the photodetector is:

$$E_e = \frac{I_e}{l^2} = \frac{I_e}{(l_0 - \Delta l)^2},$$

where I_e is the energy intensity of the light source located at point S;

$$I_e = \frac{\Phi_t}{\Delta\Omega} = \frac{\Phi_{in1,a}}{\Delta\Omega} \frac{1-k}{1+k};$$

$$E_e = \frac{\Phi_{in1,a}}{(l_0 - \Delta l)^2 \Delta\Omega} \frac{1-k}{1+k}.$$

If the photodetector's reading is the same, regardless of the presence or absence of the transparent slide between the light source and
the photodetector, it results that:

$$E_e = E_{e0};$$

$$\frac{\Phi_{in1,a}}{(l_0 - \Delta l)^2 \Delta \Omega} \frac{1-k}{1+k} = \frac{\Phi_{in1,a}}{l_0^2 \Delta \Omega};$$

$$\frac{1}{(l_0 - \Delta l)^2} \frac{1-k}{1+k} = \frac{1}{l_0^2};$$

$$\frac{l_0 - \Delta l}{l_0} = \sqrt{\frac{1-k}{1+k}}; \qquad \Delta l = \left(1 - \frac{1}{n}\right) d;$$

$$d = \frac{n}{n-1} l_0 \left(1 - \sqrt{\frac{1-k}{1+k}}\right);$$

$$k = \frac{(n-1)^2}{(n+1)^2}.$$

Problem 10. The Relativistic Doppler Effect

A source of electromagnetic radiation, S, moves with speed $\vec{v}$ relative to a fixed observer, O, so that, as Figure 10.1 indicates, the direction of the vector $\vec{v}$ does not coincide with the direction of the vector $\overrightarrow{OS}$. The source's emission of electromagnetic oscillations begins in position S_1, when the angle between vectors $\vec{v}$ and $\overrightarrow{OS_1}$ is θ, and ends when the source is in position S_2, very close to position S_1. The physical parameters and characteristics of the emitted radiation in relation to the reference system of the source S (mobile reference system) are: frequency, v_s; period, T_s; wavelength, λ_s.

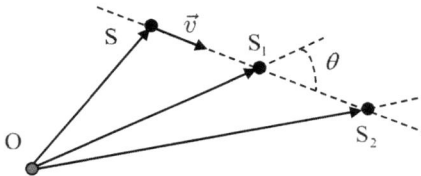

Fig. 10.1

Determine the quantities of the physical characteristics of the radiation recorded by the fixed observer O (fixed reference system): frequency, v_{obs}; period, T_{obs}; wavelength, λ_{obs}.

Particular cases: (1) $\theta = 0; \theta = \pi; \rightarrow$ longitudinal Doppler effect; (2) $\theta = \pi/2; \rightarrow$ transverse Doppler effect. We know the speed of light in a vacuum, c.

Solution

Following the drawing in Figure 10.2, the source's emission of electromagnetic oscillations begins in position S_1, when the angle between the vectors $\vec{v}$ and $\overrightarrow{OS_1}$ is θ, at time t_1' (relative to the source) and time t_1 (relative to the fixed observer). The emission of the N electromagnetic oscillations terminates in the very nearby position S_2 at time t_2' (relative to the source) and time t_2 (relative to the fixed observer), so that we have:

$$\Delta t' = t_2' - t_1'; \quad \Delta t = t_2 - t_1 \neq \Delta t';$$

$$\Delta t = \frac{\Delta t'}{\sqrt{1 - \beta^2}}.$$

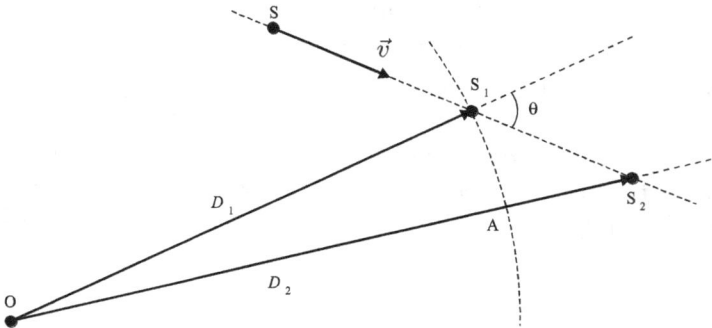

Fig. 10.2

Under these conditions, the reception of the N electromagnetic oscillations by the fixed observer will begin at the time

$$\tau_1 = t_1 + \frac{D_1}{c},$$

and it will end at the time

$$\tau_2 = t_2 + \frac{D_2}{c},$$

so that the reception duration of the N electromagnetic oscillations will be:

$$\Delta\tau = \tau_2 - \tau_1 = \Delta t + \frac{D_2 - D_1}{c};$$

$$D_2 - D_1 \approx S_1 S_2 \cos\theta = v\cos\theta\Delta t;$$

$$\Delta\tau = (1 + \beta\cos\theta)\Delta t.$$

It results that:

$$N = \nu_s\Delta t' = \nu_s\sqrt{1 - \beta^2}\Delta t;$$

$$N = \nu_{obs}\Delta\tau = \nu_{obs}(1 + \beta\cos\theta)\Delta t;$$

$$\nu_{obs} = \nu_s\frac{\sqrt{1 - \beta^2}}{1 + \beta\cos\theta};$$

$$\lambda_{obs} = \lambda_s\frac{1 + \beta\cos\theta}{\sqrt{1 - \beta^2}}.$$

The variants $\theta = 0$ and $\theta = \pi$, representing the longitudinal Doppler effect, correspond to the situations highlighted in Figure 10.3, for which we have

$$\nu_{obs} = \nu_s\frac{\sqrt{1 - \beta^2}}{1 \pm \beta}; \quad \lambda_{obs} = \lambda_s\frac{1 \pm \beta}{\sqrt{1 - \beta^2}},$$

known as the *longitudinal relativistic Doppler effect*.

$$\theta = \sphericalangle(\,\vec{OS};\vec{v}\,) = 0 \qquad\qquad \theta = \pi$$

Fig. 10.3

In addition, $\nu_{obs} \neq \nu_s$ and $\lambda_{obs} \neq \lambda_s$, and for $\theta = \pi/2$, the following relations are obtained:

$$\nu_{obs} = \nu_s\sqrt{1 - \beta^2} < \nu_s;$$

$$T_{obs} = \frac{T_s}{\sqrt{1 - \beta^2}} > T_s;$$

$$\lambda_{obs} = \frac{\lambda_s}{\sqrt{1 - \beta^2}} > \lambda_s,$$

representing the *transversal relativistic Doppler effect*, for which no classical correspondence exists. The transversal Doppler effect proves that even when the detection of light propagation toward the observer is perpendicular to the direction of movement of the source, the frequency and wavelength of the recorded radiation differ from those of the emitted radiation. However, their variations are much smaller than those corresponding to the longitudinal relativistic Doppler effect.

Problem 11. The Ives and Stilwell Experiment

Among the experiments that confirmed the existence of the relativistic Doppler effect (longitudinal and transverse), especially the transverse effect, predicted by the theory of special relativity, is the experiment proposed by Ives and Stilwell in 1938.

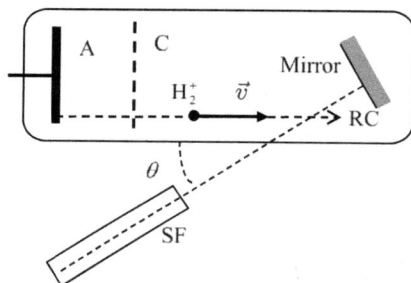

Fig. 11.1

In a *channel radiation* tube (a H_2^+ ion bundle), a plane mirror is mounted, as shown in Figure 11.1. Channel radiation is corpuscular radiation made up of positive ions that, after being accelerated in the intense electric field of an electrical discharge, propagates behind the perforated cathode of a discharge tube. It was discovered by E. Goldstein in 1886.

Using a spectrophotometer whose axis falls within the limits of the experimental curves, with the normal on the plane of the mirror, both the radiation emitted by the H_2^+, which comes directly from these ions, and that reflected by the mirror are recorded. The speed of the ions at the moment of crossing the axis of the spectrophotometer is $\vec{v}$, and the angle between the axis of the spectrophotometer and the axis of the radiation tube is θ. If the light emitted by the H_2^+ ions were observed along the RC direction (channel radiation) in the absence of mirrors, as shown in Figure 11.1, then the wavelength of the recorded radiation would correspond to the longitudinal Doppler effect. The wavelength of the emitted radiation relative to its own reference system (attached to the H_2^+ ions) is λ_s.

Determine the wavelengths λ_1 and λ_2 of the spectral lines recorded by the spectrophotometer, specifying their location relative to the spectral line λ_s. The speed of light in a vacuum, c, is known. For the values of the two wavelengths, *identify* the contribution $\Delta\lambda$ from the longitudinal Doppler effect, as well as the contribution $\Delta\lambda'$ from the transverse Doppler effect.

Solution

In the mirror shown in Figure 11.2, the radiation emitted from the source S (the H_2^+ ion) is recorded by the observer O (the spectrophotometer) following the direction of the normal to the mirror.

The part of the wave reaching the mirror seems to come towards the observer, being emitted by a virtual source S', representing the image of S in the plane mirror. Using the previous results, this wave

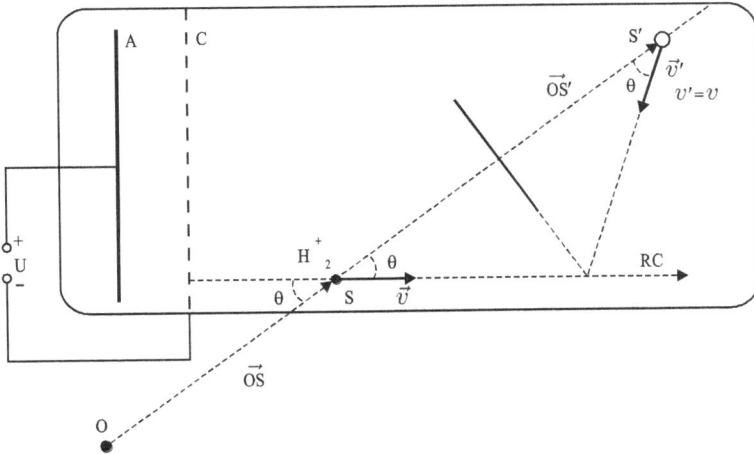

Fig. 11.2

will be recorded by the spectrophotometer as having the following wavelength:

$$\lambda'_{\text{obs}} = \lambda_1 = \lambda_s \frac{1 + \beta \cos(\overrightarrow{OS'}; \vec{v'})}{\sqrt{1 - \beta^2}};$$

$$\lambda_1 = \frac{1 + \beta \cos(\pi - \theta)}{\sqrt{1 - \beta^2}} \lambda_s;$$

$$\lambda_1 = \frac{\lambda_s(1 - \beta \cos \theta)}{\sqrt{1 - \beta^2}}.$$

This wave, received by the spectrophotometer O after reflection on the mirror, will be recorded with a variation $-\Delta\lambda$ in the wavelength, corresponding to the longitudinal Doppler effect, and with an additional variation $+\Delta\lambda'$, corresponding to the transverse Doppler effect:

$$\lambda_1 = \lambda_s - \Delta\lambda + \Delta\lambda'.$$

The part of the wave emitted directly to the observer will be registered as having the following wavelength:

$$\lambda''_{\text{obs}} = \lambda_2 = \lambda_s \frac{1 + \beta \cos(\overrightarrow{OS}; \vec{v})}{\sqrt{1 - \beta^2}};$$

$$\lambda_2 = \lambda_s \frac{1 + \beta \cos \theta}{\sqrt{1 - \beta^2}}.$$

This wave is recorded with a length variation of $+\Delta\lambda$, corresponding to the longitudinal Doppler effect, and an additional variation $+\Delta\lambda'$, corresponding to the transverse Doppler effect:

$$\lambda_2 = \lambda_s + \Delta\lambda + \Delta\lambda'.$$

As a result, in the spectrophotometer, smooth spectral lines will appear with wavelengths λ_1 and λ_2, arranged asymmetrically with respect to the corresponding division λ_s.

The variations in the wavelengths due to the longitudinal Doppler effect and transverse Doppler effect, respectively, are given by the expressions:

$$\Delta\lambda = \frac{1}{2}(\lambda_2 - \lambda_1);$$

$$\Delta\lambda' = \frac{1}{2}(\lambda_1 + \lambda_2 - 2\lambda_s).$$

Ives and Stilwell found these values to be very close to the calculated theoretical values.

Problem 12. Invariant of Lorentz Transformations

It is known that the phase of a plane electromagnetic wave is an invariant of special Lorentz transformations.

Determine the spectral displacement corresponding to the longitudinal relativistic Doppler effect.

Solution

In the origin of the fixed inertial system XYZ, represented in Figure 12.1, there is a source S of electromagnetic harmonic oscillations, occurring along the Z axis, according to the law

$$E_{\mathrm{S}} = E_{\max} \sin \omega_s t = E_{\max} \sin 2\pi\nu_s t.$$

The end-to-end transmission of these oscillations along the Y axis with speed c represents a planar (transversal) electromagnetic wave, whose equation is

$$E = E_{\max} \sin \omega_s \left(t - \frac{y}{c}\right) = E_{\max} \sin 2\pi\nu_s \left(t - \frac{y}{c}\right),$$

where y is the coordinate of the position of the wave front at time t in relation to the fixed system XYZ, and E is the instantaneous value of the intensity of the electric field of the electromagnetic wave at the considered point.

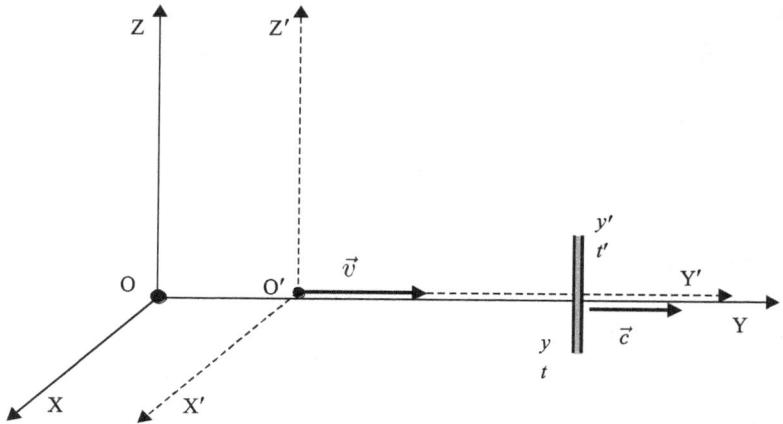

Fig. 12.1

The expression

$$\phi = 2\pi\nu_s \left(t - \frac{y}{c}\right)$$

represents the phase of the electromagnetic oscillations in the fixed system XYZ.

Corresponding to the mobile system X'Y'Z', the phase of the oscillations is

$$\phi' = 2\pi\nu_{\text{obs}} \left(t' - \frac{y'}{c}\right),$$

where y' and t' represent the spacetime coordinates of the same wavefront relative to the system X'Y'Z'.

The phase of a wave is a quantity directly proportional to the number of maxima that pass by one observer located in a certain inertial reference system.

Because counting these maxima is independent of the coordinate system, the wave's phase is an invariant of Lorentz transformations.

Using the Lorenz transformations, it results that:

$$t = \frac{t' + \frac{v}{c^2}y'}{\sqrt{1 - \beta^2}}; \quad y = \frac{y' + vt'}{\sqrt{1 - \beta^2}};$$

$$2\pi\nu_s\left(t - \frac{y}{c}\right) = 2\pi\nu_s\left(\frac{t' + \frac{v}{c^2}y'}{\sqrt{1 - \beta^2}} - \frac{y' + vt'}{c\sqrt{1 - \beta^2}}\right);$$

$$2\pi\nu_{\text{obs}}\left(t' - \frac{y'}{c}\right) = 2\pi\nu_s\left(\frac{t' + \frac{v}{c^2}y'}{\sqrt{1 - \beta^2}} - \frac{y' + vt'}{c\sqrt{1 - \beta^2}}\right);$$

$$\nu_{\text{obs}} = \nu_s\frac{1 - \beta}{\sqrt{1 - \beta^2}}.$$

Chapter 8

International Pre-Olympic Physics Contest, Zalău, Romania

Problem 1. Adventures and Gravitational Catastrophes

A. The Jump of the Polar Bear

The beginning of the 21st century. In a guide about the Spitsbergen polar ice cap (the largest island in the Svalbard Archipelago, in the Arctic Ocean), it is said that "a polar bear can move, without warning, from one frozen sector to another frozen sector, the distance between these sectors being $L = 8$ m", as shown below.

The middle of the 26th century. To populate the remote areas of the Solar System, biologists plan to send some polar bears from the land of Spitsbergen to an icy asteroid in the Kuiper belt. Physicists have warned that, through their jumps, the bears could become independent bodies of the Kuiper belt.

(a) *Estimate* the diameter that such an asteroid can have so that polar bears, brought from the land of Spitsbergen, can be comfortably placed there.

B. Gravitational Catastrophe in Our Galaxy

Let's consider that, as a result of a gravitational catastrophe occurring in a region of our galaxy, the constant of universal attraction decreases, becoming $K = \eta K_0$, where K_0 is the constant of the current gravitational attraction, and $\eta < 1$.

(b) If the current orbit of the Earth in relation to the Sun is a sky, *determine* the new shape of the Earth's orbit in relation to the Sun, as well as the elements of this orbit, after the occurrence of the gravitational catastrophe.

C. Gravitational Catastrophe in the Sun

Let's imagine that, following a catastrophe of gravitational forces, the mass of the Sun suddenly decreases, reducing to half its current value.

(c) *Determine* the new period of revolution of the Earth around the Sun.

Solution

(a) Using the notation in Figure 1.1, we determine the speed the bear has at the time of the start of the jump under the conditions offered by the Spitsbergen ice cap:

$$x_{max} = v_{0x}t = v_0 \cos \alpha \cdot 2t_u = v_0 \cos \alpha \cdot 2\frac{v_{0y}}{g} = v_0 \cos \alpha \cdot 2\frac{v_0 \sin \alpha}{g};$$

$$x_{max} = \frac{v_0^2}{g} \sin 2\alpha;$$

$$\alpha = 45°; \quad x_{max} = \frac{v_0^2}{g} = L = 8 \text{ m}; \quad g \approx 10 \, \frac{\text{m}}{\text{s}^2};$$

$$v_0 = \sqrt{gL} \approx 9 \, \frac{\text{m}}{\text{s}}.$$

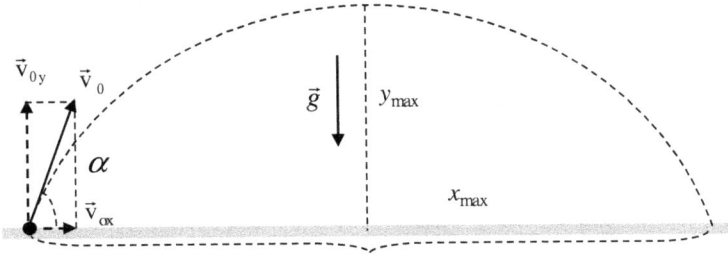

Fig. 1.1

Arriving on the asteroid's surface, the bear will execute a jump with the initial speed calculated previously. If $v_0 > v_{II}$, where v_{II} is the second cosmic velocity, under the conditions offered by the asteroid, it is obvious that the bear will leave the asteroid, which will not be at all comfortable. And if $v_0 = v_I$, where v_I is the first cosmic velocity, under the conditions offered by the asteroid, the bear will not be comfortable, for he will have to gravitate around the frozen asteroid.

As a result, for the comfort of the bear, the initial speed of the jump on the asteroid must meet the condition

$$v_0 < v_I.$$

Thus, we get:

$$v_I = \sqrt{K\frac{M}{R}},$$

where M is the mass of the asteroid, R is the asteroid's radius, and K is the constant of universal attraction;

$$M = \rho \frac{4\pi R^3}{3},$$

where ρ is the density of the asteroid (ice density);

$$v_I = D\sqrt{\frac{K\pi\rho}{3}},$$

where D is the diameter of the asteroid;

$$v_0 < D\sqrt{\frac{K\pi\rho}{3}};$$

$$\sqrt{gL} < D\sqrt{\frac{K\pi\rho}{3}}; \quad D > \sqrt{\frac{3gL}{K\pi\rho}} \approx 35\,\text{km}.$$

(b) Before the occurrence of the gravitational catastrophe, the total mechanical energy of the Earth–Sun system is:

$$E_0 = \frac{mv_0^2}{2} - K_0\frac{mM}{r_0}; \quad v_0 = \sqrt{K_0\frac{M}{r_0}}.$$

Immediately after the gravitational catastrophe, the total mechanical energy of the Earth–Sun system is:

$$E = \frac{mv_0^2}{2} - K\frac{mM}{r_0}; \quad K = \eta K_0; \quad E > E_0,$$

Thus, from that moment, the Earth's revolution around the Sun will be on an ellipse-shaped orbit, with the center of the Sun in the near focus, as indicated by Figure 1.2.

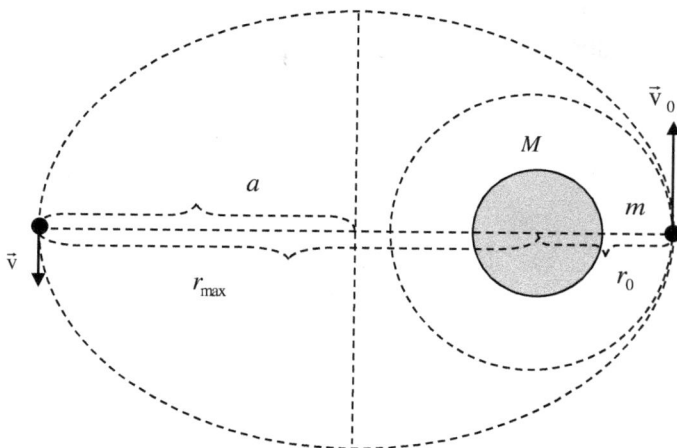

Fig. 1.2

The evolution of the Earth around the Sun, on the new orbit, occurs in accordance with the laws of conservation of mechanical energy and kinetic momentum, so that:

$$\frac{mv_0^2}{2} - K\frac{mM}{r_0} = \frac{mv_{min}^2}{2} - K\frac{mM}{r_{max}}; \quad \frac{v_0^2}{2} - K\frac{M}{r_0} = \frac{v_{min}^2}{2} - K\frac{M}{r_{max}};$$

$$r_0 m v_0 = r_{max} m v_{min}; \quad r_0 v_0 = r_{max} v_{min};$$

$$v_{min} = v_0 \frac{r_0}{r_{max}};$$

$$\frac{v_0^2}{2} - K\frac{M}{r_0} = \frac{v_0^2}{2}\frac{r_0^2}{r_{max}^2} - K\frac{M}{r_{max}};$$

$$\left(2KM - v_0^2 r_0\right) r_{max}^2 - 2KM r_0 r_{max} + v_0^2 r_0^3 = 0;$$

$$(2\eta - 1) r_{max}^2 - 2\eta r_0 r_{max} + r_0^2 = 0;$$

$$r_{max} = \frac{r_0}{2\eta - 1} > r_0;$$

$$r_{max} + r_0 = 2a; \quad a = \frac{\eta r_0}{2\eta - 1}.$$

Using the definition of the ellipse and the notation from Figure 1.3, it results that:

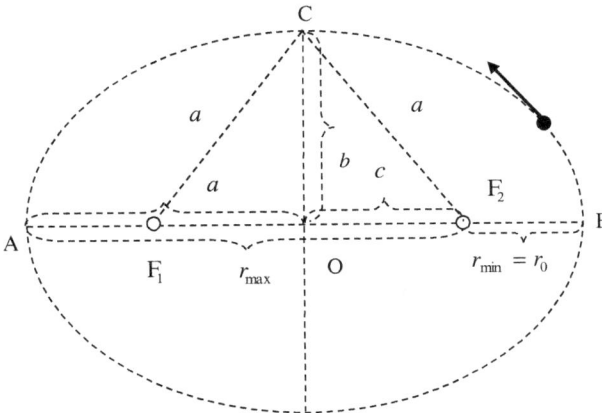

Fig. 1.3

$$r_{\min} = r_0 = a - c; \quad r_{\max} = a + c;$$

$$r_{\max} - r_0 = 2c; \quad c = \frac{1-\eta}{2\eta - 1} r_0;$$

$$b = \sqrt{a^2 - c^2} = \frac{r_0}{\sqrt{2\eta - 1}};$$

$$e = \sqrt{1 - \frac{b^2}{a^2}} = \sqrt{\left(\frac{\eta-1}{\eta}\right)^2} = \pm\frac{\eta-1}{\eta};$$

$$e = -\frac{\eta-1}{\eta} = \frac{1-\eta}{\eta};$$

$$p = a\left(1 - e^2\right) = \frac{r_0}{\eta} = b\sqrt{1 - e^2}.$$

(c) We must realize that the proposed process is hypothetical and that the laws of conservation can no longer be used to compare parameters of the Earth's movements before and after the process, because the system is not closed.

Since no other change occurs, we will consider that when the mass of the Sun decreases, it does not change the relative positions of the Sun and the Earth, nor the speed of the Earth in relation to the Sun. Accepting these postulates, in the first approximation, if the mass of the central body is reduced by half compared to the initial value, the speed of the spherical motion of the Earth becomes the speed of a parabolic motion. As a result, the Earth's orbit in relation to the Sun will become parabolic, which means that the period of the movement is infinite.

It is known, however, that on July 5, in its ellipse movement around the Sun, the Earth is close to the aphelion of its orbit, the farthest from the Sun, where its speed is lower than that required for circular movements. If the Earth was in this place at the time of the Sun's mass reduction, the Earth's speed would already be lower than that required to continue moving on a parabola. The Earth would continue moving around the Sun on an elongated elliptical orbit.

To begin with, let's establish the relationship between the speed of the Earth at the aphelion, $v_{0,\mathrm{aph}}$, on an elliptical orbit with the semi-major axis a_0 and the speed of the circular movement of the

Earth, v_0, on an orbit whose radius is $r_0 = a_0$, under the conditions of the initial Sun.

According to Kepler's second law and the law of conservation of energy, based on Figure 1.4, it results that:

$$v_{0,\text{per}}\, r_{0,\text{per}} = v_{0,\text{aph}}\, r_{0,\text{aph}};$$

$$\frac{v_{0,\text{per}}^2}{2} - K\frac{M_0}{r_{0,\text{per}}} = \frac{v_{0,\text{aph}}^2}{2} - K\frac{M_0}{r_{0,\text{aph}}};$$

$$r_{0,\min} = r_{0,\text{per}} = a_0\left(1 - e_0\right); \quad r_{0,\max} = r_{0,\text{aph}} = a_0\left(1 + e_0\right);$$

$$KM_0 = v_0^2\, r_0 = v_0^2\, a_0.$$

$$v_{0,\text{per}} = v_0\sqrt{\frac{1 + e_0}{1 - e_0}}; \quad v_{0,\text{aph}} = v_0\sqrt{\frac{1 - e_0}{1 + e_0}}.$$

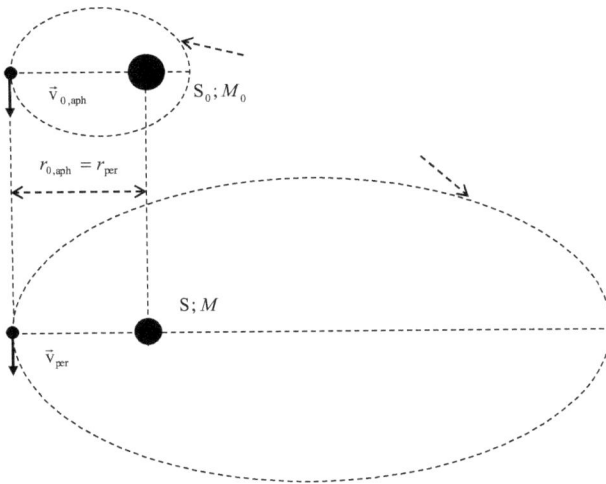

Fig. 1.4

For the new elliptical orbit of the Earth, after halving the mass of the Sun, we have:

$$r_{\text{per}} = r_{0,\text{aph}};$$

$$r_{\min} = r_{\text{per}} = a(1 - e);$$

$$a_0(1 + e_0) = a(1 - e); \quad a = a_0\frac{1 + e_0}{1 - e};$$

$$v_{\text{per}} = v_{0,\text{aph}};$$

$$v_{\text{per}} = v\sqrt{\frac{1+e}{1-e}},$$

where v is the speed of the Earth in a circular orbit whose radius is $r = a$, under the conditions of the new Sun (with the mass reduced by half, $M = M_0/2$);

$$v\sqrt{\frac{1+e}{1-e}} = v_0\sqrt{\frac{1-e_0}{1+e_0}};$$

$$v = \sqrt{K\frac{M}{r}} = \sqrt{K\frac{M_0}{2a}} = \sqrt{K\frac{M_0}{a_0}}\sqrt{\frac{a_0}{2a}} = v_0\sqrt{\frac{a_0}{2a}};$$

$$e = 1 - 2e_0; \quad a = a_0\frac{1+e_0}{2e_0}.$$

It results that:

$$T_0 = \frac{2\pi r_0}{v_0} = \frac{2\pi a_0}{v_0}; \quad T = \frac{2\pi r}{v} = \frac{2\pi a}{v};$$

$$\frac{T}{T_0} = \frac{a}{a_0}\frac{v_0}{v} = \frac{1+e_0}{2e_0}\sqrt{\frac{2a}{a_0}} = \frac{1+e_0}{2e_0}\sqrt{2}\sqrt{\frac{1+e_0}{2e_0}};$$

$$T = T_0\sqrt{2}\left(\frac{1+e_0}{2e_0}\right)^{3/2} \approx 230 \text{ years.}$$

Problem 2. Air Bubble in a Glass Sphere

A monochromatic light ray arrives on the surface of a glass sphere in the direction of one of its diameters. At a distance r from the center of the sphere, the light is scattered uniformly, in all directions, by a small air bubble embedded in the glass, as shown in Figure 2.1.

Determine the percentage of the light diffused by the air bubble that leaves the glass sphere. For the glass sphere, the following are known: radius, R; the refractive index, n. The light absorption in the glass is neglected.

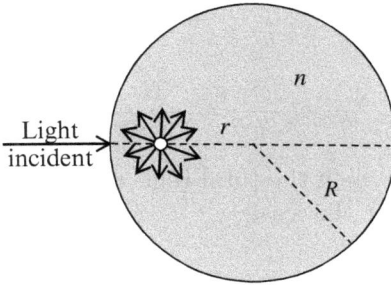

Fig. 2.1

Solution

Figure 2.2 shows the behavior of the light rays diffused by the air bubble located at point A inside the sphere when they reach the separation surface between the glass and the air.

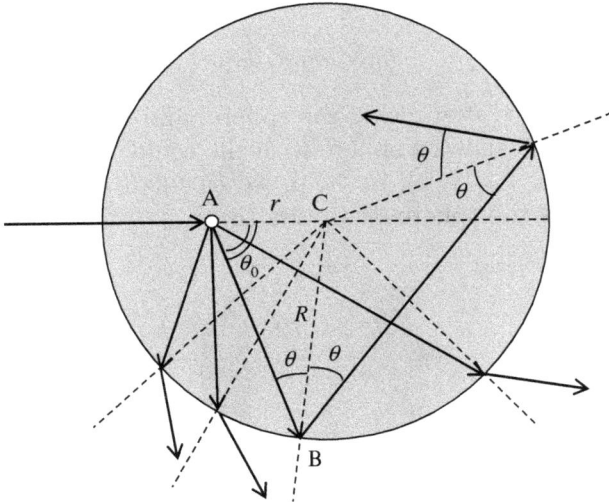

Fig. 2.2

It is observed that, depending on the value of the incident angle on the air–glass separation surface, some of the light rays leave the sphere, and the others, through total reflection, remain inside the sphere.

From the triangle ABC, writing the theorem of sines, it follows that

$$\frac{r}{\sin \theta} = \frac{R}{\sin \theta_0}.$$

The value θ_0 at which the total reflection occurs is determined by knowing the limit for this case:

$$\sin \theta = \frac{1}{n};$$

$$\sin \theta = \frac{r}{R} \sin \theta_0 = \frac{1}{n};$$

$$\sin \theta_0 = \frac{R}{nr};$$

$$\theta_{01} = \arcsin\left(\frac{R}{nr}\right); \quad \theta_{02} = \pi - \arcsin\left(\frac{R}{nr}\right).$$

The light rays are diffused by the bubble at an angle of θ_0, so that

$$\theta_{01} < \theta_0 < \theta_{02}.$$

These light rays stay inside the sphere, and the others leave the sphere. Figure 2.3 shows a sphere with the center at point A, having a certain radius, R_0. With its help, we calculate the percentage of light that leaves the sphere. It results that:

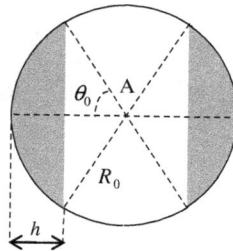

Fig. 2.3

$$\eta = \frac{2 \cdot 2\pi R_0 h}{4\pi R_0^2} = \frac{h}{R_0};$$

$$h = R_0 \left(1 - \cos\theta_0\right) = R_0 \left(1 - \sqrt{1 - \left(\frac{R}{nr}\right)^2}\right);$$

$$\eta = 1 - \sqrt{1 - \frac{R}{nr}^2}.$$

Problem 3. Flat Plate with Variable Transparency

On the main optical axis of a converging lens with focal length f at a distance $2f$ from the lens, there is a point source S, which emits monochromatic light. A very thin, non-uniformly transparent circular flat plate of radius R is placed immediately behind the lens, as shown in Figure 3.1. A screen E is placed at a distance $3f$ from the lens. The spot of light on the screen has the greatest possible illumination.

Determine the distribution law of the transparency of the plate along its radius, $T = f(r)$.

We know the radius of the lens, R_L.

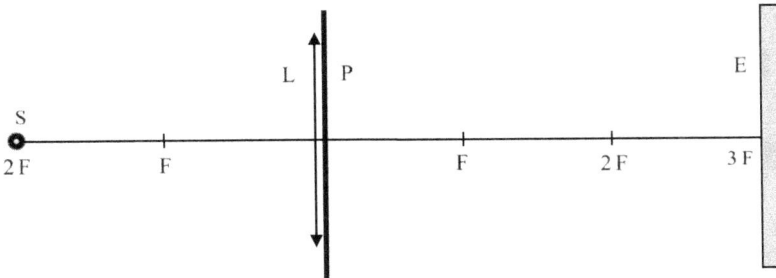

Fig. 3.1

Solution

From the lens formula, it follows that:

$$\frac{1}{d} + \frac{1}{d'} = \frac{1}{f}; \quad d = 2f; \quad d' = 2f.$$

That is, the image is formed at a f distance from the screen.

Under these conditions, if R_L is the radius of the lens, then the radius of the light spot on the screen is

$$R_\mathrm{spot} = \frac{R_\mathrm{L}}{2}.$$

At a point on the screen located at the distance r from the main optical axis ($r < R_\mathrm{spot}$), the light from the lens arrives at a distance $2r$ from the main optical axis. The illumination at this point on the screen, in the absence of the plate P, is

$$E_0 = \frac{I}{x^2} \cos i,$$

where I is the luminous intensity of the source, x is the distance from the source to the illuminated surface, and i is the angle of light incident on the respective surface,

$$E_0(r) = \frac{I}{f^2 + r^2} \cdot \frac{f}{\sqrt{f^2 + r^2}} = \frac{If}{(f^2 + r^2)^{3/2}}.$$

In the presence of plate P, as Figure 3.2 indicates, the illumination of the same point on the screen is

$$E(r) = E_0(r) \cdot T(2r) = \frac{If}{(f^2 + r^2)^{3/2}} \cdot T(2r),$$

where $T(2r)$ is the transparency of plate P at the points located at a distance of $2r$ from the main optical axis. By introducing a proportionality coefficient, k, whose value must be determined, it can be written that

$$T(2r) = k\left(f^2 + r^2\right)^{3/2}.$$

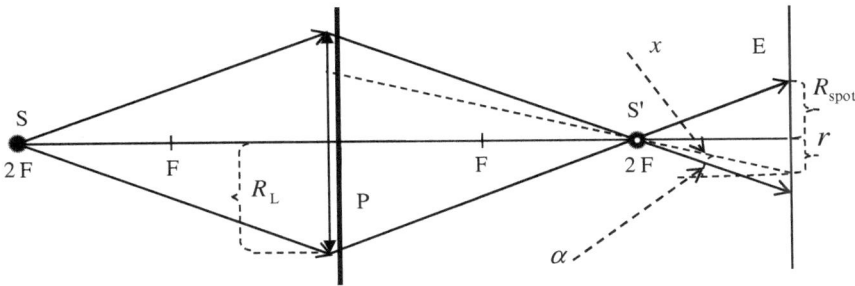

Fig. 3.2

Under these conditions, the transparency of the plate at a distance r from the main optical axis of the lens is

$$T(r) = k \left(f^2 + \frac{r^2}{4} \right)^{3/2}.$$

Since $T(r) \leq 1$, it follows that the maximum possible illumination of the spot at all points on the screen in the presence of the plate P cannot be greater than the minimum illumination of the spot in the absence of the plate. The minimum illumination of the plate P is obtained at the points on the outline of the spot, i.e., at the distance R_{spot} from the main optical axis of the lens.

It results that:

$$T(R_{\text{L}}) = 1;$$

$$k \left(f^2 + \frac{R_{\text{L}}^2}{4} \right)^{3/2} = 1; \quad k = \frac{1}{\left(f^2 + \frac{R_{\text{L}}^2}{4} \right)^{3/2}};$$

$$T(r) = \left(\frac{4f^2 + r^2}{4f^2 + R_{\text{L}}^2} \right)^{3/2}.$$

Problem 4. Light Transmitted through a Glass Cylinder

A thin opaque disk, provided with a vertical rectangular slit, covers one of the ends of a long glass cylinder, having radius R, placed on a horizontal support, as indicated in Figure 4.1. The width of the slot, d, is adjustable. A parallel beam of monochromatic light arrives at the opaque disk at an angle of incidence α, so that the illumination of the disk is E. At the other end of the cylinder is a photocell, F.

Establish the dependence on the slot width, d, of the luminous flux, Φ, recorded by the photocell, $\Phi = f(d)$. The refractive index of the glass, n, is known.

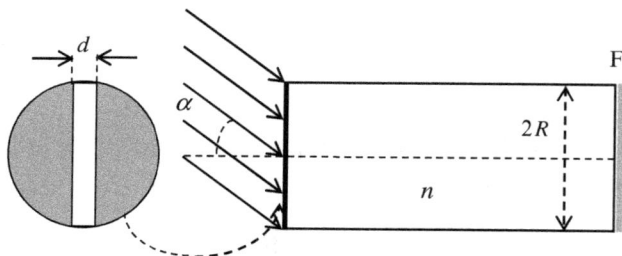

Fig. 4.1

Solution

After refraction on the face of the cylinder, as shown in Figure 4.2, the light enters the glass at an angle β with respect to the axis of the system, so that

$$\frac{\sin \alpha}{\sin \beta} = n.$$

We study the light rays located in a vertical plane at a distance x from the cylinder's axis (Figures 4.2 and 4.3). The points where all these rays arrive at the glass–air boundary are on the same side (the generator AB).

We calculate the angle γ under which the light rays fall on the boundary AB, and we compare it with the limiting angle of total internal reflection.

We study the course of the ray CD, located in the plane AA′B′B. We draw through point C a line parallel to the line AB (and to the axis of the cylinder) and delimit on it the segment CC′ = AD = L.

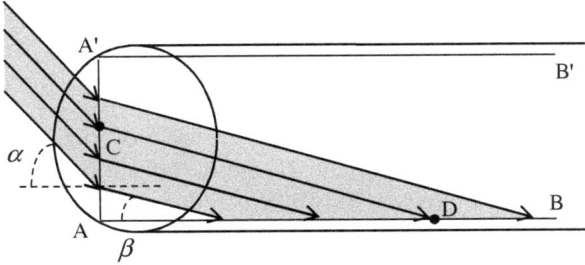

Fig. 4.2

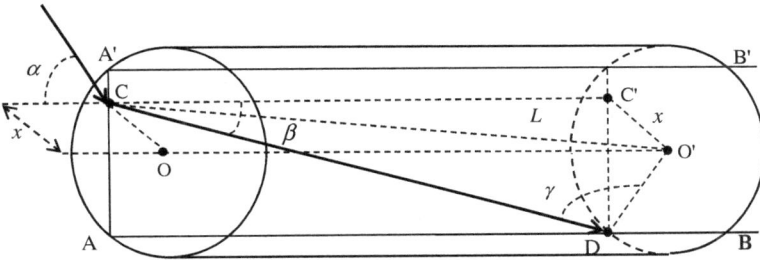

Fig. 4.3

From $\Delta CC'O'$ and from the cosine theorem written for $\Delta CO'D$, it results that:

$$(CO')^2 = L^2 + x^2;$$

$$L^2 + x^2 = \frac{L^2}{\cos^2 \beta} + R^2 - 2 \cdot \frac{LR}{\cos \beta} \cdot \cos \gamma.$$

From $\Delta C'O'D$, we get

$$L^2 \tan^2 \beta + x^2 = R^2.$$

It results that

$$\cos \gamma = \frac{\sqrt{R^2 - x^2}}{R} \cdot \frac{\sin \alpha}{n}.$$

For a certain value, $x = x_0$, the angle γ becomes equal to the corresponding limit angle for total reflection:

$$\gamma = \gamma_0 = \arcsin \left(\frac{1}{n} \right); \quad \sin \gamma_0 = \frac{1}{n}.$$

As a result,

$$x_0^2 = R^2(1 + \sin^2 \alpha - n^2)/\sin^2 \alpha,$$

so that all the light rays that fall on the side of the cylinder at the distance $x \geq x_0$ from its vertical diameter, at the first ray that falls on the side of the cylinder, fulfill the conditions for total internal reflection.

In particular, if

$$n^2 = 1 + \sin^2 \alpha,$$

the internal reflection conditions are met for all rays ($x_0 = 0$).

In this way, in the absence of the opaque plate, the light rays that fall on the entrance face arrive at the photocell in the shaded regions in Figure 4.4, for which

$$S_0 = 2R^2 \arccos\left(\frac{x_0}{R}\right) - 2x_0\sqrt{R^2 - x_0^2}.$$

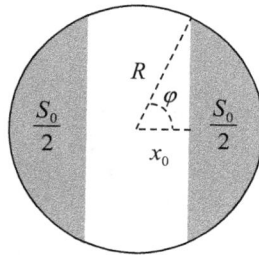

Fig. 4.4

In the presence of the opaque plate with a slit, the light rays do not reach the photocell if the slot width is $d < 2x_0$.

If

$$2x_0 < d \leq 2R,$$

the luminous flux received by the photocell is determined by the light rays that fall on the entrance face (in the shaded region in Figure 4.5), for which

$$S = S_0 - S_1 = 2R^2 \arccos\left(\frac{x_0}{R}\right) - 2x_0\sqrt{R^2 - x_0^2}$$

$$- 2R^2 \arccos\left(\frac{d}{2R}\right) + d\sqrt{R^2 - \frac{d^2}{4}}.$$

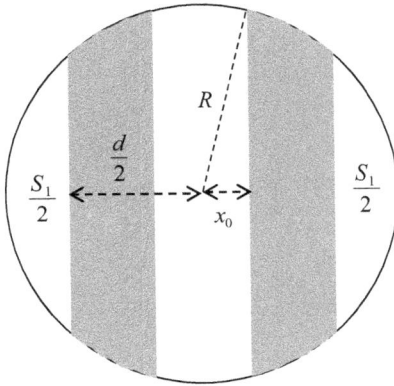

Fig. 4.5

The size of the flux is proportional to the surface area of the shaded region. Therefore, if $d \leq 2x_0$, then $\Phi = 0$, and if $2x_0 \leq d \leq 2R$, it results that:

$$\Phi = E\left[S_0 - 2R^2 \arccos\left(\frac{d}{2R}\right) + d\sqrt{R^2 - \frac{d^2}{4}}\right];$$

$$S_0 = 2R^2 \arccos\left(\frac{x_0}{R}\right) - 2x_0\sqrt{R^2 - x_0^2}.$$

Chapter 9

International Pre-Olympic Physics Contest, Cluj-Napoca, Romania

Problem 1. The Maximum Radial Speed on an Elliptical Orbit

The ecliptic orbit of star σ around the star Σ is represented in Figure 1.1, where the observation line of the system is also indicated.

(a) *Determine* the maximum radial speed of the star σ. *We know*: the semi-axes of the ellipse, a and b; K, the constant of universal attraction; and M, the mass of the star Σ.

(b) *Determine* the time interval, τ_1, in which the radial component of star σ's velocity increases from zero to the maximum value, as well as the time interval, τ_2, in which this component decreases from the maximum value to the zero value if $\tau_1/\tau_2 = n$.

Solution

(a) For a star σ, which moves with speed $\vec{v}$ relative to an observer O, as shown in Figure 1.2, we define the radial speed, $\vec{v}_{rad}$, as being the component of the speed oriented along the direction of observation of the star, and the tangential speed, $\vec{v}_{tang}$, as being the component of the speed perpendicular to the direction of vision.

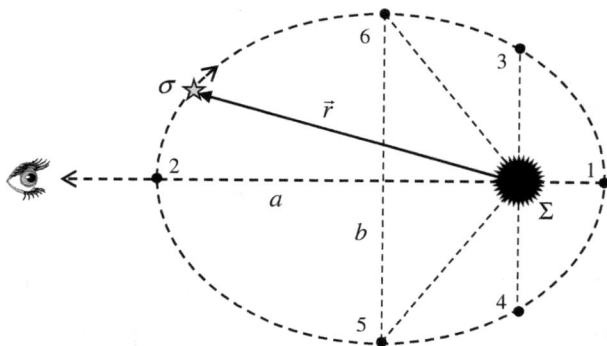

Fig. 1.1

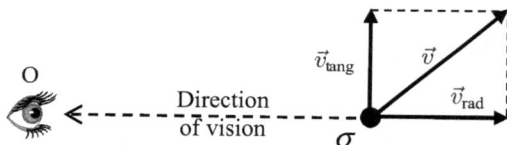

Fig. 1.2

In these conditions, as shown in Figure 1.2, it results that:

$$v_1 = v_{\max} = v_{\text{tang. max}}; \quad v_2 = v_{\min} = v_{\text{tang. min}};$$

$$v_{\text{rad},1} = v_{\text{rad},2} = 0;$$

$$v_5 = v_6 = v = v_{\text{rad}}.$$

The acceleration of star σ due to the central gravitational force, $\vec{a} = \frac{\vec{F}}{m}$, exerted on it by star Σ can be decomposed into a tangential component and a radial component, as indicated by Figure 1.3:

$$\vec{a} = \vec{a}_{\text{rad}} + \vec{a}_{\text{tang}},$$

of which only the radial component, $\vec{a}_{\text{rad}}$, determines the variation in the radial component, $\vec{v}_{\text{rad}}$, of star σ's velocity.

Figure 1.4 shows the evolution of the radial and tangential components of the velocity and acceleration vectors in relation to the observer when the line of sight is in the plane of the orbit.

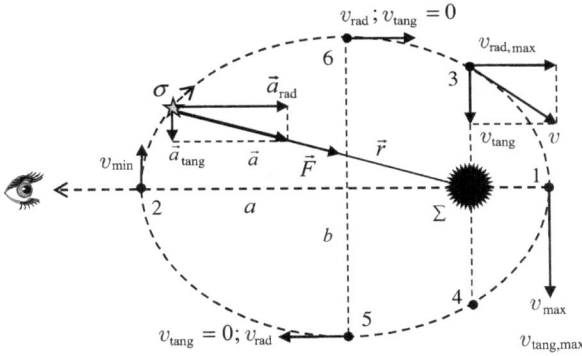

Fig. 1.3

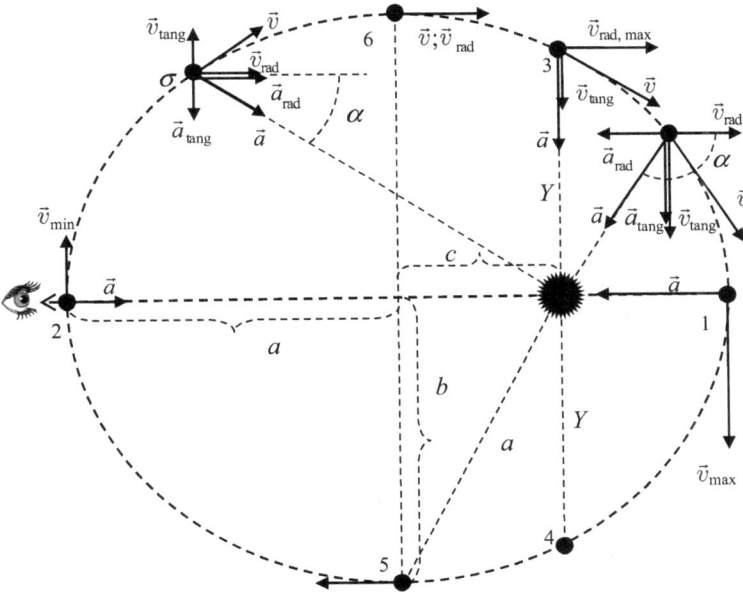

Fig. 1.4

On the 2–6–3 sector of the ellipse, when the angle α between the acceleration vector, $\vec{a}$, and the radial velocity vector, $\vec{v}_{\text{rad}}$, is

$$0 \le \alpha \le 90°,$$

the effect of acceleration (the effect of the force of gravitational attraction) on the radial component of the speed is an accelerating effect. As a result, on this sector, the radial component of the velocity is

$$0 \leq v_{\text{rad}} \leq v_{\text{rad,max}}.$$

On sector 3–1 of the ellipse, when the angle α between the acceleration vector, $\vec{a}$, and the radial velocity vector, $\vec{v}_{\text{rad}}$, is

$$90 \leq \alpha \leq 180°,$$

the accelerating effect (the effect of the gravitational attraction force) on the radial component of the velocity is a decelerating effect. As a result, on this sector, the radial component of the velocity is

$$0 \leq v_{\text{rad}} \leq v_{\text{rad,max}}.$$

Conclusions:

- for $\alpha = 90°$, when the star σ is in position 3, the radial component of its velocity has the maximum value;
- on sector 2–3 of the ellipse, the radial effect of the gravitational attraction force is an accelerating effect, and on sector 3–1 of the ellipse, the radial effect of the gravitational attraction force is a decelerating effect;
- on sector 2–3 of the ellipse, the tangential effect of the gravitational attraction force is a decelerating effect, and on sector 3–1 of the ellipse, the tangential effect of the gravitational attraction force is an accelerating effect;
- on sector 2–3 of the ellipse, the vectors $\vec{v}_{\text{rad}}$ and $\vec{a}_{\text{rad}}$ have identical orientations, and on sector 3–1, the vectors $\vec{v}_{\text{rad}}$ and $\vec{a}_{\text{rad}}$ have opposite orientations;
- at point 3, where the direction of the vector $\vec{a}_{\text{rad}}$ changes, the vector $\vec{v}_{\text{rad}}$ has an extreme (maximum) value.

At some point on the ellipse, the kinetic moment of the star σ is:

$$\vec{L} = \vec{r} \times m\vec{v}; \quad L = rmv \cdot \sin(\vec{r}; \vec{v}).$$

Thus, according to the law of conservation of kinetic momentum, it results that:

$$\vec{L} = \text{constant}; \quad L = rmv \cdot \sin(\vec{r}; \vec{v}) = r_{\text{min}} m v_{\text{max}} = r_{\text{max}} m v_{\text{min}};$$

$$r_{\min} = \frac{p}{1+e}; \quad v_{\max} = \sqrt{\frac{KM(1+e)}{1-e}}; \quad r_{\min} = \frac{p}{1-e};$$

$$v_{\min} = \sqrt{\frac{KM(1-e)}{1+e}};$$

$$p = a(1-e^2); \quad p = b\sqrt{1-e^2};$$

$$L = m\sqrt{\frac{KMa^2(1-e)^2(1+e)}{a(1-e)}} = mb\sqrt{\frac{KM}{a}}.$$

Corresponding to position 3 of star σ, when its kinetic moment is

$$L = Y m v_{\mathrm{rad,max}},$$

it results that:

$$Y m v_{\mathrm{rad,max}} = mb\sqrt{\frac{KM}{a}}; \quad v_{\mathrm{rad,max}} = \frac{b}{Y} \cdot \sqrt{\frac{KM}{a}};$$

$$\frac{x^2}{a^2} + \frac{y^2}{b^2} = 1;$$

$$x = c; \quad y = Y; \quad \frac{c^2}{a^2} + \frac{Y^2}{b^2} = 1; \quad c^2 = a^2 - b^2; \quad Y = \frac{b^2}{a};$$

$$v_{\mathrm{rad,max}} = \frac{a}{b} \cdot \sqrt{\frac{KM}{a}}.$$

(b)

$$\Delta v_{\mathrm{rad},23} = v_{\mathrm{rad,max}} - 0 = v_{\mathrm{rad,max}}; \quad \Delta t_{23} = \tau_1;$$

$$\Delta v_{\mathrm{rad},31} = 0 - v_{\mathrm{rad,max}} = -v_{\mathrm{rad,max}}; \quad \Delta t_{31} = \tau_2;$$

$$\frac{\Delta S_1}{\tau_1} = \frac{\Delta S_2}{\tau_2} = \frac{S_{\mathrm{elipse}}}{T} = \frac{\pi ab}{T}; \quad \Delta S_1 = \pi ab \cdot \frac{\tau_1}{T}; \quad \Delta S_2 = \pi ab \cdot \frac{\tau_2}{T};$$

$$\Delta S_1 + \Delta S_2 = \frac{1}{2} \cdot S_{\mathrm{elipse}}; \quad \pi ab \cdot \frac{\tau_1}{T} + \pi ab \cdot \frac{\tau_2}{T} = \frac{1}{2} \cdot \pi ab;$$

$$\tau_1 + \tau_2 = \frac{1}{2}T; \quad \frac{\tau_1}{\tau_2} = n;$$

$$\tau_1 = \frac{nT}{2(n+1)}; \quad \tau_2 = \frac{T}{2(n+1)}; \quad T = 2\pi\sqrt{\frac{a^3}{KM}}.$$

Problem 2. A Satellite in the Upper Atmosphere

A satellite moves around the Earth in an approximately circular orbit, with speed v. The change in the satellite's orbit is determined by the fact that the air in the upper atmosphere exerts on it a resistance force $F_r = qv^\alpha$, where q is a known constant.

    Determine the constant α, knowing that the radius of the satellite's orbit varies very slowly.

Solution

If after the time Δt, the radius of the satellite's orbit decreases by the amount $\Delta R \ll R$, then the variation of the total mechanical energy of the satellite–Earth system is:

$$W_R = -K\frac{mM}{R} + \frac{mv^2}{2}; \quad K\frac{mM}{R^2} = \frac{mv^2}{R};$$

$$W_R = -\frac{1}{2}K\frac{mM}{R};$$

$$W_{R-\Delta R} \approx -K\frac{mM}{R-\Delta R} + \frac{mv^2}{2}; \quad K\frac{mM}{R-\Delta R} = \frac{mv^2}{R-\Delta R};$$

$$W_{R-\Delta R} = -\frac{1}{2}K\frac{mM}{R-\Delta R};$$

$$\Delta W = W_{R-\Delta R} - W_R = -\frac{1}{2}KmM\frac{\Delta R}{R(R-\Delta R)};$$

$$\Delta W \approx -\frac{1}{2}KmM\frac{\Delta R}{R^2}.$$

During this time, Δt, the mechanical work of the air resistance force is:

$$\Delta L = F_r v \Delta t = qv^{\alpha+1}\Delta t; \quad v = \sqrt{\frac{KM}{R}}.$$

According to the theorem of the variation in the total mechanical energy of a system, it follows that:

$$\Delta W = -\Delta L;$$

$$\frac{1}{2}KmM\frac{\Delta R}{R^2} = q\left(\frac{KM}{R}\right)^{\frac{\alpha+1}{2}}\Delta t;$$

$$q\left(\frac{KM}{R}\right)^{\frac{\alpha+1}{2}} = \frac{1}{2}\frac{KmM}{R^2}\frac{\Delta R}{\Delta t},$$

where $\frac{\Delta R}{\Delta t}$, representing the speed of the satellite's approach to the Earth, is a constant quantity;

$$q\frac{(KM)^{\frac{\alpha+1}{2}}}{R^{\frac{\alpha+1}{2}}} = \frac{1}{2}\frac{KmM}{R^2}\frac{\Delta R}{\Delta t}.$$

Because the previous equality must be true regardless of the value of R, it follows that:

$$\frac{\alpha+1}{2} = 2; \quad \alpha = 3.$$

Problem 3. An Abandoned Satellite

A satellite abandoned in a circular orbit at $H_0 \ll R_0$, where R_0 is the radius of the Earth, is braked in the upper layers of the atmosphere. The angular acceleration of the satellite is ε.

Determine the satellite's altitude after time t. The acceleration is known according to the terrestrial gravity on the ground, g_0.

Solution

The resistance to the forward movement of the atmosphere reduces the satellite's orbiting altitude. In this way, the satellite advances towards regions where the gravitational acceleration is higher. The movement of the satellite is accelerated, so its speed is increasing (paradox of satellites). This result is in contradiction to what we would have expected to happen, in general, because of friction.

Under these conditions, it can be considered that the angular speed of the satellite increases over time according to the law of uniformly accelerated circular motion,

$$\omega = \omega_0 + \varepsilon\, t,$$

from which it results that:

$$\omega_0 = \frac{v_0}{R_0 + H_0} = \frac{1}{R_0 + H_0}\sqrt{\frac{KM}{R_0 + H_0}} = \frac{R_0}{R_0 + H_0}\sqrt{\frac{g_0}{R_0 + H_0}};$$

$$\omega = \frac{v}{R_0 + H} = \frac{1}{R_0 + H}\sqrt{\frac{KM}{R_0 + H}} = \frac{R_0}{R_0 + H}\sqrt{\frac{g_0}{R_0 + H}};$$

$$\frac{R_0}{R_0 + H}\sqrt{\frac{g_0}{R_0 + H}} = \frac{R_0}{R_0 + H_0}\sqrt{\frac{g_0}{R_0 + H_0}} + \varepsilon t;$$

$$\frac{R_0^2 g_0}{(R_0 + H)^3} = \frac{R_0^2 g_0}{(R_0 + H_0)^3} + 2\frac{R_0 \varepsilon t}{R_0 + H_0}\sqrt{\frac{g_0}{R_0 + H_0}} + \varepsilon^2 t^2;$$

$$\varepsilon^2 t^2 \to 0;$$

$$\frac{R_0^2 g_0}{(R_0 + H)^3} \approx \frac{R_0^2 g_0}{(R_0 + H_0)^3} + 2\frac{R_0 \varepsilon t}{R_0 + H_0}\sqrt{\frac{g_0}{R_0 + H_0}};$$

$$\frac{R_0^2 g_0}{R_0^3\left(1 + \frac{H}{R_0}\right)^3} = \frac{R_0^2 g_0}{R_0^3\left(1 + \frac{H_0}{R_0}\right)^3} + 2\frac{R_0 \varepsilon t}{R_0 + H_0}\sqrt{\frac{g_0}{R_0 + H_0}};$$

$$\frac{g_0}{R_0}\left(1 - 3\frac{H}{R_0}\right) = \frac{g_0}{R_0}\left(1 - 3\frac{H_0}{R_0}\right) + 2\frac{R_0 \varepsilon t}{R_0 + H_0}\sqrt{\frac{g_0}{R_0 + H_0}};$$

$$H = H_0 - \frac{2}{3}\frac{R_0^3 \varepsilon t}{g_0(R_0 + H_0)}\sqrt{\frac{g_0}{R_0 + H_0}}.$$

Problem 4. Satellite in a Rarefied Atmosphere

A satellite of the Earth with mass m, moving in the upper atmosphere in a circular orbit with radius r, encounters a resistance force F_r from the rarefied air.

Determine the satellite's speed variation after a rotation around the Earth. The altitude of the satellite's orbit is considered small compared to the radius of the Earth. *We know:* M, the mass of the Earth, and K, the constant of gravitational attraction.

Solution

In a circular orbit with radius r, the total mechanical energy of the satellite–Earth system is

$$E = \frac{mv^2}{2} - K\frac{mM}{r} = -\frac{1}{2}K\frac{mM}{r}.$$

After a rotation, when the radius of the orbit has decreased and is $r' = r - \Delta r$, so that $\Delta r \ll r$, the total mechanical energy of the satellite–Earth system has decreased and is

$$E' = \frac{mv'^2}{2} - K\frac{mM}{r - \Delta r} = -\frac{1}{2}K\frac{mM}{r - \Delta r}.$$

The variation in the total mechanical energy of the system due to air resistance is

$$\Delta E = E' - E = -\frac{1}{2}KmM\frac{\Delta r}{r(r - \Delta r)}.$$

According to the theorem of the variation in the total mechanical energy of a system, it results that:

$$\Delta E = L_{\mathrm{r}}; \quad -\frac{1}{2}KmM\frac{\Delta r}{r(r - \Delta r)} = -2\pi r F_{\mathrm{r}};$$

$$r - \Delta r \approx r; \quad \Delta r = \frac{4\pi r^3 F_{\mathrm{r}}}{KmM}.$$

As a result of the action of the air resistance force, the radius of the satellite's orbit decreases, and the speed of the satellite increases (paradox of satellites). The kinetic energies of the satellite in orbits with radii r and $(r - \Delta r)$ are, respectively:

$$E_{\mathrm{c}} = \frac{mv^2}{2} = K\frac{mM}{2r}; \quad E'_{\mathrm{c}} = \frac{mv'^2}{2} = K\frac{mM}{2(r - \Delta r)},$$

so that the variation in the kinetic energy of the satellite after a complete revolution around the Earth is:

$$\Delta E_{\mathrm{c}} = K\frac{mM\Delta r}{2r^2} = 2\pi r F_{\mathrm{r}};$$

$$\Delta E_{\mathrm{c}} = \frac{m}{2}\left(v'^2 - v^2\right) = \frac{m}{2}\left(v' - v\right)\left(v' + v\right)$$

$$= \frac{m}{2}\left(2v + \Delta v\right)\Delta v \approx mv\Delta v;$$

$$\Delta v = \frac{2\pi r F_{\mathrm{r}}}{mv} = \frac{2\pi r F_{\mathrm{r}}}{m}\sqrt{\frac{r}{KM}}.$$

Problem 5. Gyroscopic Disk

The ends of a horizontal axis with length $2l$, represented in Figure 5.1, in the middle of which is fixed a gyroscopic disk with mass m and with the moment of inertia compared to its own axis I, are inserted securely into the inner rings of some spherical ball bearings, while the outer rings of the bearings rest on the tops of the vertical supports S_1 and S_2.

    The disk rotates uniformly around its own axis with angular velocity $\vec{\omega}$, and the support S_2 rotates uniformly around the support S_1 with angular velocity $\vec{\Omega}$.

    Determine the value of Ω for which the system continues to rotate uniformly around the support S_1 and, after removing the support S_2, the axis of the disk rests in the horizontal plane.

    It is known that $\Omega \ll \omega$.

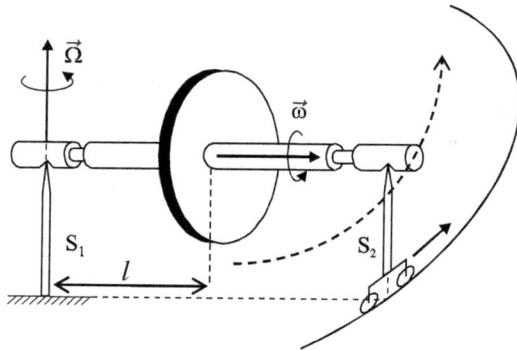

Fig. 5.1

Solution

If the axis of the disk were fixed in space and the disk rotated around it with angular velocity $\vec{\omega}$, then the kinetic moment of the disk, $\vec{J}_{rot} = I\vec{\omega}$, would be oriented along the axis of the disk.

When the axis of the disk also rotates around the vertical support S_1, then an additional angular momentum, $\vec{j}_{rot}$, occurs, so that the total angular momentum, $\vec{J}_{rot} + \vec{j}_{rot}$, is no longer oriented along the disk's axis.

If $\Omega \ll \omega$, then $j_{rot} \ll J_{rot}$ and, as a result, j_{rot} can be neglected in relation to J_{rot}, so that in these conditions, the total kinetic moment of the system is identified with the $\vec{J}_{rot}$ component and is permanently oriented along the disk's axis, rotating in space "solidary" with it, as shown in Figure 5.2.

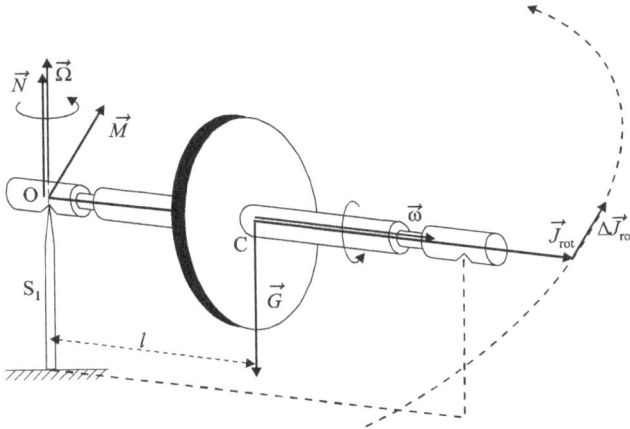

Fig. 5.2

If $\vec{J}_{rot}$ is the kinetic momentum vector of the system at time $t_0 = 0$, with the origin at point O (the sprue point of the disk axis on top of the support S_1) and with its direction along the disk axis, then, after the time Δt, the kinetic moment of the system is $\vec{J}'_{rot}$, a vector with its origin at point O and modulus $J'_{rot} = J_{rot}$, also oriented along the axis of the disk, but in its new position, rotated in the horizontal plane with an angle $\Delta\varphi$, as shown in Figure 5.3. From this, it results that the peak of the angular momentum vector of the system during the interval of time Δt is represented by the vector:

$$\Delta\vec{J}_{\text{rot}} = \vec{J}'_{\text{rot}} - \vec{J}_{\text{rot}};$$

$$\Delta J_{\text{rot}} \approx J_{\text{rot}}\Delta\varphi,$$

whose orientation, for $\Delta\varphi$, is very small, $\Delta\vec{J}_{\text{rot}} \perp \vec{J}_{\text{rot}}$.

It is known, however, that any variation in the kinetic moment of a mechanical system is caused by a non-zero resultant moment of the external forces acting on the system. The external forces acting on the system are the normal reaction of the top of the support, $\vec{N}$, and the weight, $\vec{G}$. Their resulting moment is reduced to

$$\vec{M} = \overrightarrow{OC} \times \vec{G},$$

a horizontal vector perpendicular to the disk axis ($\vec{M} \perp \vec{J}_{\text{rot}}$; $\vec{M}//\Delta\vec{J}_{\text{rot}}$), having the modulus

$$M = mgl.$$

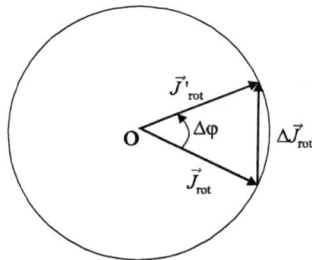

Fig. 5.3

According to the kinetic momentum variation theorem, it follows that:

$$\vec{M} = \frac{\Delta\vec{J}_{\text{rot}}}{\Delta t};$$

$$M = \frac{\Delta J_{\text{rot}}}{\Delta t};$$

$$mgl = \frac{J_{\text{rot}}\Delta\varphi}{\Delta t}; \quad \frac{\Delta\varphi}{\Delta t} = \Omega;$$

$$\Omega = \frac{mgl}{J_{\text{rot}}}; \quad \Omega = \frac{mgl}{I\omega}.$$

This represents the angular velocity with which the system must rotate around the vertical of the support S_1 so that, in the absence of the support S_2, the system continues its movement, i.e., while the disk rotates around its axis, this axis rotates around the vertical point O, remaining permanently in the same horizontal plane. This movement of the disk axis is called *precession movement*.

Since, during its precession movement, the disk axis remains permanently in the horizontal plane of points O and C, the resultant of the external forces acting on the system is

$$\vec{N} + \vec{G} = 0,$$

from which it follows that there must be a pressure $\vec{F}$, oriented vertically downwards, exerted by the end of the disk axis on the top of the support S_1, equal to the weight of the system ($\vec{F} = \vec{G}$), to which a normal reaction should correspond:

$$\vec{N} = -\vec{F} = -\vec{G}.$$

Indeed, the force $\vec{F}$ exists, and it is the result of the gyroscopic effect. If on the free end of the disk axis, when the axis performs the precession movement in the horizontal plane of the support point, a small impulse is applied for a short time vertically downwards/upwards, then, simultaneously with the horizontal precession, the axis will oscillate vertically up and down, representing a *nutation movement*.

Problem 6. Collision of Disks

Two disks (A and B) with identical masses and thicknesses but with different radii (R_A and R_B, respectively) can move over the flat and horizontal surface of a table with an air cushion, sliding without friction, as shown in Figure 6.1.

Disk A moves through horizontal plane translation with speed $\vec{v}$, parallel to the axis OX, at a distance b from it ($R_B < b < R_A + R_B$) and perfectly elastically collides with disk B at rest (its center being at the origin O of the XY axis system from the horizontal plane of the table), so that the distance from its center to the direction of $\vec{v}$ is b, as indicated in Figure 6.1.

(a) *Determine* the velocities of the two disks after their perfectly elastic collision if there is no relative movement of one disk in relation to the other disk at the contact point of the disks. Special case: $b = 0$.

(b) *Determine* the directions in which the two disks move after the collision.

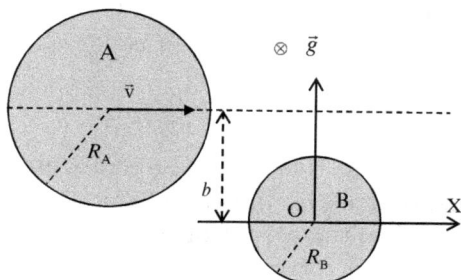

Fig. 6.1

Solution

(a) Corresponding to the moment of collision of the disks, at their point of contact, let $\vec{n}$ and $\vec{\tau}$, respectively, be versors of the line of the centers and of the common tangent at the point of contact and the angle α between the direction of the centers (at the moment of collision) and the axis OX, as shown in Figure 6.2.

If $\vec{u}_A$ and $\vec{u}_B$ are the velocities of the centers of the two disks after the collision, as shown in Figure 6.3, in accordance with the law of conservation of momentum, maintaining the direction of the centers of the disks, with the versors $\vec{n}$ and $\vec{\tau}$, we get

$$m\, u_{A,n} + m\, u_{B,n} = mv \cos \alpha. \tag{9.1}$$

In the projection on the tangent of the point of contact of the two disks, where equal and opposite frictional forces act on the disks, the variations in the impulses of the two disks are equal and in opposite directions:

$$\frac{\Delta \vec{p_1}}{\Delta t} = -\frac{\Delta \vec{p}}{\Delta t}; \quad \Delta p_1 = \Delta p_2;$$

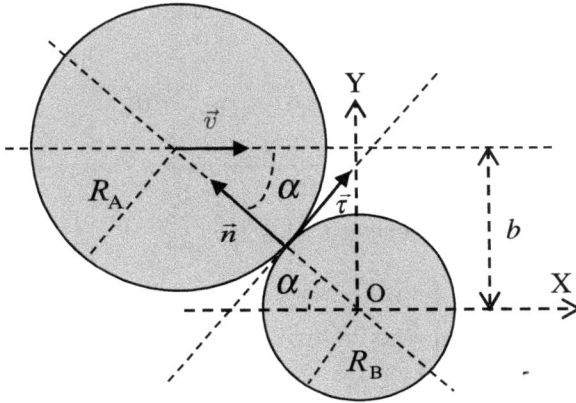

Fig. 6.2

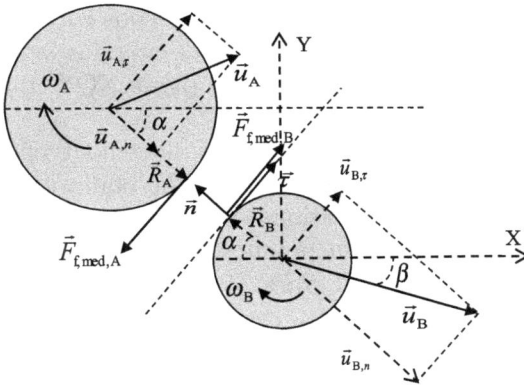

Fig. 6.3

$$\Delta p_1 = mv \sin \alpha - mu_{A,\tau}; \quad \Delta p_2 = mu_{B,\tau};$$

$$-mu_{A,\tau} + mv \sin \alpha = mu_{B,\tau};$$

$$mu_{A,\tau} + mu_{B,\tau} = mv \sin \alpha,$$

where m is the mass of a disk.

From the definition of the coefficient of restitution for the collision of the two disks, it follows that:

$$k = \frac{u_{B,n} - u_{A,n}}{v_{A,n} - v_{B,n}},$$

where $v_{A,n}$ and $v_{B,n}$ are the components along $\vec{n}$ of the disk velocities before the collision;

$$v_{A,n} = v \cos \alpha; \quad v_{B,n} = 0;$$

$$k = \frac{u_{B,n} - u_{A,n}}{v \cos \alpha} = 1,$$

because the collision is perfectly elastic;

$$\begin{cases} u_{B,n} - u_{A,n} = v \cos \alpha; \\ u_{A,n} + u_{B,n} = v \cos \alpha; \end{cases}$$

$$u_{A,n} = 0; \quad u_{B,n} = v \cos \alpha.$$

Due to the frictional forces from the contact of the disks (forces in the direction of $\vec{\tau}$, equal in mode and of opposite directions), they will also acquire after the collision an instantaneous rotational movement around the axes of the centers, with the angular velocities $\vec{\omega}_A$ and $\vec{\omega}_B$, respectively, perpendicular to the plane XOY and of the same direction.

In this way, after the collision, the movement of each disk is a parallel plane movement (the result of the composition of a translational movement and a rotational movement). Because there is no relative motion at the contact points of those two disks, the resultant instantaneous tangential velocities of the contact points on the two disks, as shown in Figure 6.4, are equal in mode and of the same

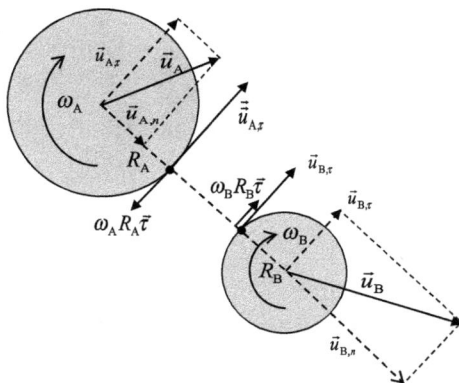

Fig. 6.4

direction, so that

$$u_{A,\tau} - \omega_A R_A = u_{B,\tau} + \omega_B R_B.$$

Since the directions of the instantaneous rotations at the contact points of the disks are opposite, the sliding friction forces that act on each disk in the direction of the common tangent will be opposite. The kinetic moment of disk B, as a result of the collision, is: Paragraph Font result A. On the other hand, from the theorem of the variation of angular momentum, for disk B, we have:

$$\vec{R}_B \times m\vec{u}_B = \vec{R}_B \times \vec{P}_{int,med} + \vec{R}_B \times \vec{P}_{ext,B};$$

$$P_{ext,B} \ll P_{int,med};$$

$$\vec{P}_{int,med} = \vec{F}_{f,\,B;med}\,\Delta t;$$

$$\vec{F}_{f,B;med} = \frac{1}{\Delta t} \int_{t_0}^{t_c} \vec{F}_{f,B}dt.$$

It results that:

$$R_B m u_{B,\tau} = R_B P_{int,med} = R_B F_{f,B;med}\Delta t;\quad F_{f,med,B} = \frac{m u_{B,\tau}}{\Delta t}.$$

On the other hand, from the theorem of kinetic momentum variation, for disk B, we have:

$$\frac{\Delta \vec{J}_{rot,B}}{\Delta t} = \vec{M}_B = \vec{R}_B \times \vec{F}_{f,med,B};$$

$$I_B \omega_B = R_B F_{f,med,B}\Delta t;$$

$$R_B m\, u_{B,\tau} = I_B \omega_B.$$

From the theorem of kinetic momentum variation, for disk A, we have:

$$\vec{R}_A \times m\vec{u}_A = \vec{R}_A \times \vec{P}_{int,med} + \vec{R}_A \times \vec{P}_{ext,A};$$

$$P_{ext,A} \ll P_{int,med};$$

$$\vec{P}_{int,med} = \vec{F}_{f,\,A;med}\,\Delta t;$$

$$\vec{F}_{f,\,A;med} = \frac{1}{\Delta t} \int_{t_0}^{t_c} \vec{F}_{f,A}dt;$$

$$R_A m u_{A,\tau} = R_A P_{int,med} = R_A F_{f,A;med}\Delta t;\quad F_{f,med,A} = \frac{m u_{A,\tau}}{\Delta t};$$

$$u_{A,\tau} = \omega_A R_A + u_{B,\tau} + \omega_B R_B;$$

$$u_{A,\tau} = \omega_A R_A + u_{B,\tau} + \omega_B R_B;$$

$$\frac{\Delta \vec{J}_{\text{rot,A}}}{\Delta t} = \vec{M}_A = \vec{R}_A \times \vec{F}_{f,\text{med,A}}.$$

The collision of the two disks complies with the law of conservation of kinetic momentum. If I_A and I_B are the moments of inertia of the two disks in relation to the axes of symmetry perpendicular to the plane XOY, it results that:

$$\begin{cases} mu_{A,\tau}(R_A + R_B) + I_A\omega_A + I_B\omega_B = mvb; \\ mu_{B,\tau}R_B = I_B\omega_B; \\ u_{A,\tau} + u_{B,\tau} = v\sin\alpha; \\ u_{A,\tau} = \omega_A R_A + u_{B,\tau} + \omega_B R_B; \end{cases}$$

$$u_{A,\tau} - u_{B,\tau} = \omega_A R_A + \omega_B R_B;$$

$$\omega_A = \frac{u_{A,\tau} - u_{B,\tau} - \omega_B R_B}{R_A};$$

$$\omega_B = \frac{mu_{B,\tau}R_B}{I_B};$$

$$\omega_A = \frac{u_{A,\tau} - u_{B,\tau} - \frac{mu_{B,\tau}R_B^2}{I_B}}{R_A};$$

$$mvb = mu_{A,\tau}(R_A + R_B) + I_A\frac{u_{A,\tau} - u_{B,\tau} - \frac{m u_{B,\tau}R_B^2}{I_B}}{R_A}$$
$$+ mu_{B,\tau}R_B;$$

$$mvb = mu_{A,\tau}(R_A + R_B) + I_A\frac{(u_{A,\tau} - u_{B,\tau})I_B - m u_{B,\tau}R_B^2}{I_B R_A}$$
$$+ mu_{B,\tau}R_B;$$

$$mvb = u_{A,\tau}\left[m(R_A + R_B) + \frac{I_A}{R_A}\right]$$
$$+ u_{B,\tau}\left(mR_B - \frac{I_A}{R_A} - \frac{mR_B^2 I_A}{I_B R_A}\right);$$

$$I_A = \frac{1}{2}mR_A^2; \quad I_B = \frac{1}{2}mR_B^2;$$

$$mvb = u_{A,\tau}\left[m(R_A + R_B) + \frac{1}{2}mR_A\right]$$

$$+ u_{B,\tau}\left(mR_B - \frac{1}{2}mR_A - mR_A\right);$$

$$\begin{cases} vb = u_{A,\tau}\left(R_B + \frac{3}{2}R_A\right) + u_{B,\tau}\left(R_B - \frac{3}{2}R_A\right); \\ u_{A,\tau} + u_{B,\tau} = v\sin\alpha; \end{cases}$$

$$vb = u_{A,\tau}\left(R_B + \frac{3}{2}R_A\right) + (v\sin\alpha - u_{A,\tau})\left(R_B - \frac{3}{2}R_A\right);$$

$$vb = u_{A,\tau}\left(R_B + \frac{3}{2}R_A\right) + (v\sin\alpha)\left(R_B - \frac{3}{2}R_A\right)$$

$$- u_{A,\tau}\left(R_B - \frac{3}{2}R_A\right);$$

$$vb = u_{A,\tau}\left(\frac{3}{2}R_A\right) + (v\sin\alpha)\left(R_B - \frac{3}{2}R_A\right) + u_{A,\tau}\left(\frac{3}{2}R_A\right);$$

$$3R_A u_{A,\tau} = vb - v\sin\alpha\left(R_B - \frac{3}{2}R_A\right);$$

$$b = (R_A + R_B)\sin\alpha;$$

$$u_{A,\tau} = \frac{5}{6}v\sin\alpha; \quad u_{B,\tau} = \frac{1}{6}v\sin\alpha;$$

$$\omega_A = \frac{u_{A,\tau} - u_{B,\tau} - \frac{mu_{B,\tau}R_B^2}{I_B}}{R_A};$$

$$\omega_A = \frac{v\sin\alpha}{3R_A}; \quad \omega_B = \frac{v\sin\alpha}{3R_B};$$

$$\vec{u}_A = \vec{u}_{A,\tau} + \vec{u}_{A,n}; \quad \vec{u}_A = \vec{u}_{A,\tau}; \quad u_A = \frac{5}{6}\frac{vb}{R_A + R_B};$$

$$\vec{u}_B = \vec{u}_{B,\tau} + \vec{u}_{B,n}; \quad u_{B,\tau} = \frac{1}{6} v \sin \alpha; \quad u_{B,n} = v \cos \alpha \,;$$

$$u_B = \sqrt{u_{B,\tau}^2 + u_{B,n}^2} = \frac{v \sqrt{36 \left(R_A + R_B \right)^2 - 35 b^2}}{6 \left(R_A + R_B \right)} \,;$$

$$b = 0; \quad u_A = 0; \quad u_B = v.$$

(b)
$$\vec{u}_B = \vec{u}_{B,x} + \vec{u}_{B,y};$$

$$u_{B,x} = u_{B,\tau} \sin \alpha + u_{B,n} \cos \alpha = v \left[1 - \frac{5}{6} \frac{b^2}{\left(R_A + R_B \right)} \right];$$

$$u_{B,y} = u_{B,\tau} \cos \alpha - u_{B,n} \sin \alpha = -\frac{5 v b}{6} \frac{\sqrt{\left(R_A + R_B \right)^2 - b^2}}{\left(R_A + R_B \right)^2} \,;$$

$$\mathrm{tg}\,\beta = \frac{u_{B,y}}{u_{B,x}},$$

where $\beta = \angle(\vec{u}_B; \mathrm{OX})$.

Problem 7. Rolling Disk

A massy and homogeneous cylindrical disk with radius r, rotating around its own axis with angular velocity ω, rolls without slipping on a horizontal support plane, the plane of the disk being inclined to the support plane at an angle α, as indicated in Figure 7.1. It describes on the support plane a circle with the radius $R \ll r$ during the time T.

Determine T and R. We know the gravitational acceleration, g.

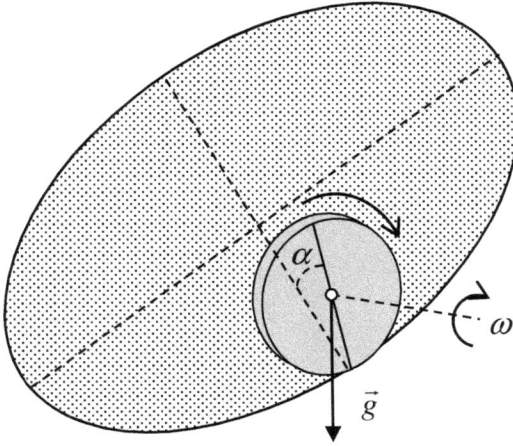

Fig. 7.1

Solution

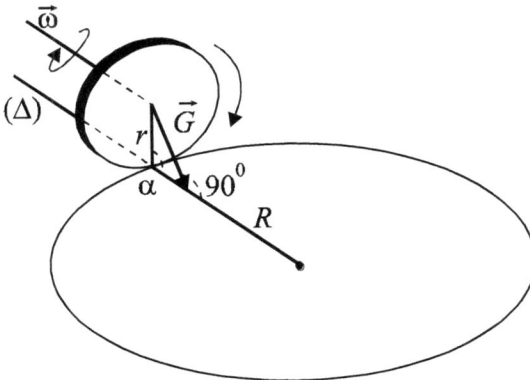

Fig. 7.2

Using the theorem of the variation of kinetic momentum, the Steiner theorem and the drawing in Figure 7.2, it results that:

$$\dot{\vec{J}}_{\text{rot}} = \vec{\Omega} \times \vec{J}_{\text{rot}} = \vec{M};$$

$$\Omega J_{\text{rot}} \sin \alpha = mgr \sin(90 - \alpha);$$

$$J_{\text{rot}} = I_\Delta \omega = (I_0 + mr^2)\omega;$$

$$I_0 = \frac{1}{2}mr^2;$$

$$\frac{3}{2}\Omega \omega r \sin \alpha = g \cos \alpha;$$

$$T = \frac{3\pi \omega r}{g} \operatorname{tg} \alpha;$$

$$\omega r = \Omega R;$$

$$R = \frac{3}{2}\frac{\omega^2 r^2}{g} \operatorname{tg} \alpha.$$

Chapter 10

International Pre-Olympic Physics Contest 2017, Satu Mare, Romania

Problem 1. Bicycle Wheel

An experimenter, at rest on a rotating chair, as shown in Figure 1.1, is handed the vertical axis of a bicycle wheel, which is in rapid rotation with angular velocity $\vec{\Omega}$, and is additionally uniformly loaded with a large axial moment of inertia, I.

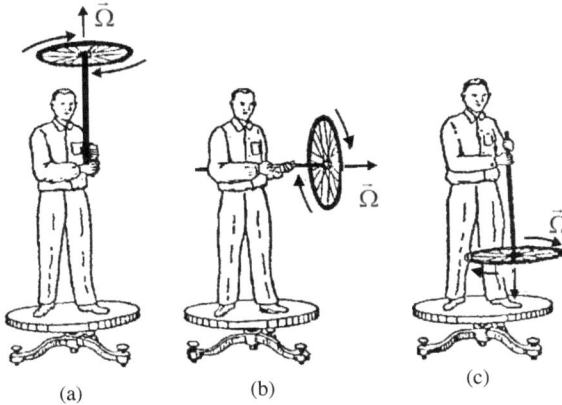

Fig. 1.1

(a) *Determine* the angular speed of the man–chair assembly, $\vec{\omega}$, when the experimenter tilts the wheel axis at an angle α with respect to the vertical. *We know*: I_0, the axial moment of inertia of the man–chair assembly, and I, the axial moment of inertia of the wheel.

Particular cases: $\alpha = 90^0$, $\alpha = 180^0$.

Specify what the experimenter must do to be able to easily hand over the bicycle wheel axle.

(b) *Analyze* the case in which the experimenter receives the entire axis of the bicycle as in scenario b, considering that the chair is initially at rest.

Determine the angular velocity, $\vec{\omega}$, of the man–chair assembly, when, after receiving the wheel axle as in scenario b, the experimenter raises the bicycle wheel axle to a vertical position, so that the wheel is above his head.

Specify what the experimenter must do to be able to easily hand over the bicycle wheel axle.

(c) *Analyze* the case in which the experimenter receives the bicycle wheel axle as in scenario c, considering that the chair is initially at rest.

Determine the angular speed of the man–chair assembly, $\vec{\omega}$, when, after receiving the wheel axle as in scenario c, the experimenter raises the axis of the bicycle wheel to a vertical position, so that the wheel is above his head.

Specify what the experimenter must do to be able to easily hand over the bicycle wheel axle.

Note that: In each step of the described experiment, the position of the experimenter's body will be considered to remain vertical.

Solution

(a) The kinetic moment of the wheel in relation to the vertical axis of symmetry of the whole system is $\vec{J} = I\vec{\Omega}$, directed vertically upward, as indicated in drawing a of Figure 1.2. At the initial moment, this is the total kinetic moment of the man–chair–wheel system.

The experimenter tilts the wheel axis, forming an angle α with the vertical so that in the new position, the kinetic moment of the wheel in relation to its axis is the same, $\vec{J} = I\vec{\Omega}$, as indicated by Figures 1.1 and 1.2, respectively.

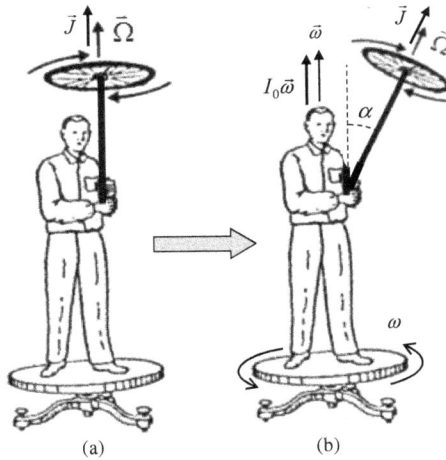

Fig. 1.2

In the new situation, when the axis of the bicycle wheel is inclined to the vertical by an angle α, the component of its kinetic moment along the vertical axis ($\perp$) of the man–chair system is:

$$J_{\perp,\text{wheel}} = J \cos \alpha = I\Omega \cos \alpha;$$

$$\vec{J}_{\perp,\text{wheel}} = I\vec{\Omega} \cos \alpha.$$

This implies a variation in the vertical component ($\perp$) of the wheel's kinetic moment:

$$(\Delta \vec{J})_{\perp,\text{wheel}} = \vec{J}_{\perp,\text{wheel}} - \vec{J} = I\vec{\Omega} \cos \alpha - I\vec{\Omega};$$

$$(\Delta \vec{J})_{\perp,\text{ wheel}} = -I\vec{\Omega}(1 - \cos \alpha).$$

Its orientation is represented in Figure 1.3.

According to the law of conservation of kinetic momentum, the vertical component ($\perp$) of the total kinetic momentum of the system (man–chair–wheel) will remain constant ($\vec{J}$) if, after tilting the wheel axis, the vertical component ($\perp$) of the varying kinetic moment of the man–chair system is

$$(\Delta \vec{J})_{\perp,\text{man–chair}} = -(\Delta \vec{J})_{\perp,\text{wheel}},$$

which corresponds to a rotation around the vertical of the man–chair assembly with the angular velocity $\vec{\omega}$, orientated along the ascending

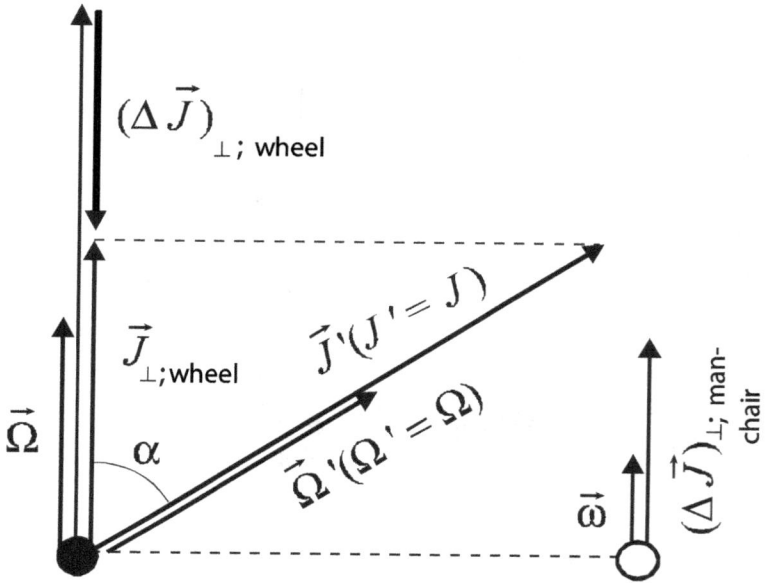

Fig. 1.3

vertical so that

$$I_0 \vec{\omega} = (\Delta \vec{J})_{\perp, \text{man-chair}},$$

where I_0 is the axial moment of inertia of the man–chair assembly;

$$\omega = \frac{I}{I_0}\Omega(1 - \cos \alpha).$$

For $\alpha = 90°$, as shown in Figure 1.4, the angular velocity of the man–chair assembly is calculated as follows:

$$I \vec{\Omega} = I_0 \vec{\omega}';$$

$$\omega' = \frac{I}{I_0}\Omega.$$

Alternatively:

$$\alpha = 90°; \quad \cos 90° = 0; \quad \omega' = \frac{I}{I_0}\Omega.$$

For $\alpha = 180°$, as Figure 1.5 indicates, the angular velocity of the man–chair assembly is calculated as follows:

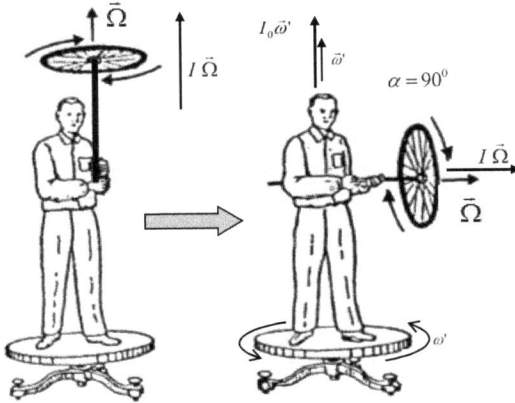

Fig. 1.4

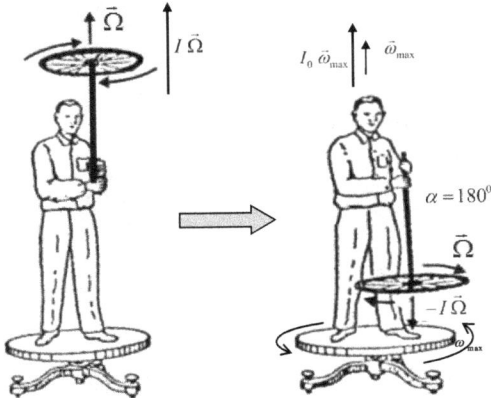

Fig. 1.5

$$I\,\vec{\Omega} = I_0\,\vec{\omega}_{\text{max}} + (-I\,\vec{\Omega});$$

$$2I\,\vec{\Omega} = I_0\,\vec{\omega}_{\text{max}}; \quad 2I\,\Omega = I_0\,\omega_{\text{max}};$$

$$\omega_{\text{max}} = 2\frac{I}{I_0}\Omega.$$

Alternatively:

$$\alpha = 180°; \quad \cos 180° = -1; \quad \omega_{max} = 2\frac{I}{I_0}\Omega.$$

At the end of each of the two particular cases, we discover that, for the man–chair–wheel assembly, the total kinetic moment is $\vec{J}$.

Returning the wheel to its position in Figure 1.1 a will bring the man–chair assembly to a rest again, and the experimenter will be able to hand over the wheel without difficulty.

(b) Using the drawing from Figure 1.6, in accordance with the law of conservation of axial kinetic moment, it results that:

$$0 = I\vec{\Omega} + I_0\vec{\omega}';$$

$$\vec{\omega}' = -\frac{I}{I_0}\vec{\Omega}; \quad \omega' = \frac{I}{I_0}\Omega.$$

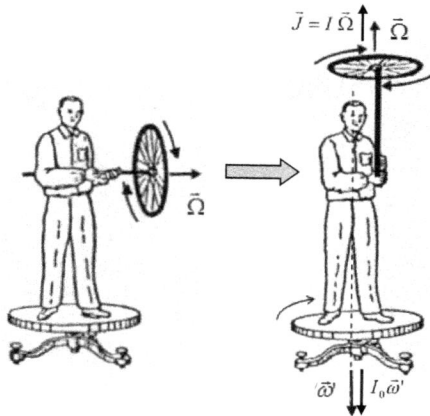

Fig. 1.6

(c) Using the diagram in Figure 1.7, in accordance with the law of conservation of axial kinetic moment, it results that:

$$I\vec{\Omega} = I_0\vec{\omega}_{max} + (-I\vec{\Omega});$$

$$2I\vec{\Omega} = I_0\vec{\omega}_{max}; \quad \vec{\omega}_{max} = 2\frac{I}{I_0}\vec{\Omega}; \quad \omega_{max} = 2\frac{I}{I_0}\Omega.$$

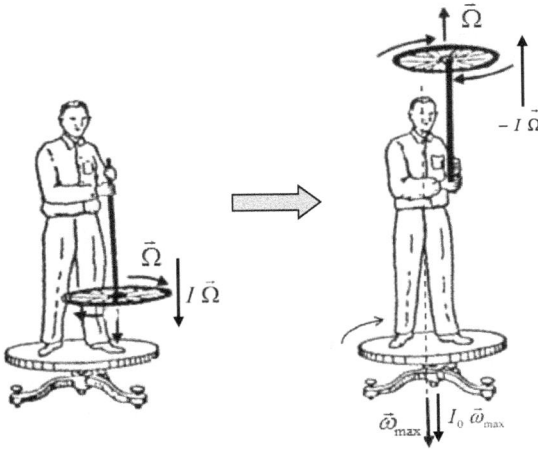

Fig. 1.7

Problem 2. A Spinning Egg

A boiled egg (a solid body with longitudinal axial symmetry) located in a special device, which provides it with a rotational movement around the longitudinal axis of symmetry, with a high angular velocity, $\vec{\omega}$, is released on a horizontal support, as shown in drawings a and b in Figure 2.1.

In each scenario, the initial value of the angle between the axis of longitudinal symmetry of the egg and the vertical is the same, α.

Justify the evolution of the value of the angle α after releasing the egg on the horizontal support in each of the two scenarios.

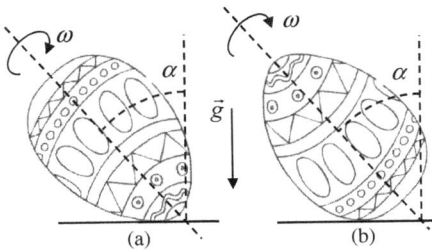

Fig. 2.1

Solution

The egg, a solid body with axial symmetry, whose longitudinal cross-section is shown in Figure 2.2, rotates with angular velocity $\vec{\omega}$ around its longitudinal axis and is placed with its pointed end on a horizontal support.

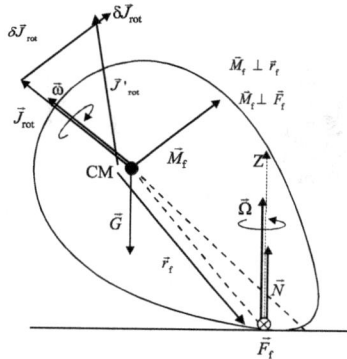

Fig. 2.2

Due to the shape of the egg, its axis of rotation, on which its center of mass CM is located, does not pass through the point of contact of the egg with the support.

The external forces acting on the egg, represented in the drawing, are the weight $(\vec{G})$, the normal reaction of the support $(\vec{N})$, and the frictional force $(\vec{F}_f,$ perpendicular to the plane of the diagram, i.e., entering the diagram).

The moment of the external forces relative to the point of contact, which, according to the kinetic momentum variation theorem, determines the emergence of the gyroscopic effect, materializing in the egg's precessional motion with angular velocity $\vec{\Omega}$ around the vertical point of contact, is

$$\vec{M} = I\,\vec{\Omega} \times \vec{\omega},$$

where the angular velocity of precession, $\vec{\Omega}$, represents the angular velocity with which the kinetic moment vector $(\vec{J}_{\rm rot})$ rotates around the vertical point of contact.

The moment of the frictional force relative to the center of mass (CM) of the egg, $\vec{M}_f$, whose orientation is represented in the diagram, causes the kinetic moment of the egg to vary by the magnitude $\delta \vec{J}_{rot}$ in a very small time interval, so that $\vec{J}_{rot}$ is brought to a vertical position, and the egg rises at the pointed end. For the situation represented in Figure 2.3, when the spinning egg, which rotates around its axis with the angular velocity $\vec{\omega}$, is placed on the horizontal support on its the round end, the moment of the external forces governs the precession movement with the angular velocity $\vec{\Omega}$ around the vertical point of contact, as in the previous case, while the moment of the frictional force relative to the center of mass, $\vec{M}_f$, determines, in a very short time interval, a variation $\delta \vec{J}_{rot}$ in the kinetic moment, so that $\vec{J}_{rot}$ is brought to a horizontal position and then inverted. Thus, the egg continues its movement, inverting its position to rest on its pointed end, before evolving according to the first scenario analyzed.

Conclusion: The rotation of the spinning egg, supported by its pointed end, gives it stability during rotation.

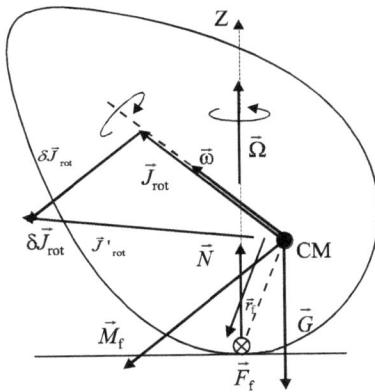

Fig. 2.3

Problem 3. Intergalactic Cloud

At the time of its formation, an intergalactic, spherical, and homogenous cloud rotates around its own axis with the angular velocity ω_{cloud}. The cloud is made of rarefied gas and cosmic dust. It has a radius R and a mass m.

Due to internal gravitational effects, the cloud substance has accumulated in two massive spheres of different radii. Thus, a binary stellar system is created, and its components rotate rapidly around their common center of mass.

(a) *Determine* the rotation period (T_{rot}) of the two stars formed in the initial cloud around the common center of mass, if the stars' masses are m_1 and $m_2 < m_1$, respectively. The gravitational attraction constant, K, is known.

(b) *Demonstrate* that the two stars can not only rotate around the center of mass, but also oscillate along the line of their centers. *Determine* the period (T_{osc}) of the "small" oscillations of the two stars.

(c) The astronomical observations made by an interstellar cosmic laboratory on this binary stellar system have proved that the oscillation period of its components uniformly decreases over time. *Specify* the physical meaning of these observations and *determine* the minimum value of the system's oscillation period. *Estimate* the evolution of the system after the moment when the minimum rotation period is reached.

Solution

(a) According to the notation in Figure 3.1, if $\vec{L}_{cloud}$ is the initial angular moment of the cosmic cloud, and $\vec{L}$ is the total angular moment of the binary stellar system, the entire evolution of the system complies with the law of conservation of angular momentum. Thus, it results that:

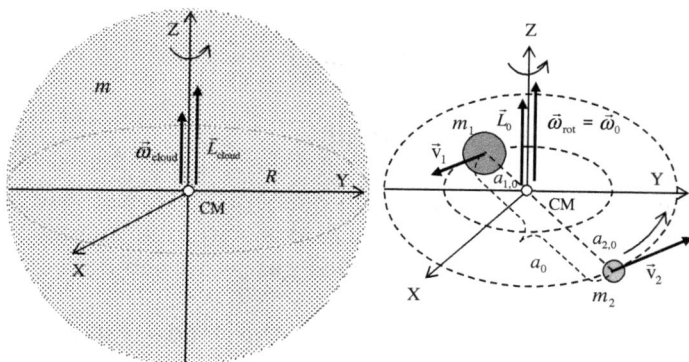

Fig. 3.1

$$L_{\text{cloud}} = I_{\text{cloud}}\omega_{\text{cloud}} = \frac{2}{5}mR^2\omega_{\text{cloud}};$$

$$L_0 = m_1\,v_1\,a_{1,0} + m_2\,v_2\,a_{2,0} = m_1\omega_{\text{rot}}a_{1,0}^2 + m_2\omega_{\text{rot}}a_{2,0}^2$$

$$= \omega_{\text{rot}}(m_1a_{1,0}^2 + m_2a_{2,0}^2);$$

$$m_1a_{1,0} = m_2a_{2,0}; \quad a_{1,0} + a_{2,0} = a_0; \quad m_1 + m_2 = m;$$

$$a_{1,0} = \frac{m_2a_0}{m_1 + m_2}; \quad a_{2,0} = \frac{m_1a_0}{m_1 + m_2}; \quad a_{1,0} = \frac{m_2}{m_1}a_{2,0};$$

$$L_0 = \omega_{\text{rot}}(m_1a_{1,0}^2 + m_2a_{2,0}^2);$$

$$L_0 = \omega_{\text{rot}}\left(m_1\frac{m_2^2}{m_1^2}a_{2,0}^2 + m_2a_{2,0}^2\right) = \omega_{\text{rot}}m_2\left(1 + \frac{m_2}{m_1}\right)a_{2,0}^2;$$

$$L_0 = L_{\text{cloud}};$$

$$\omega_{\text{rot}} = \omega_0 = \frac{L_{\text{cloud}}}{m_2\left(1 + \frac{m_2}{m_1}\right)a_{2,0}^2};$$

$$F_{\text{cp},2} = m_2\omega_{\text{rot}}^2 a_{2,0} = m_2\frac{L_{\text{cloud}}^2}{m_2^2\left(1 + \frac{m_2}{m_1}\right)^2 a_{2,0}^4}a_{2,0} = \frac{L_{\text{cloud}}^2}{m_2\left(1 + \frac{m_2}{m_1}\right)^2 a_{2,0}^3};$$

$$F_{\text{g}} = K\frac{m_1m_2}{a_0^2}; \quad a_0 = \frac{a_{2,0}(m_1 + m_2)}{m_1};$$

$$F_{\text{g}} = K\frac{m_1m_2}{\frac{a_{2,0}^2(m_1+m_2)^2}{m_1^2}} = K\frac{m_1^3m_2}{(m_1 + m_2)^2a_{2,0}^2} = K\frac{m_1^3m_2}{m_1^2\left(1 + \frac{m_2}{m_1}\right)^2 a_{2,0}^2};$$

$$F_{\text{g}} = K\frac{m_1m_2}{\left(1 + \frac{m_2}{m_1}\right)^2 a_{2,0}^2};$$

$$F_{\text{cp},2} = F_{\text{g}};$$

$$\frac{L_{\text{cloud}}^2}{m_2\left(1 + \frac{m_2}{m_1}\right)^2 a_{2,0}^3} = K\frac{m_1m_2}{\left(1 + \frac{m_2}{m_1}\right)^2 a_{2,0}^2};$$

$$\frac{L_{\text{cloud}}^2}{m_2a_{2,0}} = Km_1m_2;$$

$$a_{2,0} = \frac{L_{\text{cloud}}^2}{Km_1m_2^2}; \quad m_1a_{1,0} = m_2a_{2,0}; \quad a_{2,0} = \frac{m_1}{m_2}a_{1,0};$$

$$a_{1,0} = \frac{L_{\text{cloud}}^2}{Km_1^2m_2};$$

$$a_{1,0} + a_{2,0} = a_0;$$

$$a_0 = \frac{L_{\text{cloud}}^2}{Km_1^2m_2} + \frac{L_{\text{cloud}}^2}{Km_1m_2^2} = \frac{L_{\text{cloud}}^2}{Km_1m_2}\left(\frac{1}{m_1} + \frac{1}{m_2}\right);$$

$$a_0 = \frac{L_{\text{cloud}}^2(m_1 + m_2)}{Km_1^2m_2^2};$$

$$L_{\text{cloud}} = \frac{2}{5}mR^2\omega_{\text{cloud}} = \frac{2}{5}(m_1 + m_2)R^2\omega_{\text{cloud}};$$

$$\omega_{\text{rot}} = \omega_0 = \frac{L_{\text{cloud}}}{m_2\left(1 + \frac{m_2}{m_1}\right)a_{2,0}^2}; \quad a_{2,0} = \frac{m_1a_0}{m_1 + m_2};$$

$$\omega_{\text{rot}} = \omega_0 = \frac{L_{\text{cloud}}}{m_2\left(\frac{m_1+m_2}{m_1}\right)\frac{m_1^2a_0^2}{(m_1+m_2)^2}} = \frac{L_{\text{cloud}}(m_1 + m_2)}{m_1m_2a_0^2} = \frac{2\pi}{T_{\text{rot}}};$$

$$a_0 = \frac{L_{\text{cloud}}^2(m_1 + m_2)}{Km_1^2m_2^2};$$

$$T_{\text{rot}} = \frac{2\pi m_1m_2}{L_{\text{cloud}}(m_1 + m_2)}a_0^2 = \frac{2\pi m_1m_2}{L_{\text{cloud}}(m_1 + m_2)}\frac{L_{\text{cloud}}^4(m_1 + m_2)^2}{K^2m_1^4m_2^4};$$

$$T_{\text{rot}} = \frac{2\pi(m_1 + m_2)L_{\text{cloud}}^3}{K^2m_1^3m_2^3}; \quad L_{\text{cloud}} = \frac{2}{5}mR^2\omega_{\text{cloud}}.$$

(b) According to the notation in Figures 3.2 and 3.3, it results that:

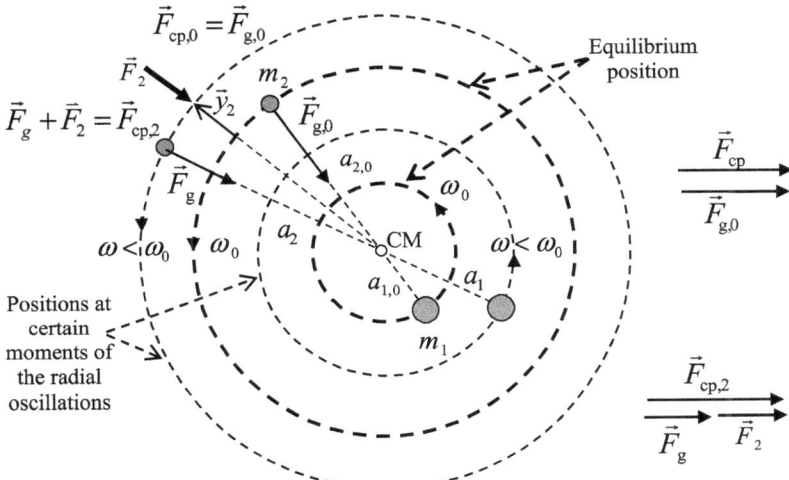

Fig. 3.2

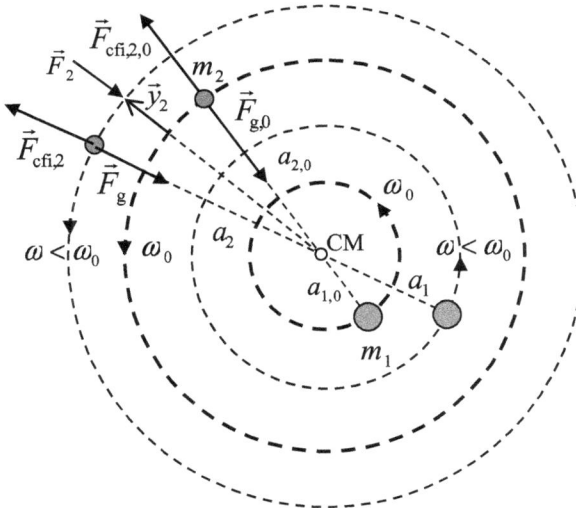

Fig. 3.3

$$\vec{F}_{g,0} = \vec{F}_{cp,0}; \quad F_{g,0} = F_{cp,0};$$

$$\vec{F}_g + \vec{F}_2 = \vec{F}_{cp}; \quad F_g + F_2 = F_{cp}; \quad F_2 = F_{cp} - F_g;$$

$$\vec{F}_{2,0} = \vec{F}_{cfi,2,0} + \vec{F}_{g,0} = 0; \quad F_{cfi,2,0} = F_{g,0};$$

$$F_{g,0} = K\frac{m_1 m_2}{a_0^2}; \quad a_0 = \frac{a_{2,0}(m_1 + m_2)}{m_1}; \quad F_{g,0} = K\frac{m_1 m_2}{\left(1 + \frac{m_2}{m_1}\right)^2 a_{2,0}^2};$$

$$F_{cfi,2,0} = m_2 \omega_0^2 a_{2,0};$$

$$m_2 \omega_0^2 a_{2,0} = K\frac{m_1 m_2}{\left(1 + \frac{m_2}{m_1}\right)^2 a_{2,0}^2}; \quad \omega_0^2 a_{2,0} = K\frac{m_1}{\left(1 + \frac{m_2}{m_1}\right)^2 a_{2,0}^2};$$

$$\omega_0^2 = K\frac{m_1}{\left(1 + \frac{m_2}{m_1}\right)^2 a_{2,0}^3};$$

$$\frac{L_{cloud}^2}{m_2 a_{2,0}} = Km_1 m_2; \quad L_{cloud}^2 = Km_1 m_2^2 a_{2,0};$$

$$\omega_0^2 = K\frac{m_1 m_2^2 a_{2,0}}{m_2^2 \left(1 + \frac{m_2}{m_1}\right)^2 a_{2,0}^4}; \quad \omega_0^2 = \frac{L_{cloud}^2}{m_2^2 \left(1 + \frac{m_2}{m_1}\right)^2 a_{2,0}^4};$$

$$\omega_0 = \frac{L_{cloud}}{m_2 \left(1 + \frac{m_2}{m_1}\right) a_{2,0}^2};$$

$$F_{cfi,2,0} = m_2 \frac{L_{cloud}^2}{m_2^2 \left(1 + \frac{m_2}{m_1}\right)^2 a_{2,0}^4} \cdot a_{2,0};$$

$$F_{cfi,2,0} = \frac{L_{cloud}^2}{m_2 \left(1 + \frac{m_2}{m_1}\right)^2 a_{2,0}^3};$$

$$F_{g,0} = K\frac{m_1 m_2}{\left(1 + \frac{m_2}{m_1}\right)^2 a_{2,0}^2};$$

$$F_{\mathrm{cfi},2,0} = F_{\mathrm{g},0};$$

$$\frac{L_{\mathrm{cloud}}^2}{m_2 \left(1 + \frac{m_2}{m_1}\right)^2 a_{2,0}^3} = K \frac{m_1 m_2}{\left(1 + \frac{m_2}{m_1}\right)^2 a_{2,0}^2};$$

$$\vec{F}_2 = \vec{F}_{\mathrm{g}} + \vec{F}_{\mathrm{cfi},2} \neq 0;$$

$$F_2 = F_{\mathrm{g}} - F_{\mathrm{cfi},2} \neq 0;$$

$$F_{\mathrm{g}} = K\frac{m_1 m_2}{a^2}; \quad a = \frac{a_2(m_1 + m_2)}{m_1}; \quad F_{\mathrm{g}} = K\frac{m_1 m_2}{\left(1 + \frac{m_2}{m_1}\right)^2 a_2^2};$$

$$F_{\mathrm{cfi},2} = m_2 \omega^2 a_2;$$

$$\omega_{\mathrm{rot}} = \frac{L_{\mathrm{cloud}}}{m_2 \left(1 + \frac{m_2}{m_1}\right) a_2^2}; \quad \omega = \frac{L_{\mathrm{cloud}}}{m_2 \left(1 + \frac{m_2}{m_1}\right) a_2^2};$$

$$F_{\mathrm{cfi},2} = m_2 \cdot \frac{L_{\mathrm{cloud}}^2}{m_2^2 \left(1 + \frac{m_2}{m_1}\right)^2 a_2^4} a_2;$$

$$F_{\mathrm{cfi},2} = \frac{L_{\mathrm{cloud}}^2}{m_2 \left(1 + \frac{m_2}{m_1}\right)^2 a_2^3};$$

$$L_0 = L;$$

$$I_0 \omega_0 = I\omega; \quad \omega = \frac{I_0}{I}\omega_0 = \frac{m_1 a_{1,0}^2 + m_2 a_{2,0}^2}{m_1 a_1^2 + m_2 a_2^2}\omega_0;$$

$$a_{1,0} = \frac{m_2 a_0}{m_1 + m_2}; \quad a_{2,0} = \frac{m_1 a_0}{m_1 + m_2};$$

$$a_1 = \frac{m_2 a}{m_1 + m_2}; \quad a_2 = \frac{m_1 a}{m_1 + m_2};$$

$$\omega = \frac{m_1 \frac{m_2^2 a_0^2}{(m_1+m_2)^2} + m_2 \frac{m_1^2 a_0^2}{(m_1+m_2)^2}}{m_1 \frac{m_2^2 a^2}{(m_1+m_2)^2} + m_2 \frac{m_1^2 a^2}{(m_1+m_2)^2}}\omega_0 = \left(\frac{a_0}{a}\right)^2 \omega_0;$$

$$a > a_0; \rightarrow \omega < \omega_0;$$

$$F_{\text{cfi},2,0} = \frac{L_{\text{cloud}}^2}{m_2 \left(1 + \frac{m_2}{m_1}\right)^2 a_{2,0}^3}; \quad F_{\text{cfi},2} = \frac{L_{\text{cloud}}^2}{m_2 \left(1 + \frac{m_2}{m_1}\right)^2 a_2^3};$$

$$a_2 > a_{2,0}; \rightarrow F_{\text{cfi},2} < F_{\text{cfi},2,0};$$

$$F_{\text{g},0} = K \frac{m_1 m_2}{\left(1 + \frac{m_2}{m_1}\right)^2 a_{2,0}^2}; \quad F_{\text{g}} = K \frac{m_1 m_2}{\left(1 + \frac{m_2}{m_1}\right)^2 a_2^2};$$

$$F_{\text{g}} < F_{\text{g},0};$$

$$F_2 = F_{\text{g}} - F_{\text{cfi},2} = K \frac{m_1 m_2}{\left(1 + \frac{m_2}{m_1}\right)^2 a_2^2} - \frac{L_{\text{cloud}}^2}{m_2 \left(1 + \frac{m_2}{m_1}\right)^2 a_2^3};$$

$$a_2 = a_{2,0} + y_2; \quad y_2 \ll a_{2,0};$$

$$F_2 = K \frac{m_1 m_2}{\left(1 + \frac{m_2}{m_1}\right)^2 (a_{2,0} + y_2)^2} - \frac{L_{\text{cloud}}^2}{m_2 \left(1 + \frac{m_2}{m_1}\right)^2 (a_{2,0} + y_2)^3};$$

$$F_2 = K \frac{m_1 m_2}{\left(1 + \frac{m_2}{m_1}\right)^2 a_{2,0}^2} \left(1 + \frac{y_2}{a_{2,0}}\right)^{-2}$$

$$- \frac{L_{\text{cloud}}^2}{m_2 \left(1 + \frac{m_2}{m_1}\right)^2 a_{2,0}^3} \left(1 + \frac{y_2}{a_{2,0}}\right)^{-3};$$

$$F_2 = K \frac{m_1 m_2}{\left(1 + \frac{m_2}{m_1}\right)^2 a_{2,0}^2} \left(1 - 2\frac{y_2}{a_{2,0}}\right)$$

$$- \frac{L_{\text{cloud}}^2}{m_2 \left(1 + \frac{m_2}{m_1}\right)^2 a_{2,0}^3} \left(1 - 3\frac{y_2}{a_{2,0}}\right);$$

$$F_2 = K \frac{m_1 m_2}{\left(1 + \frac{m_2}{m_1}\right)^2 a_{2,0}^2} - 2K \frac{m_1 m_2}{\left(1 + \frac{m_2}{m_1}\right)^2 a_{2,0}^3} y_2$$

$$- \frac{L_{\text{cloud}}^2}{m_2 \left(1 + \frac{m_2}{m_1}\right)^2 a_{2,0}^3} + \frac{3 L_{\text{cloud}}^2}{m_2 \left(1 + \frac{m_2}{m_1}\right)^2 a_{2,0}^4} y_2;$$

$$F_{g,0} = K \frac{m_1 m_2}{\left(1 + \frac{m_2}{m_1}\right)^2 a_{2,0}^2}; \quad F_{\text{cfi},2,0} = \frac{L_{\text{cloud}}^2}{m_2 \left(1 + \frac{m_2}{m_1}\right)^2 a_{2,0}^3};$$

$$F_{\text{cfi},2,0} = F_{g,0};$$

$$F_2 = -2K \frac{m_1 m_2}{\left(1 + \frac{m_2}{m_1}\right)^2 a_{2,0}^3} y_2 + \frac{3L_{\text{cloud}}^2}{m_2 \left(1 + \frac{m_2}{m_1}\right)^2 a_{2,0}^4} y_2;$$

$$F_2 = \frac{1}{\left(1 + \frac{m_2}{m_1}\right)^2 a_{2,0}^3} \left(\frac{3L_{\text{cloud}}^2}{m_2 a_{2,0}} - 2K m_1 m_2\right) y_2;$$

$$\frac{L_{\text{cloud}}^2}{m_2 a_{2,0}} = K m_1 m_2;$$

$$F_2 = \frac{1}{\left(1 + \frac{m_2}{m_1}\right)^2 a_{2,0}^3} (3K m_1 m_2 - 2K m_1 m_2) y_2;$$

$$F_2 = \frac{K m_1 m_2}{\left(1 + \frac{m_2}{m_1}\right)^2 a_{2,0}^3} y_2; \quad k = \frac{K m_1 m_2}{\left(1 + \frac{m_2}{m_1}\right)^2 a_{2,0}^3};$$

$$a_{2,0} = \frac{m_1 a_0}{m_1 + m_2}; \quad a_0 = \frac{L_{\text{cloud}}^2 (m_1 + m_2)}{K m_1^2 m_2^2};$$

$$k = \frac{K m_1 m_2}{\left(1 + \frac{m_2}{m_1}\right)^2 \frac{m_1^3 a_0^3}{(m_1 + m_2)^3}} = \frac{K m_1 m_2}{\frac{(m_1 + m_2)^2}{m_1^2} \frac{m_1^3}{(m_1 + m_2)^3} \frac{L_{\text{cloud}}^6 (m_1 + m_2)^3}{K^3 m_1^6 m_2^6}};$$

$$k = \frac{K^4 m_1^6 m_2^7}{L_{\text{cloud}}^6 (m_1 + m_2)^2} = m_2 \omega_{\text{osc}}^2 = m_2 \frac{4\pi^2}{T_{\text{osc}}^2};$$

$$F_2 = k y_2; \quad \vec{F}_2 = -k \vec{y}_2;$$

$$T_{\text{osc}} = 2\pi \frac{L_{\text{cloud}}^3 (m_1 + m_2)}{K^2 m_1^3 m_2^3}.$$

(c)
$$\frac{K^4 m_1^6 m_2^6}{L_{\text{cloud}}^6 (m_1 + m_2)^2} = \omega_{\text{osc}}^2;$$

$$\omega_{\text{osc}} = \frac{K^2 m_1^3 m_2^3}{L_{\text{cloud}}^3 (m_1 + m_2)};$$

$$T_{\text{osc}} \downarrow \Rightarrow \omega_{\text{osc}} \uparrow \Rightarrow \omega_{\text{osc,max}};$$

$$m_1 + m_2 = \text{constant};$$

$$\omega_{\text{osc}} = \omega_{\text{osc,max}}, \quad \text{when } m_1 = m_2 = \frac{m}{2};$$

$$\omega_{\text{osc,max}} = \frac{K^2 m^5}{64 L_{\text{cloud}}^3} = \frac{4\pi^2}{T_{\text{osc,min}}^2}; \quad T_{\text{vibr,min}} = 16\pi \frac{L_{\text{cloud}}}{K m^2} \sqrt{\frac{L_{\text{cloud}}}{m}}.$$

$$T_{\text{osc}} = 2\pi \frac{L_{\text{cloud}}^3 (m_1 + m_2)}{K^2 m_1^3 m_2^3}.$$

Other Books by Mihail Sandu

1. **PROBLEME DE FIZICĂ PENTRU GIMNAZIU**, *Editura Didactică şi Pedagogică*, Bucharest, 1977, 327 p.

2. **FIZICĂ, MANUAL PENTRU CLASA a VI-a**, coauthor Emanuel Nichita, *Editura Didactică şi Pedagogică*, Bucharest, 1978, 160 p.

3. **PROBLEME DE FIZICĂ PENTRU GIMNAZIU**, coauthors Emanuel Nichita and Tudorel Ştefan, *Editura Didactică şi Pedagogicş*, Bucharest, 1982, 223 p.

4. **CAIET METODIC CU PROBLEME DE FIZICĂ PENTRU GIMNAZIU**, *Casa Personalului Didactic*, Rm. Vâlcea, 1983, 181 p.

5. **GHIDUL PROFESORULUI ŞI AL ELEVULUI PENTRU CERCURILE DE FIZICĂ, VOL. I**, *Casa de Cultură a Ştiinţei şi Tehnicii pentru tineret*, Rm. Vâlcea, 1984, 498 p.

6. **GHIDUL PROFESORULUI ŞI AL ELEVULUI PENTRU CERCURILE DE FIZICĂ, VOL. II**, *Casa de Cultură a Ştiinţei şi Tehnicii pentru tineret*, Rm. Vâlcea, 1984, 405 p.

7. **CULEGERE DE PROBLEME DE FIZICĂ, VOL. I**, *Societatea de Ştiinţe Fizice şi Chimice din România*, Bucharest, 1986, 338 p.

8. **CULEGERE DE PROBLEME DE FIZICĂ, VOL. II**, *Societatea de Ştiinţe Fizice şi Chimice din România*, Bucharest, 1986, 118 p.

9. **PROBLEME DE FIZICĂ**, *Editura Scrisul Românesc*, Craiova, 1988, 277 p.

10. **GHID PENTRU CERCURILE DE FIZICĂ**, *Editura Academiei Române*, Bucharest, 1991, 325 p.

11. **PROBLEME DE FIZICĂ PENTRU GIMNAZIU**, *Editura Didactică şi Pedagogică*, Bucharest, 1991, 208 p.

12. **500 PROBLEME DE FIZICĂ**, *Editura Tehnică*, Bucharest, 1991, 283 p.

13. **PROBLEME DE PERFORMANŢĂ ÎN FIZICĂ**, *Editura Tehnică*, Bucharest, 1992, 340 p.

14. **PROBLEME DE FIZICĂ PENTRU GIMNAZIU**, *Editura Lumina*, Chişinău, Republic of Moldova, 1993, 272 p.

15. **CAIET METODIC CU PROBLEME DE FIZICĂ PENTRU GIMNAZIU**, *Editura Lumina*, Chişinău, Republic of Moldova, 1993, 179 p.

16. **TEME ŞI PROBLEME PENTRU CERCURILE DE FIZICĂ**, *Editura Hyperion XXI*, Bucharest, 1993, 254 p.

17. **PROBLEME DE FIZICĂ DIN REVISTA "KVANT", VOL. I**, *Editura Didactică şi Pedagogică*, Bucharest, 1993, 182 p.

18. **PROBLEME DE FIZICĂ PENTRU LICEU**, *Editura Ex Libris*, Rm. Vâlcea, 1993, 378 p.

19. **PROBLEME DE FIZICĂ DIN REVISTA "KVANT", VOL. II**, *Editura Didactică şi Pedagogică*, Bucharest, 1994, 194 p.

20. **CULEGERE DE PROBLEME DE FIZICĂ, VOL. I**, *Editura TipCim*, Chişinău, Republic of Moldova, 1995, 338 p.

21. **CULEGERE DE PROBLEME DE FIZICĂ, VOL. II**, *Editura TipCim*, Chişinău, Republic of Moldova, 1995, 117 p.

22. **GHIDUL PROFESORULUI ŞI AL ELEVULUI PENTRU CERCURILE DE FIZICĂ, VOL. I**, *Editura TipCim, Chişinău*, Republic of Moldova, 1995, 498 p.

23. **PROBLEME DE FIZICĂ PENTRU GIMNAZIU**, *Editura Didactică şi Pedagogică*, Bucharest, 1995, 202 p.

24. **GHIDUL PROFESORULUI ŞI AL ELEVULUI PENTRU CERCURILE DE FIZICĂ, VOL. II**, *Editura TipCim, Chişinău*, Republic of Moldova, 1996, 400 p.

25. **PROBLEME DE FIZICĂ PENTRU GIMNAZIU**, *Editura ALL*, Bucharest, 1996, 1st Edition, 503 p.

26. **PROBLEME DE FIZICĂ DIN REVISTA "KVANT", VOL. III**, *Editura Didactică şi Pedagogică*, Bucharest, 1996, 208 p.

27. **PROBLEME DE FIZICĂ DIN REVISTA "KVANT", VOL. IV**, *Editura Didactică şi Pedagogică*, Bucharest, 1996, 229 p.

28. **PROBLEME DE FIZICĂ PENTRU GIMNAZIU**, *Editura ALL*, Bucharest, 1997, 2nd Edition, 503 p.

29. **PROBLEME DE FIZICĂ – CONCURSURI ŞI OLIMPIADE**, *Editura Petrion*, Bucharest, 1998, 335 p.

30. **PROBLEME DE FIZICĂ PENTRU GIMNAZIU**, *Editura ALL*, Bucharest, 1998, 3rd Edition, 503 p.

31. **MECANICĂ FIZICĂ** (university course), *Editura Didactică şi Pedagogică*, Bucharest, 2002, 462 p.

32. **MECANICĂ TEORETICĂ** (university course), *Editura Didactică şi Pedagogică*, Bucharest, 2002, 476 p.

33. **ASTRONOMIE** (university course), *Editura Didactică şi Pedagogică*, Bucharest, 2003, 400 p.

34. **FIZICĂ. ÎNVĂȚĂMÂNTUL LICEAL ȘI CERC-ETAREA ȘTIINȚIFICĂ. MONOGRAFIA JUDEȚ-ULUI VÂLCEA**, *Editura Conphys*, Rm. Vâlcea, 2004, 406 p.

35. **IN MEMORIAM – PROFESORUL LIVIU TĂTAR**, *Editura Conphys*, Rm. Vâlcea, 2004.

36. **PROBLEME DE FIZICĂ PENTRU GIMNAZIU**, *Editura ALL*, Bucharest, 2005, 4th Edition, 416 p.

37. **TEORIA RELATIVITĂȚII** (university course), *Editura Didactică și Pedagogică*, Bucharest, 2005, 341 p.

38. **EVRIKA! PROBLEME DE FIZICĂ**, *Editura Didactică și Pedagogică*, Bucharest, 2005, 600 p.

39. **CONCURSUL DE SELECȚIE PENTRU OLIMPIADA INTERNAȚIONALĂ DE FIZICĂ – 2005**, coauthor, *Editura Conphys*, Rm. Vâlcea, 2006, 262 p.

40. **PROBLEME DE FIZICĂ PENTRU GIMNAZIU**, *Editura ALL*, Bucharest, 2007, 5th Edition, 416 p.

41. **ACTIVITĂȚI OLIMPICE LA C.P.P.P. CĂLIMĂ-NEȘTI-CĂCIULATA, 2004 – 2005 – 2006, ASTRO-NOMIE ȘI ȘTIINȚE**, *Editura Conphys*, Rm. Vâlcea, 2007, 611 p.

42. **ACTIVITĂȚI OLIMPICE LA C.P.P.P. CĂLIMĂ-NEȘTI-CĂCIULATA, 2004 – 2005 – 2006, FIZICĂ**, *Editura Conphys*, Rm. Vâlcea, 2007, 530 p.

43. **TOP-FIZ, PROBLEME DE FIZICĂ**, *Editura Didactică și Pedagogică*, Bucharest, 2009, 833 p.

44. **LUCRĂRI EXPERIMENTALE DE FIZICĂ**, *Editura Conphys*, Rm. Vâlcea, 2009, 156 p.

45. **ȘTIINȚE PENTRU JUNIORI. CHIMIE – FIZICĂ – BIOLOGIE**, coauthor, *Editura MISTRAL*, Bucharest, 2009, 358 p.

46. **CONCURSUL INTERJUDEŢEAN DE FIZICĂ "LIVIU TĂTAR"**, coauthor, *Editura Universitaria*, Craiova, 2011, 236 p.

47. **ŞTIINŢE PENTRU JUNIORI. CHIMIE – FIZICĂ – BIOLOGIE**, coauthor, *Editura MISTRAL*, 2nd Edition, Bucharest, 2011, 358 p.

48. **PROBLEME ŞI EXPERIMENTE DE FIZICĂ**, *Editura Didactică şi Pedagogică*, Bucharest, 2012, 820 p.

49. **INTERNATIONAL OLYMPIAD ON ASTRONOMY AND ASTROPHYSICS, PROBLEMS, 2007–2013, SOLUTIONS, COMMENTS, DETAILS, EXTENSIONS**, *Editura CYGNUS*, Suceava, 2014, 600 p.

50. **OLIMPIADA INTERNAŢIONALĂ DE ASTRONOMIE Ş ASTROFIZICĂ, PROBLEME, VOL. 1**, *Editura CYGNUS*, Suceava, 2015, 400 p.

51. **OLIMPIADA INTERNAŢIONALĂ DE ASTRONOMIE ŞI ASTROFIZICĂ, PROBLEME, VOL. 2**, *Editura CYGNUS*, Suceava, 2015, 460 p.

52. **CONCURSUL INTERJUDEŢEAN DE FIZICĂ "LIVIU TĂTAR"**, coauthor, *Editura else*, Craiova, 2016, 280 p.

53. **INTERNATIONAL OLYMPIAD ON ASTRONOMY AND ASTROPHYSICS, PROBLEMS, VOL. I**, *Trisula Adisakti*, Jakarta, Indonesia; Amazon, USA, 2016, 500 p.

54. **INTERNATIONAL OLYMPIAD ON ASTRONOMY AND ASTROPHYSICS, PROBLEMS, VOL. II**, *Trisula Adisakti*, Jakarta, Indonesia; Amazon, USA, 450 p.

55. **ASTRONOMIE Ş ASTROFIZICĂ – EVENIMENTE OLIMPICE – 1996–2010, PROBLEME**, *Editura Didactică şi Pedagogică*, Bucharest, 2018, 972 p.

56. **EXPERIMENTE DE FIZICĂ**, *Editura Didactică şi Pedagogică*, Bucharest, 2018, 476 p.

57. **ASTRONOMIE ŞI ASTROFIZICĂ. TEME ŞI PROBLEME PENTRU OLIMPIADE Ę CONCURSURI**, *Editura Didactică şi Pedagogică*, Bucharest, 2019.

58. **ASTRONOMY AND ASTROPHYSICS. THEMES AND PROBLEMS FOR OLYMPIADS AND COMPETITIONS**, *Editura Didactică şi Pedagogică*, Bucharest, 2019.

59. **ASTRONOMIE ŞI ASTROFIZICĂ – EVENIMENTE OLIMPICE – 2011–2013, PROBLEME, VOL. II**, *Editura Didactică şi Pedagogică*, Bucharest, 2020, 500 p.

60. **ASTRONOMIE ŞI ASTROFIZICĂ – EVENIMENTE OLIMPICE – 2013–2015, PROBLEME, VOL. III**, *Editura Didactică şi Pedagogică*, Bucharest, 2020, 500 p.

61. **ASTRONOMIE ŞI ASTROFIZICĂ – EVENIMENTE OLIMPICE – 2016–2017, PROBLEME, VOL. IV**, *Editura Didactică şi Pedagogică*, Bucharest, 2020, 500 p.

62. **ASTRONOMIE ŞI ASTROFIZICĂ – EVENIMENTE OLIMPICE – 2017–2018, PROBLEME, VOL. V**, *Editura Didactică şi Pedagogică*, Bucharest, 2020, 500 p.

63. **ASTRONOMIE ŞI ASTROFIZICĂ – EVENIMENTE OLIMPICE – 2019, PROBLEME, VOL. VI**, *Editura Didactică şi Pedagogică*, Bucharest, 2020, 548 p.

64. **INTERNATIONAL ASTRONOMY OLYMPIAD, VOL. I**, *Editura CYGNUS*, Suceava, 2021, 464 p.

65. **INTERNATIONAL ASTRONOMY OLYMPIAD, VOL. II**, *Editura CYGNUS*, Suceava, Romania, 2021, 484 p.

66. **INTERNATIONAL OLYMPIAD ON ASTRONOMY AND ASTROPHYSICS, VOL. III**, *Editura CYGNUS*, Suceava, Romania, 2021, 589 p.

67. **ASTRONOMY AND ASTROPHYSICS. ALL MY PROBLEMS FOR INTERNATIONAL OLYMPIADS, VOL. I**, *Editura CYGNUS*, Suceava, Romania, 2021, 344 p.

68. **ASTRONOMY AND ASTROPHYSICS. ALL MY PROBLEMS for INTERNATIONAL OLYMPIADS, VOL. II,** *Editura CYGNUS,* Suceava, Romania, 2021, 314 p.

69. **SUB CERUL BUCOVINEI, VOL. I, CONCURSUL INTERJUDEŢEAN DE FIZICĂ CYGNUS, CONCURSUL NAŢIONAL DE FIZICĂ MARIN DACIAN BICA, Clasele VI–IX, 2015–2019,** *Editura CYGNUS,* Suceava, 2021, 300 p.

70. **SUB CERUL BUCOVINEI, Vol. II, CONCURSUL INTERJUDEŢEAN DE FIZICĂ CYGNUS, CONCURSUL NAŢIONAL DE FIZICĂ MARIN DACIAN BICA, CLASELE X–XII, 2015–2019,** *Editura CYGNUS,* Suceava, 2021, 326 p.

71. **SUB CERUL BUCOVINEI, VOL. III, CONCURSUL NAŢIONAL DE ASTRONOMIE ŞI ASTROFIZICĂ MARIN DACIAN BICA FIZICĂ, 2015–2019, CONCURSUL INTERDISCIPLINAR ŞTIINŢELE PĂMÂNTULUI, OLIMPIADA NAŢIONALĂ DE ASTRONOMIE ŞI ASTROFIZICĂ,** *Editura CYGNUS,* Suceava, 2021, 388 p.

72. **PROBLEME DE FIZICĂ, VOL. I, EVENIMENTE OLIMPICE, SELECŢII PENTRU IPhO 1995–2009,** *Editura Didactică şi Pedagogică,* Bucharest, 2021, 340 p.

73. **PROBLEME DE FIZICĂ, VOL. II, EVENIMENTE OLIMPICE, SELECŢII PENTRU IPhO 2010–2021,** *Editura Didactică şi Pedagogică,* Bucharest, 2021, 340 p.

74. **ASTRONOMIE ŞI ASTROFIZICĂ, EVENIMENTE OLIMPICE, 2020–2021, PROBLEME, VOL. VII,** *Editura Didactică şi Pedagogică,* Bucharest, 2022, 600 p.

75. **PROBLEME DE PERFORMANŢĂ ÎN FIZICĂ, VOL. I,** *Editura Didactică şi Pedagogică,* Bucureşti, 2023, 664 p.

76. **ASTRONOMIE ŞI ASTROFIZICĂ, EVENIMENTE OLIMPICE, 2020–2022, PROBLEME, VOL. VIII,** *Editura Didactică şi Pedagogică,* Bucharest, 2023, 595 p.

77. **ASTRONOMY AND ASTROPHYSICS – MY NEW PROBLEMS, VOL. I** (for IOAA 2023, Edition XVI, Poland), *Editura Didactică și Pedagogică*, București, 2023, 395 p.

78. **ASTRONOMY AND ASTROPHYSICS – MY NEW PROBLEMS, VOL. II** (for IOAA 2023, Edition XVI, Poland), *Editura Didactică și Pedagogică*, București, 2023, 460 p.

79. **ASTRONOMY AND ASTROPHYSICS – MY NEW PROBLEMS, SECOND ED., VOL. I** (for IOAA Juniors 2023, Edition II, Greece), *Editura Didactică și Pedagogică*, Bucharest, 2023, 395 p.

80. **ASTRONOMY AND ASTROPHYSICS – MY NEW PROBLEMS, SECOND ED.,VOL. II** (for IOAA Juniors 2023, Edition II, Greece), *Editura Didactică și Pedagogică*, Bucharest, 2023, 460 p.

81. **ASTRONOMY AND ASTROPHYSICS – MY NEW PROBLEMS, THIRD ED., VOL. I** (for IAO 2023, Edition XXVII, China), *Editura Didactică și Pedagogică*, Bucharest, 2023, 460 p.

82. **ASTRONOMY AND ASTROPHYSICS – MY NEW PROBLEMS, THIRD ED., VOL. II** (for IAO 2023, Edition XXVII, China), *Editura Didactică și Pedagogică*, Bucharest, 2023, 460 p.

83. **PROBLEME DE PERFORMANȚĂ ÎN FIZICĂ, VOL. II,** *Editura Didactică și Pedagogică,* Bucharest, 2023, 500 p.

84. **PROBLEME DE PERFORMANȚĂ ÎN FIZICĂ, VOL. III,** *Editura Didactică și Pedagogică,* Bucharest, 2024, 500 p.

85. **INTERNATIONAL PRE-OLYMPIC CONTEST OF PHYSICS, ROMANIA, HUNGARY, MOLDOVA,** *Editura Didactică și Pedagogică*, Bucharest, 2024, 350 p.

86. **ȘTIINȚE PENTRU JUNIORI ÎN ROMÂNIA 2004–2023, PROBLEMELE MELE DE FIZICĂ,** *Editura Didactică și Pedagogică*, Bucharest, 2024, 480 p.

87. **SCIENCE FOR JUNIORS IN ROMANIA 2004–2023, MY PROBLEMS OF PHYSICS,** *Editura Didactică şi Pedagogică*, Bucharest, 2024, 480 p.

88. **ŞTIINŢE PENTRU JUNIORI ÎN ROMÂNIA, 2024, PROBLEMELE MELE DE FIZICĂ,** *Editura Didactică şi Pedagogică*, Bucharest, 2024, 200 p.

89. **SCIENCE FOR JUNIORS IN ROMANIA, 2024, MY PROBLEMS OF PHYSICS,** *Editura Didactică şi Pedagogică*, Bucharest, 2024, 200 p.

90. **CONCURSUL NAŢIONAL DE FIZICĂ ŞI CHIMIE IMPULS PERPETUUM! AL ELEVILOR DE GIMNAZIU DIN MEDIUL RURAL, PROBLEMELE MELE DE FIZICĂ, 2011, 2013–2020,** *Editura CONPHYS*, Rm. Vâlcea, 2025, 200 p.

91. **PROBLEMELE MELE DE FIZICĂ PENTRU EVENIMENTE OLIMPICE, VOL. I, 2022–2023,** *Editura Didactică şi Pedagogică*, Bucharest, 2025, 300 p.

92. **PROBLEMELE MELE DE FIZICĂ PENTRU EVENIMENTE OLIMPICE, VOL. II, 2023–2024,** *Editura Didactică şi Pedagogică*, Bucharest, 2025, 350 p.

93. **ASTRONOMIE ŞI ASTROFIZICĂ, PENTRU ELEVI OLIMPICI, STUDENŢI ŞI PROFESORI**, *Editura Didactică şi Pedagogică*, Bucharest, 2025, 450 p.

94. **ASTRONOMY AND ASTROPHYSICS OLYMPIAD: TOPICS, PROBLEMS AND SOLUTIONS, VOL. I**, *World Scientific*, London, 2025, 596 p.

95. **ASTRONOMY AND ASTROPHYSICS OLYMPIAD: PROBLEMS AND SOLUTIONS, VOL. 2,** *World Scientific*, London, 2025, 432 p.

96. **ASTRONOMY AND ASTROPHYSICS: PROBLEMS AND SOLUTIONS, VOL. 3,** *World Scientific*, London, 2025, 456 p.

97. **PHYSICS OLYMPIAD: PROBLEMS AND SOLUTIONS**, *World Scientific*, London, 2025, 396 p.

www.ingramcontent.com/pod-product-compliance
Lightning Source LLC
Chambersburg PA
CBHW061616220326
41598CB00026BA/3788